THE LAW AND ECOLOGY OF PESTICIDES AND PEST MANAGEMENT

Ecology and Law in Modern Society

Series Editors:
Richard O. Brooks
Professor Emeritus, Vermont Law School, USA
and
Ross A. Virginia
Environmental Studies Program, Dartmouth College, USA

This series presents a legal and ecological perspective on important environmental issues such as declining biodiversity and the ecological significance of endangered species, the effects of pollution on natural and managed systems, the ecology of fragile ecosystems (mountains, grazed lands), and pollution and its effects on ecosystem services. The scientific basis for understanding important areas of environmental law will be presented by experts in their fields. The formulating of environmental law to influence human activities will be analyzed by leading legal scholars. The central legal cases and challenges will be explored and the resulting success of the statutes considered. It is the intent of the series to provide a balance in the treatment of legal and ecological material not traditionally found in environmental law texts.

Also in the series

Mountain Resorts
Ecology and the Law
Edited by Janet E. Milne, Julia LeMense and Ross A. Virginia
ISBN 978-0-7546-2315-1 (Hb)

Law and Ecology
The Rise of the Ecosystem Regime
Richard O. Brooks, Ross Jones and Ross A. Virginia
ISBN 978-0-7546-2038-9 (Hb) and 978-0-7546-2316-8 (Pb)

The Law and Ecology of Pesticides and Pest Management

MARY JANE ANGELO
University of Florida Levin College of Law, USA

ASHGATE

Published by
Ashgate Publishing Limited
Wey Court East
Union Road
Farnham
Surrey, GU9 7PT
England

Ashgate Publishing Company
110 Cherry Street
Suite 3-1
Burlington, VT 05401-3818
USA

www.ashgate.com

British Library Cataloguing in Publication Data
Angelo, Mary Jane.
The law and ecology of pesticides and pest management. --
(Ecology and law in modern society)
1. Pesticides--Law and legislation--United States.
2. Pesticides--Environmental aspects--United States.
3. Agricultural pests--Control--United States--History.
4. Agricultural pests--Control--Environmental aspects.
5. Agricultural pests--Integrated control.
6. Environmental law--United States.
I. Title II. Series
344.7'3046334-dc23

The Library of Congress has cataloged the printed edition as follows:
Angelo, Mary Jane.
The law and ecology of pesticides and pest management / by Mary Jane Angelo.
pages cm. -- (Ecology and law in modern society)
Includes bibliographical references and index.
ISBN 978-0-7546-7431-3 (hardback) -- ISBN 978-0-7546-9128-0 (ebook) -- ISBN 978-1-4094-7479-1 (epub) 1. Pesticides--Law and legislation--United States. 2. Insect pests--Control--Law and legislation--United States. I. Title.
KF3959.A96 2013
344.7304'6334--dc23

2012043459

ISBN 9780754674313 (hbk)
ISBN 9780754691280 (ebk – PDF)
ISBN 9781409474791 (ebk – ePUB)

MIX
Paper from responsible sources
FSC® C013985
FSC www.fsc.org

Printed in the United Kingdom by Henry Ling Limited,
at the Dorset Press, Dorchester, DT1 1HD

Contents

List of Figures

List of Photos

List of Bridge Documents

Series Editors' Preface
A Decade of Change in the Concept of the Ecosystem Legal Regime

Richard O. Brooks and Ross A. Virginia

Introduction

When Rachel Carson published *The Silent Spring* in 1962, she revealed the serious environmental damages from the uses of DDT and, in doing so, she offered a primer in ecology, describing the ways in which this pesticide coursed through that complex system of interacting biotic and abiotic elements we call "the ecosystem." Rachel Carson's fundamental insights helped to prompt the gradual adoption of environmental laws that, at least in part, incorporate the theories and conclusions of the ecological sciences. In the following 40 years, a field of scholarship emerged which explored the complex relations between law and ecology. In 2002, *Law and Ecology: The Rise of the Ecosystem Regime* was published in which we defined the outlines of these two disciplines, traced the history of their relationship, and found the emergence of "ecosystem regimes": i.e., new social institutions designed to properly govern ecosystems (Brooks, Virginia, and Jones, 2002). This book was the first in a series that explored the relations of law and ecology in specific settings. In the next work, *Mountain Resorts: Ecology and Law* (2009), Janet Milne et al. explored the legal and environmental relationships of mountain resorts to mountain ecology. Now, Professor Mary Jane Angelo's *The Law and Ecology of Pesticides and Pest Management* offers a fine-grained analysis of the "application" of specific ecological concepts to the difficult task of formulating a law for the use and management of pesticides.

Since *The Law and Ecology of Pesticides and Pest Management* is part of a series, it is useful to compare the manner of its preparation to the other books in the series. In each of these works, interdisciplinarity is achieved in different ways. In the foundation to the series, *Law and Ecology*, a small group of authors—one lawyer (Brooks), one ecologist (Virginia) and one lawyer–ecologist (Jones)—prepared the work. In the second volume, one legal scholar (Milne) assembled, supervised and, with another lawyer (LeMense), and an ecologist (Virginia), edited and synthesized the work of an interdisciplinary team of lawyers, ecologists, planners, and economists to prepare a series of careful case studies. Professor

Angelo's book is the product of one author with academic qualifications both in the fields of ecology and the law.

Professor Angelo's study is a haunting return to "the scene of the crime" of uncontrolled pesticides which Rachel Carson described 50 years ago. But, unlike Carson's situation, the use of pesticides has multiplied, better knowledge of their benefits and harms has been gathered, our ecological knowledge has improved, and many new environmental laws have been adopted. These laws have to be applied in the context of an industrialized agricultural sector which has been a central contributor to the overall economic growth and prosperity of the United States, and to global food security, but which has also produced significant environmental harms and promises to produce more damage in the immediate future.

In this context, Professor Angelo's study documents this new and complicated reality in the effort to control pesticide production and use. Professor Angelo has created a framework and synthesis that provides the first in-depth ecological perspective on the fields of pesticide law and regulation. The result is a critique of the development of our absolute dependence on pesticides in conventional agriculture. The legal issues become complex because, unlike much of contaminant law, where compounds are by-products or inadvertent releases from industrial processes and the goal is to reduce emissions, pesticides are deliberately applied toxins where efficacy is often dose dependent. Pesticides can also move from targeted sites by air and water, and bio-magnify up the food chain with serious non-target effects on other organisms, including humans. Professor Angelo shows that an understanding of pesticides and their use requires knowledge about the evolution of pesticide resistance, degradation pathways and persistence in the environment, and the complicated food web interactions connecting life in soils with plants and animals.

The strength of Professor Angelo's approach comes from her balanced emphasis on the ecological and legal dimensions of pesticide law. Using a historical approach she traces the evolution of the modern frameworks from Integrated Pest Management (IPM) to Ecologically Based Pest Management (EBPM). By example, she shows how a pesticide regime could be imbedded within a sustainable agriculture. A variety of instruments would need to interact to minimize pesticide use and harm. These range from ecosystem services payments to new statutory schemes that monitor pesticide use and promote best practices for adaptive management, including a shift to more localized decision making to match with site-specific ecological conditions. Angelo makes a compelling case for important regulatory changes in the aging foundational laws and programs that now regulate pesticides. In the end Angelo's work builds a framework for a pesticide regime, deeply based in ecological science, to manage agricultural pests through a new integration of pesticide laws with economic incentives and current science to move today's agriculture toward sustainable agriculture, ushering in the new field of "agro-ecology."

Professor Angelo's approach, however, must be viewed as part of a much needed comprehensive reform of our entire environmental law and its political economy (Kysar, 2003; Speth, 2008). Such a reform seems unlikely at the moment when public concerns for environmental issues are preempted by serious economic issues confronting the United States, Europe, Brazil, India, and China. It is likely that such reforms will take place only as the public realizes the growing worldwide impacts of global warming on natural and managed ecosystems (agriculture). To be sure, there are other serious possible catastrophes facing our world (Posner, 2004), including, but not limited to, the threats of terrorism, asteroids, and the global spread of disease; but global warming has pervasive, long-term, and in many cases irreversible environmental effects which will seriously disrupt our ordinary way of life and our wellbeing. If the predicted frequency of extreme weather events such as drought, flooding, hurricanes, and severe storms continues to increase as predicted by IPCC (2012), then a shift in public culture will focus attention to the environment itself and, in doing so, we predict that society will demand an integrated and ecological view of environmental problems. This recognition will galvanize a new awareness of the ecosystems affected by such warming and, in turn, make us more viscerally aware of our global ecosystem—in short a new public culture will emerge. This shift in public awareness and new expectations from science, law, and policy institutions will lead to the increased formation and strengthening of ecosystem regimes to prevent further global warming increases and mitigate and adapt to the effects of previous global warming. But these regimes will extend beyond global warming to govern the other detrimental effects of an "autonomous nature" (Heyd, 2007).

The Public Culture for Ecosystems in the Twenty-first Century

In *Law and Ecology*, we projected an optimistic history of the emergence of ecosystem regimes during the second half of the twentieth century. However, we cautioned that such an emergence depended heavily upon the emergence of a supportive public culture. In the first decade of the new millennium, there are sobering developments. Two ecologically destructive wars took place, in Iraq and Afghanistan, and war clouds are hanging over the Middle East. The rapid growth of China has taken place at the expense of her own environment and massive contributions to global warming. An economic crisis in Europe and the US has resulted in a retreat of public concern about the environment in general, and climate change, in particular. The retreat is marked by the relative public indifference to the BP Deepwater Horizon Gulf Oil Spill in April 2010 and to the increasingly ominous signs of the impacts of global warming. These ongoing symptoms of "environmental crisis fatigue" are echoed in many of the controversial arguments of Nordhaus and Shellenberger who, in *The Death of Environmentalism* (2004), argued that our current "insecure affluence" would not lead to values supportive of new environmental measures.

If the history of the early twenty-first century simply provoked a lessening of environmental protection enthusiasm, such events could be marked up to the winds of political change. But the challenge is deeper. *There is now a profound skepticism towards the very science supporting environmentalism.* Part of this skepticism is due to recognition of the complexity of the ecological sciences; the topic of complexity, both in general and as applied to natural systems and environmental law has received much attention in the last 10 years. But this skepticism about the applications of ecology is also reinforced by the admittedly limited accomplishments of those major place-based ecosystem regimes relying heavily upon science to promote restoration in systems such as the Florida Everglades, the Chesapeake Bay, and the California Bay Delta. In addition, the focus of the public skepticism is directed at the scientific models supporting predictions of global warming and its effects, despite the overwhelming number of scientists who accept the findings of global warming science (Anderegg et al., 2010). Part of the public resistance to global warming science is its limited ability to establish linkages between the science and particular weather events. Since the scientific support for evidences of global warming is firm, while the public resists such evidence, it is probable that only when the more severe effects of global warming are felt will the public culture turn to embrace a recognition of global warming. Similarly, only when the detrimental effects of the functioning of large-scale ecosystems are experienced will there be support for such ecosystem regimes to control those effects. Whatever the ultimate fate of large-scale place-based ecosystem management regimes, the influence of ecology upon environmental law and policy continues (although not in a dramatic way), and new developments have taken place in the reformulation of our understanding of law, our science of ecology, and the establishments of linkages between law and ecology.

The Reformulation of Environmental Law

In *Law and Ecology*, we listed the different ways in which ecosystems might be "framed" in law and traced the history of the National Environmental Policy Act and other laws as examples of legal adjustments to the insights of ecology. In Janet Milne's *Mountain Resorts: Ecology and the Law*, she carefully identified the vegetation, wildlife, and water impacts of resort development. More importantly, she structured a comparative examination of the regulation of four separate mountain ecosystems including a Canadian system. In addition, her analysis did not limit itself to obvious ecosystem law, such as the contributions of NEPA to structuring an ecosystem regime but she also identified how other federal (forestry), state (mini-impact laws), and provincial environmental and land use planning laws served to protect montane ecological functions and services. In her final chapter, she and her colleagues set forth a possible ecosystem model for mountain ecosystems. Since these volumes were written, other legal scholars have both praised and criticized NEPA and other laws as devices for protecting ecosystems. In this volume, Mary Jane Angelo has documented and critiqued their current role pertaining to the effective governance of pesticides.

The second major reformulation of environmental law to accommodate ecological science was the adoption of ecosystem regimes, whether in the form of coastal, river, forest, mountain, and lake management programs. As described in *Law and Ecology*, the law involved not merely statutes and regulations, but a complex mix of these along with plans, memorandums of understanding, guidance documents, and other "bridge documents," all of which made up an ecosystem regime. Since we wrote, scholarship has caught up with action and new accounts of this law are being set forth (e.g., Karkkainen, 2002), Professor Angelo herself has made significant contributions to this field (Angelo, 2006).

One of the reasons for seeking new appropriate legal vehicles is the apparent change in ecological theories and conclusions over time. Thus, for example, in past theories, the "climax" community was an acceptable concept; now it is either largely rejected or heavily modified to incorporate newer concepts of succession and resilience (Flournoy, 2009). Recent biological theory has questioned the notion of the "species" and hence, the laws which depend upon that concept, e.g., the Endangered Species Act (ESA), have become contested (Frankham et al., 2012). Similarly, the concept of "biodiversity" has been better defined and measured (Maguran and McGill, 2011; McGill, 2008), enabling the refinement of biodiversity protections (Parenteau, 2012).

Perhaps the most controversial example of changes in ecology is the adoption by some ecologists of a non-equilibrium theory of ecology (Botkin, 1990). This adoption raises questions about conventional environmental laws which assume a natural equilibrium in order to establish ecological baselines for management targets; consequently, some legal scholars (Profeta, 1996) argue for more adaptive and flexible legal regimes *in light of the uncertainty created by a non-equilibrium environment.*

In addition to the adjustment of existing laws to use ecological knowledge and accommodate ecosystem management, some legal scholars have gone beyond to suggest that the entire notion of environmental law must be reformulated. Eric Orts (1995) suggests that environmental law be viewed exemplifying "reflexive law" in which the limits of direct legal regulation are recognized, and more informal modes of regulation through the use of other institutions, such as modified markets or self-auditing, be adopted. Dan Tarlock (2004) offers similar insights by reviewing the three jurisprudential bases of environmental law—positivism, sociological jurisprudence and "revolutionary law,"—all of which he finds to be inadequate. Instead, he proposes to view environmental law as the securing of bounded decision processes. In another approach, Daniel Farber in *Eco-Pragmatism* (1991) and Brian Norton in *Sustainability* (2002), have suggested a pragmatic rationale for environmental regulation. Without going into the details of these and other approaches to reformulating environmental law, it is important to recognize that such legal scholarship is continuing to ask foundational questions about the nature of the law itself in order to understand how ecological knowledge and law might be usefully understood.

Rather than adopt a jurisprudence or philosophy of law and ecology, some recent authors have concentrated upon political theory, which describes and prescribes the ideal kind of society and institutions necessary to guide an ecologically oriented law. Thus, in *Law and Ecology*, we called attention to the work of John Dryzek (1987) who outlined how a democratic regime might be shaped in light of ecological concerns. More recently, other authors (e.g., Speth, *supra*; Guth, 2008; Kysar, *supra*) have sought to outline a broader political ideal which accommodates ecological insights and regulations, but also envisages large scale reforms of our present polity. Although such approaches may seem idealistic, they serve the purpose of supplying a longer-term road map and highlighting the radical conditions which might be needed to meet the looming challenge of global warming. In doing so, they provide visions of the ideal societal conditions to support ecological management in light of this crisis. Hopefully, they will not detract from more prosaic incremental environmental measures for fixing our malfunctioning ecosystems (Johnson, 1978).

Changes in Ecology since 2002

The disciplines of population, community, and ecosystem ecology have experienced incremental advances since 2002, due to improvements in instrumentation, computing power, and more human capital invested in the environmental sciences seen in the growth of university programs in sustainability science, environmental health, and global change (NAS, 2011). In 2002, in *Law and Ecology*, we identified four areas of ecology that had gained special importance to the environmental law through funding priorities established by environmental management agencies and basic funding from agencies such as NSF, NASA, and NIH. These areas were biogeochemistry of entire ecosystems, conservation biology, restoration ecology, and the interplay of ecology and the social sciences (ecological economics, urban ecology etc.).

The first area, the interdisciplinary study of the earth's biogeochemical cycles, has rapidly advanced from new technology in remote sensing, spatial analysis, and mapping, and from improved computer simulation of climate and terrestrial, and marine productivity. Since *Law and Ecology*, the fourth Assessment Report of the Intergovernmental Panel on Climate Change (IPCC) was published, in 2007, and elements of the fifth report are in draft form. The climate science is clear: the IPCC concluded that warming of the earth's climate is unequivocal, and that the probability that most of this warming is due to anthropogenic release of greenhouse gases is greater than 90 percent (IPCC, 2007). Despite the strong scientific consensus, agreement at the international level stalled over issues such as setting of emission caps acceptable to developed and developing countries, and general uncertainty over the economic consequences of controls on carbon emission and mitigation strategies. These issues prevented collective action at the UN Copenhagen Climate Conference (COP-15) in December 2009. Despite greater scientific certainty about the causes and consequences of global climate change, the international climate regime has made minor progress since *Law and Ecology* was published. However, programs operating at national, state and regional scales have proliferated (Selin and VanDeveer, 2011). US states have entered into regional cooperation, and in some cases cooperation with Canadian provenances, to develop a capacity to share energy policies and align as much as possible, capture economic efficiency gains in a shared regional marketplace, and exercise environmental leadership to their constituents. In 2012 the US EPA released the first comprehensive greenhouse gas emissions data on a state-by-state by sector and facility basis using 2010 information under its Greenhouse Gas Reporting Program. This data is essential to any future comprehensive US greenhouse gas policy and/or a US commitment to a binding international agreement on greenhouse gases. Climate change remains an area of international law where progress in developing the best available science has not clearly driven the agreement process forward.

For the other three areas we listed above, ecosystem-based management remains the central paradigm animating much research. There has been steady pressure for this management effort to deal with issues of climate change and the newer concepts of managing for resilience and sustainability. For example, Craig and Ruhl (2010) review the pressing environmental problems facing coastal and estuarine ecosystems. These ecosystems hold unusual value in ecological currency of biodiversity, productivity, and nutrient cycling, and direct economic value from fisheries and high land values at the coastal margin. These authors propose that sustainable governance of the coasts is required and argue that long-held principles such as the polluter pays principle, the use of best available science, the application of the precautionary principle for setting burdens of proof, may be recast into sustainable governance systems, equivalent to the regimes discussed in *Law and Ecology* (2002). Each of these principles operate as a driver for the proper adoption of ecological research.

Moreover, these management systems employ a place-based management combined with resilience science and adaptive techniques, which require the proper sequences of ecological research. Important to this management process is the resolution of ecosystem scale issues that link integrated coastal zone management to watershed management (both of which may have different scales), along with the use of economic incentives such as market-based water pollution trading to sustain the ecosystem services of water quality and biodiversity in the coastal zone. Another critical element in ecological science is the recognition that, under climate change and rising sea level management, the goals of ecological management themselves must be flexible and adaptable in light of the changing science.

Underlying the perceived need for flexibility, and hence an adaptive management system, lies the limitation of complete knowledge of the ecosystem and its components. This lack of knowledge was anticipated by systems theorists (Braybrooke and Lindblom, 1963) and environmental law scholars (Tarlock, 2004).

Such a lack of knowledge, in turn creates problems both for the adoption of environmental legislation and the courts (Bosselman, 2002). These problems of adequate knowledge are illustrated by recent court decisions (*In Re* Polar Bear Endangered Species Act Listing and Section 4 (d) Rule Litigation 1993).

Securing the adequacy of ecological information for ecological management is not limited to natural science data. Since 2002, the interplay between ecology and the social sciences has also grown closer. One example is the development of ecological economics and new approaches for valuation of ecosystem services including participatory decision making and discourse-based valuation methods, which are incorporated into conservation biology as well as ecosystem management and protection plans (Zografos and Howarth, 2008). Overall, the steady gains in the bridging of economics with ecology and the continued growth of ecological economics as a discipline has created more information, new techniques, and new conceptual frameworks from which incremental changes in administrative environmental law can be derived (Kysar, *supra*).

Reconceiving Ecosystem Regimes

In *Law and Ecology*, we argued that there was a historical rise in the formation of ecological regimes—systems of environmental law which are based upon the conclusions of ecology which recognize and govern ecosystems. Since 2002, there has been an increase in the formation of such regimes throughout the world and within the United States. Equally important, there has been a systematic effort to think through the implications of an effort to manage an ecosystem commons. Two of the early leaders in the practical effort have been the Chesapeake Bay (Ernst, 2003) and Yellowstone Park programs (Pritchard, 1999), followed by the comprehensive effort to restore the Florida Everglades and the California Bay Delta. We are now blessed with careful studies of these efforts to employ environmental laws to manage these extensive ecosystems and there continues to be a process of ongoing ecological research pertaining to each of these ecosystem regimes.

The most systematic effort to apply ecological concepts of ecosystem management has been the Florida Comprehensive Everglades Restoration Program. This program sought to systematically apply an adaptive ecosystem approach to the restoration of the Everglades ecosystem (USGS, USACE, 2004). In tandem with such efforts, even more comprehensive and expanded notions of ecosystem management have been advanced under the ambitious banner of "panarchy," which seeks to unify attention to both ecological and social systems. Whether such an expansion of such comprehensive efforts will be successful and fruitful, for both science and environmental law, remains to be seen. Perhaps a note of caution is in order.

In *Law and Ecology*, we summarized the early history of the Everglades program. The program has been monitored by the National Research Council in the past 10 years. The fourth report (NRC, 2012), concluded that despite a myriad of projects on the periphery of the glades in the 12 years of the project, little progress had been made on restoring the hydrology of the historical Everglades ecosystem, and it endorsed a more focused Central Everglades Planning Project to restore water quality and quantity. *The experience of the Everglades project, as well as other programs of a large scope raises questions about the feasibility and desirability of such projects.* Such experience seems to echo the early failure of comprehensive ecological research of Van Dyne in the mid-century (described in *Law and Ecology*), and seems to confirm the critique of comprehensive approaches by Braybrooke and Lindblom (which is also described in *Law And Ecology*).

Further light on the feasibility of such large scale approaches has been shed by the work of Oran Young. At the time of *Law and Ecology*, Young was the principal founder of regime theory. In 2002, Young issued a thoughtful new book, *The Institutional Dimensions of Environmental Change: Fit, Interplay and Scale.* In this book, Young identifies two quite distinct approaches to regime theory—a collective action model and a social practice model—and he identifies three "cutting-edge themes" of fit, interplay, and scale. The question of fit is the question of the congruence of the institution designed to manage the system and the nature of the ecosystem itself. The interplay issue is the question of the relations between the management of an ecosystem with other vertical and horizontal arrangements. The issue of scale is the question as to what level of systems should be selected for study, comparison, and action. Young identifies the sources of these problems to be due to imperfect knowledge, institutional constraints, and "rent seeking." Young explores these issues in considerable detail, including an assessment of their relevance to design principles for ecosystem regimes. These issues are familiar to environmental lawyers, since they confront the issues of setting regulatory boundaries ("fit"), of defining the relationship among other environmental laws and programs within the same jurisdiction, and of setting the relationships of management among lower or higher jurisdictions ("interplay"). Lawyers also help to legislate and litigate to determine the appropriate reach of regulation ("scale"). Hence, the theoretical discussions of fit, interplay, and scale have direct relevance to the law.

A second major contributor to regime theory is the late Nobel Prize winner, Elinor Ostrom, who has advanced the theoretical study of "the governing of the commons." In the last decade she set forth the role of rules in the management of "common pool resources" in *Understanding Institutional Diversity* (2009), by describing and classifying different kinds of rules and exploring their role in the management of the commons, and elaborated upon the "design principles" for regimes to govern the commons. This classification of rules and their roles offers different ways of understanding environmental laws. It remains for ecologically oriented lawyers to employ the Ostrom insights to understand the workings of environmental law. Despite these theoretical linkages, including the insights and conclusions of Young and Ostrom, which are particularly influential in the scholarly consideration of the governance of discrete resources, e.g., water, wildlife, and lumber, their scholarship has yet to demonstrate its usefulness to a fine-grained application to the complex web of ecosystem relations. An outgrowth of the ecosystem regime approach is illustrated by the more "down-to-earth" notion of ecosystem management. At about the same time that *Law and Ecology* was published, John Nagle and J.B. Ruhl authored *The Law of Biodiversity and Ecosystem Management* (2002). This book set forth the efforts to legally adopt the notion of ecosystem management at the federal level, as well as identifying some of the obstacles to such an effort. The authors also examined a number of case studies of the management of forest, grassland, freshwater, coastal and marine, and "extreme" (both fragile (desert, coral and island) and human dominated (urban and suburban)) systems. Since this work, there has been extensive scholarship, both legal and non-legal about ecosystem management. One of the major changes in ecosystem management has been the development of adaptive management in which measures undertaken to govern ecosystems would be monitored, evaluated, and adjusted in light of changing conditions. As we pointed out in *Law and Ecology*, the recognition of a variety of difficulties in implementing comprehensive or "synoptic plans" leads by default to the less ambitious adoption of various forms of incremental planning and action. "Adaptive management" is another form of such incremental planning and implementation. Part of such adaptive management is a robust system of continuous evaluation (NRC, 2003).

One of the leaders in such evaluative efforts is the National Academy of Science and its National Research Council. Not only is the Council's evaluative research built into the Florida Everglades program, but the Council has also evaluated the use of science within that program (NRC, 2003; *see also* Halvorsen and Davis, 1996). Despite the commendable work of the NRC, there has been a vigorous critique of the NRC assessment process (Fein, 2011).

Within the ambit of the management of some specific ecosystems, such monitoring and evaluation has taken place. As a consequence, there has also been legal and ecological scholarship based upon that monitoring and evaluation (see for example, Richardson, 2008). The legal and environmental literature is festooned with evaluations of ecosystem management (Ernst, 2003).

The Dimensions of Sustainability and Ecosystem Services

Another recent dimension of the study of ecological regimes is the serious attention paid to the securing of sustainability of ecosystems. Sustainability can refer to the resilience and robustness of ecosystems, irrespective of any human intervention; it can also refer to the ensuring of a continuous supply of a given stock, e.g., fish, lumber, or a service within a system. But the contemporary meaning of sustainability extends to an intervention into nature in such a way that the overall natural capital of the ecosystem may be maintained for future generations. In *Law and Ecology*, we traced the origin of the notion of sustainability and the exemplification of its use in the US Forest Service Rules. In the second book in our series, Janet Milne's

Mountain Resorts: Ecology and the Law, a discussion of sustainability science as applied to mountain developments was set forth. More recently, there have been a variety of economic works exploring sustainability, as well as a journal devoted to the topic. There have also been extensive interdisciplinary explorations of the concept of sustainability, including Norton's classic work, *Sustainability*, which sought to explore the balance between economic and ecological concerns. Unlike many of the other ecological terms, "sustainable" and its various definitions are found in many statutes and regulations. As a consequence, there is much litigation regarding sustainability, some of which we reviewed in *Law and Ecology*. Similarly there are a variety of law books and a myriad of law journals exploring its ramifications.

A related concept to the sustainability of ecosystem regime theory and practice is the adoption of the notion of "ecosystem services" (mentioned above), as part of the developing field of ecological economics (NAS, 2004). Recent environmental texts have identified ecosystem services (Friedman et al., 2012), and economics works have sought to measure the economic value of such services (Daly and Farley, 2004). Texts that trace the implications of ecosystems services have been prepared, and the extent to which the existing law can accommodate ecosystem services has been explored (Ruhl, *supra*). Since the turn of the twenty-first century, two major statutes have incorporated direct reference to "ecosystem services" (Ocean Act of 2000; Ocean and Atmospheric Research and Development Program of 2011), and more than 140 administrative regulations bearing upon "ecosystem services" have been promulgated. With the advent of the use of "ecosystem services," there have been a variety of efforts to define their meaning and scope within existing and proposed laws and regulations (Doyle, 2011; Ruhl, 2010).

The Establishment of Linkages between Public Culture, Law, Ecology, and Ecosystem Regimes

In 2002, when *Law and Ecology* first appeared, we suggested several "conceptual frameworks" in which these two disciplines might be harnessed: (1) as part of a more comprehensive framework (e.g., "law and science"); (2) re-conceptualizing one or both disciplines (e.g., "evolutionary law"); (3) becoming part of an applied theory (e.g. ecosystem management); (4) settling within a third mediating framework (e.g. "ecosystem regimes"); (5) linking through the expansion of a third discipline (e.g., ecological economics); (6) turning to the adoption of new concepts (e.g., sustainability); and (7) viewing the effort of ecosystem management within an ethical framework to accommodate ecology and law (e.g., ecological integrity). All of these avenues have been pursued in practice or scholarship since 2002. In this preface, we have already identified some of these avenues of research, including reconceptualizing law (e.g., "reflexive law" above), relying upon an applied theory (e.g., environmental management), looking to the new discipline of ecological economics, or turning to ecological concepts such as sustainability. Other ways to link ecology to environmental law which have not been mentioned above are: (1) the development of "linkage persons"—persons who have a demonstrated knowledge of both ecology and law; (2) the formulation of "bridge documents"—legal documents with low visibility to the public, e.g., regulations and guidelines, which embrace both ecological content and legal implications; (3) the migrating of specific ecological concepts into environmental laws and scholarship; and (4) the adoption of an ecologically based ethical framework. First, with the spread of environmental studies programs with some environmental law teachers, and environmental law programs with increasing offerings of ecology, the educational ground for establishing a supply of such linkage persons is now established. The author of this book exemplifies a "linkage person," since Professor Angelo has benefitted from both environmental science and law training. Second, another way of linking the fields is through bridge documents; Professor Angelo herself sets forth a list of such documents pertaining to both the ecology and law of pesticides. Third, it is the migration of ecological concepts into the law that constitutes the greatest opportunity for linking ecology and law. In our *Law and Ecology*, we set forth a list of 50 important ecological concepts ranging from habitat and trophic level to habitat and materials cycling. *Professor Angelo has also included a specific list of ecological concepts either found within the law of pesticides or necessary to understand that law.* The migration of ecological concepts into the law is exemplified by the movement of the concept of ecosystem itself into recent federal legislation. Within the last couple of years alone, a myriad of laws have identified, defined, and employed the notion of ecosystem within their law; these include laws such as the Southwest Forest Health and Wildlife Protection Act, the Water Resources Development Act, the Water Supply, Reliability and Environmental Improvement Act, the Great Lakes-St. Lawrence River Basin Compact, and the Energy Independence Act and many more. There has also been the recent migration of many other ecological concepts. For example, a quick check of federal regulations found 214 regulations in the past five years that treated predator–prey relations seriously, and 117 regulations in the past 10 years that treated the concept of trophic levels. One of the editors of this series has traced the effort to employ the concept of "carrying capacity" in the land use planning of the Florida Keys (Brooks and Virginia, 2006). "Resilience" is also a commonly invoked concept (Flournoy, *supra*) in ecological management, especially in the efforts to restore ecosystems. The virtue of the slow and steady migration of ecological concepts into the workings of the law is that their migration does not require the erection of organizations. Instead, such migrations can take place through informal usage as well as the low visibility amendment of existing laws and regulations, i.e., "bridge documents."

A fourth and final way of linking the fields of ecology and law is through an ethical framework which draws upon ecology and is adopted within the environmental laws. One example is Laura Westra's proposal for an "ethical integrity principle" which would embrace and guide our actions in relationship to the environment (Westra, 1994). A critique of the applicability of this concept to the planning for national parks has been explored by Shaun Fluker (2010).

Conclusion

In the third volume of this *Ecology and Law in Modern Society* series, Professor Angelo's *The Law and Ecology of Pesticides and Pest Management* captures many of the themes of the first decade of the twenty-first century discussed above. She demonstrates the relevance of many ecological concepts—resilience, biodiversity, food webs—and others to an understanding of the role of "pests" in the environment. She draws upon a variety of ecologically related tools— conservation biology, ecosystem management, adaptive management, ecological risk assessment, ecosystem services payments—which have witnessed development in the past few years. She has also invoked ecological ethical considerations such as an ecological integrity baseline. These concepts and disciplines are not treated in the abstract, but in the context of the past development of agriculture in the United States, the growth of industrialized agriculture, and the public policy responses to that growth including the farm subsidy legislation. She has documented the ecological impacts of the use of pesticides, and both the history and present status of efforts to control the manufacture and use of pesticides. This context is necessary to an understanding and evaluation of the current legal response to pesticides, its conflicts with wildlife protection law, and its international dimensions. Finally, she sets forth the outline of an ecologically based pest management, and the legal and policy proposals necessary to establish such a management regime.

Professor Angelo's work, and the work of the past decade, is indeed impressive. Given the events of the past 10 years and the present mood of the nation, it remains necessary for there to be a significant change in the public culture that would support that work. As we have suggested above, perhaps the advent of the looming problems of global warming will provoke such a change. In the meantime, the next step is the education of scientists, lawyers, and others who will seek to make these changes in the public culture to better support the application of ecology, its concepts, methods, and conclusions, to public policy and the laws which implement that policy. That public policy must be viewed also in light of the more recent conclusions of environmental ethics which can hopefully cope with intergenerational problems, resolve the proper use of cost/benefit analysis, adopt strategies for scientific uncertainty, and deal with collective action problems (Gardiner, 2011).

References

Anderegg, W.R.L., Prall, J.W., Harold, J. and Schneider, S.H., "Expert Credibility in Climate Change" 107 Proceedings of the National Academy of Sciences (PNAS) 12107 (2010).

Angelo, M.J., "Embracing Uncertainty, Complexity, and Change: An Eco-Pragmatic Reinvention of a First Generation Environmental Law" 33 Ecology Law Quarterly 105 (2006).

Bosselman, F., "What Law Makers Can Learn From Large Scale Ecology" 17 JLUEL (2002).

Botkin, D., *Discordant Harmonies: A New Ecology for the Twenty-first Century* (1990).

Braybrooke, D. and Lindblom, C.E., *A Strategy of Decision: Policy Evaluation as a Social Process* (1963).

Brooks, R. and Virginia, R.A., "The Great Migration of Ecological Ideas in Law" (Discussion Draft, Society of Human Ecology, October (2006)).

Brooks, R.O., Virginia, R.A. and Jones, R., *Law and Ecology, The Rise of the Ecosystem Regime* (2002).

Craig, R.K. and Ruhl, J.B. "Governing for Sustainable Coasts: Complexity, Climate Change, and Coastal Ecosystem Protection" 2 Sustainability 1361 (2010).

Daly, H.E. and Farley, J., *Ecological Economics: Principles and Applications* (2004).

Doyle, M.W., "Evolving Law and Policy for Freshwater Ecosystem Service Markets" 36 William and Mary Law and Policy Review 153 (2011).

Dryzek, J., *Rational Ecology* (1987).

Ernst, H.B., *Chesapeake Bay Blues, Science, Politics and the Struggle to Save the Bay* (2003).

Farber, D.A., *Eco-Pragmatism* (1991).

Fein, I., "Reassessing the Role of the National Research Council: Peer Review, Political Tool, or Science Court?" 99 California Law Review 465 (2011).

Flournoy, A., "Protecting a Natural Resource Legacy while Promoting Resilience: Can it be Done?" 87 Nebraska Law Review 1008 (2009).

Fluker, S., "Ecological Integrity in Canada's National Parks: The False Promise of Law" 29 Windsor Rev. of Legal and Social Issues (2010).

Frankham, R. et al., "Implications of Different Species Concepts for Conserving Biodiversity" 153 Biological Conservation 25 (2012).

Friedman, A. et al., *Environmental Science: Foundations and Application* (2012).

Gardiner, S., *A Perfect Moral Storm: The Ethical Tragedy of Climate Change* (2011).

Guth, J., "Law for the Ecological Age" 9 Vermont Journal of Environmental Law 431 (2008).

Halvorsen, W. and Davis, G., *Science, Ecosystem and Management in Parks* (1996).

Heyd, T., *Encountering Nature: Toward an Environmental Culture* (2007).

IPCC, "Summary for Policymakers", in *Climate Change 2007: The Physical Science Basis. Contribution of Working Group I to the Fourth Assessment Report of the Intergovernmental Panel on Climate Change*, S. Solomon et al., eds (2007).

IPCC, "Summary for Policymakers", in *Managing the Risks of Extreme Events and Disasters to Advance Climate Change Adaptation. A Special Report of Working Groups I and II of the Intergovernmental Panel on Climate Change*, C.B. Field et al., eds (2012).

Johnson, W., *Muddling Toward Frugality* (1978).

Karkkainen, B.C., "Collaborative Ecosystem Governance: Scale, Complexity, and Dynamism" 21 Virginia Environmental Law Journal 189 (2002).

Kysar, D.A. "Law, Environment, and Vision" 97 Northwestern University Law Review 675 (2003).

Kysar, D.A., "Breakthrough: From the Death of Environmentalism to the Politics of Possibility by Ted Nordhaus and Micael Shellenberger" 121 Harvard Law Review 2041 (2008).

Maguran, A. and McGill, B., *Biological Diversity, Frontiers of Management and Assessment* (2011).

Milne, J., LeMense, J. and Virginia, R.A., *Mountain Resorts: Ecology and the Law* (2009).

Nagle, J.C. and Ruhl, J.B., *The Law of Biodiversity and Ecosystem Management* (2002).

NAS, *A Data-Based Assessment of Research-Doctorate Programs in the United States* (2011).

Nordhaus, T. and Shellenberger, M., *The Death of Environmentalism* (2004).

Norton, B.G. *Sustainability: A Philosophy of Adaptive Ecosystem Management* (2005).

NRC, *Adaptive Monitoring and Assessment for the Comprehensive Everglades Restoration Plan* (2003).

NRC, *Progress toward Restoring the Ecosystem* (2012).

NRC, *Science and the Greater Everglades Ecosystem Restoration* (2003).

NRC, NAS, *Valuing Ecosystem Services: Toward Better Environmental Decision-Making* (2004).

Orts, E., "Reflexive Environmental Law" 89 Northwestern University Law Review 1227 (1995).

Ostrom, E., *Understanding Institutional Diversity* (2009).

Parenteau, P., "Species and Ecosystem Impacts", Chapter 10 in M. Gerrard and K. Kuh, eds, *The Law of Adaptation to Climate Change* 307 (2012).

Posner, R., *Catastrophe, Risk and Response* (2004).

Pritchard, J.A. *Preserving Yellowstone's Natural Conditions: Science and the Perception of Nature* (1999).

Profeta, T.H., "Managing Without Balance, Environmental Regulation in Light of Ecological Advances" 7 Duke Envtl L. & Policy Forum 71 (1996).

Richardson, C., *The Everglades Experiments: Lessons for Ecosystem Restoration* (2008).

Ruhl, J.B., "Ecosystem Services and the Clean Water Act: Strategies for Fitting New Science Into Old Law" 40 Environmental Law 1381 (2010).

Selin, H. and VanDeveer, S.D., "Climate Change Regionalism in North America" 28(3) Review of Policy Research 295–304 (2011).

Speth, J.G., *The Bridge at the Edge of the World: Capitalism, the Environment and Crossing From Crisis to Sustainability* (2008).

Stanford Encyclopedia of Philosophy, "Species" (July 4, 2002, rev. January 27, 2010).

Tarlock, A.D., "Is there a there there in Environmental Law?" 19 Journal of Land Use and Environmental Law 213 (2004).

USGS, USACE, *First National Conference on Ecosystem Restoration: Sustainable Ecosystem Restoration Through Integration of Science, Planning and Policy* (2004).

Westra, L., *An Environmental Proposal for Ethics: The Principle of Integrity* (1994).

Young, O., *The Institutional Dimensions of Environmental Change: Fit, Interplay and Scale* (2002).

Zografos, C. and Howarth, R.B., eds, *Deliberative Ecological Economics* (2008).

Preface

This book represents a culmination of my work over the course of my career thus far. My professional interest in the relationship between pest ecology, pest management and the law governing pesticides began many years ago while studying biological sciences and entomology for my undergraduate and graduate degrees. Specifically, my graduate work focused on the ecology of insect pest species. However, my personal interest in these matters dates back to childhood. As a child growing up in the 1960s, I was fascinated by the newly emerging environmental movement. I am not sure how old I was when I first read a book by Rachel Carson, but I remember how her ideas, particularly with regard to the impact of pesticides on the environment, grabbed hold of me and became permanently fixed in my mind. During this same time, I was intrigued to hear the stories told by my mother, Mary L. Angelo, of her work in the pesticide research laboratory at Rohm and Haas, a major early pesticide manufacturer during World War II. These childhood experiences coupled with my love of nature led me to study biology and entomology.

Years later, after making the switch from entomology to law, issues of pests and pesticide ecology continued to haunt me. As a lawyer for the US Environmental Protection Agency (EPA), I worked closely with EPA's Office of Pesticide Programs on a variety of pesticide-related matters and was part of EPA's early involvement in development regulations for pesticidal genetically modified organisms. Throughout this time, I could not help but see the many shortcomings in our pesticide regulatory programs. I saw a great disconnect between the innovative research on integrated pest management and other approaches to ecologically based pest management that I had seen as a graduate student, and the type of synthetic chemical based pest control that continued to dominate our agricultural system. I continued to think that there must be a better way to use the ecological knowledge being developed in our research institutions to make smarter and more environmentally friendly decisions regarding pest management; and it seemed that our legal and policy choices should be able to do a better job helping to facilitate these changes. Yet, I was disheartened to learn how poor our laws and policies were at encouraging more ecologically based approaches. I also became acutely aware of the serious conflicts that continue to exist between pesticide law and wildlife protection law.

Later, as an attorney for the St. Johns River Water Management District in Florida, I saw firsthand how the legacy of past pesticide use could continue to cause ecological tragedy. Decades after the US banned DDT and its organochlorine relatives, residues of these pesticides in the soil of former farmlands caused a significant bird die-off, which included the deaths of endangered birds, during an environmental restoration effort. More recently, as a law professor, my scholarship has been dominated by research on how we can better integrate ecology into environmental law, particularly pesticide law.

This is the first book to explore the connection between the evolution of environmental law and policy relating to pesticides and development of our ecological understanding of pests and pest control. The interdisciplinary character of this book should make it of interest to lawyers, scientists, and environmental and agricultural professionals. As part of the *Ecology and Law in Modern Society* series, this book seeks to explore the desirability and feasibility of placing ecological science at the center of an understanding of pesticide law and policy. This book is the first to explore the ecological and legal landscapes related to pesticide use in an interdisciplinary fashion. Although concerns over the ecological impacts of pesticides such as DDT fueled the environmental movement of the late 1960s and early 1970s and led to the creation of the field of environmental law, the ecological impacts of pesticide use have been largely ignored by the law and by legal scholars for more than 30 years. Despite the substantial impacts of pesticides on the environment, most environmental law texts touch only briefly, and in a cursory manner, on pesticide issues. Similarly, ecological texts dealing with pesticide impacts largely ignore the role of the law in addressing these concerns. This book is the first to provide a serious treatment of the significance of pesticide issues in environmental law and the first to provide an ecological perspective on the legal issues.

As was set forth in the first book of this series, *Law and Ecology: The Rise of the Ecosystem Regime*,[1] four major factors have influenced the manner in which environmental effects, including those of pesticide use, are addressed by society: "(1) 'ecology,' i.e., the developing science of ecology; (2) 'public culture,' i.e., the emergence of a public culture which increasingly embraces an ethos of nature and sustainability; (3) 'public policy,' i.e., local, state, and federal policies, largely the result of the arousal of vigorous environmental movements and the emergence of 'ecosystem regimes' as central players in the environmental policy debate; and (4) 'environmental law,' i.e., environmental constitutional provisions, legislation, regulations, court rulings, and international agreements, laws and treaties."[2] Although each of these influences constitutes its own academic discipline, they are inextricably intertwined. Historically, each academic discipline, if it addresses the other influences at all, has only given scant treatment to them. This book strives to integrate all four influences in a more balanced and interdisciplinary manner.

Although this book addresses issues relating to a wide range of pests and pesticides, its focus is primarily on agricultural pests and pesticides. This is in large part because agricultural pesticide use accounts for approximately 78 percent

---

1 Richard O. Brooks, Ross Jones and Ross A. Virginia, Law and Ecology: The Rise of the Ecosystem Regime (2002).

2 *Id.* at xii.

of all pesticide use in the US, while industrial, commercial, and government uses only make up roughly 12 percent, and home and garden uses comprise only about 8 percent.[3] Agricultural pesticides and pest management also provide a useful lens through which to explore the complex relationships between pesticides and the impact of pesticides and pest management systems law and policy.

3 EPA Pesticide Industry Sales and Usage 2006/2007, at 12, *available at* http://www.epa.gov/op0001/07pestsales/usage2007_table3_4.htm (last visited May 26, 2011).

Acknowledgements

I would like to acknowledge and thank the many people who have contributed to this book in many different ways. Most importantly, perhaps, I would like to acknowledge the person who inspired my life-long interest in pesticide law and ecology, the late Rachel Carson, who fearlessly spoke out on environmental concerns years before it became popular to do so and whose work paved the way for many of the environmental laws that continue to help to protect human health and natural resources. I would like extend special thanks to Richard O. Brooks at the Vermont Law School and Ross A. Virginia at Dartmouth University for recognizing my work on pesticide law and suggesting that I undertake this project, and for providing insightful and helpful comments on an early draft of this book. I would like to extend my sincerest appreciation to the many outstanding research assistants, all former students at the University of Florida College of Law, who assisted me in researching various aspects of this book or assisted me in previous research projects that have led to the writing of this book. These research assistants include Brenda Appledorn, Ryan Feinberg, Meredith Fields, Seth Hennes, Jocelyn Ho, Drew Melville, Melissa Newmans, Joanna Reilly-Brown, Kevin Shuler, and Christina Storz.

I would also like to thank *Ecology Law Quarterly*, the *George Mason Law Review*, the *Nebraska Law Review*, the *Harvard Environmental Law Review*, and the *Wake Forest Law Review* for permitting me to use portions of articles I published previously in their journals in this book. Numerous colleagues and friends have helped me to formulate my ideas, including Alyson Flournoy, Michael Wolf, Christine Klein, Elizabeth Rowe, Dr. Norman Leppla, and Dr. Danielle Treadwell.

Finally, I would like to thank my parents, Mary L. Angelo and the late Joseph F. Angelo, and my husband, Reed Pedlow, for their unwavering support and encouragement that has allowed me to pursue my passions.

Introduction

Modern industrialized agriculture with its concomitant reliance on chemical pesticide inputs contributes to substantial harms to both human health and the environment. Harm to the environment is best understood through both ecological concepts and related management approaches. These ecological concepts and management tools can inform our understanding of the ecological risks that result from pesticide use and pest management, and can reveal the uncertainties associated with these risks to facilitate well-informed and reasoned policy choices. A variety of past legal and policy efforts to address the risks associated with pesticide use, both at the national and international level, have fallen short, at least in part due to their failure to incorporate ecological concepts and tools. Not only do most existing laws and policies fail to make appropriate use of ecological science, most tend to encourage pest management practices that are not ecologically sound. Consequently, this book argues that ecologically based pest management is warranted, and suggests a variety of legal and policy proposals to promote such an approach.

Ecological impacts from pesticides differ from other types of environmental harm in that pesticides, by definition, are intended to kill organisms in the environment, disrupt ecological systems, and reduce biodiversity.[1] In other words, pesticides are used to "tilt the balance in favor of crops and humans rather than to encourage the diversity of pests which compete for these crops."[2] Moreover, pesticides are intentionally released into the environment in large quantities on a continual basis. In other areas of environmental law, the offending substance is a byproduct or waste of manufacturing, energy production, or other some other process designed to produce a product or provide a service. Pesticide law, on the other hand, must directly address the "product"—i.e., the pesticide—which by design has ecologically harmful qualities. Therefore, while most environmental laws are designed to reduce the quantities of pollutants or hazardous substances released into the environment, pesticide law is inherently different. With pesticides, the substance that causes the harm is the commercial product itself. Eliminating or greatly reducing the quantity of a pesticide into the environment defeats the purpose of the pesticide—i.e., killing and disrupting

1 William Vorley & Dennis Keeney, *The Greening of Industry versus Greenwash: Introducing a Case Study*, *in* Bugs in the System: Redesigning the Pesticide Industry for Sustainable Agriculture 1, 2 (William Vorley & Dennis Kenney, eds, 1998).

2 *Id.* at 2.

living organisms. Accordingly, pesticide law has evolved in a very different way to other areas of environmental law.

Early pesticide laws were primarily consumer protection laws designed to protect consumers from false claims about products that did not work. Even later attempts to incorporate human health and environmental concerns were extremely limited due to the standards, scope, and mechanisms provided in these laws. Environmental regulators have struggled to find ways to use these limited pesticide laws to address the ecological risks posed by the ever-changing suite of pesticides in commercialization.

At the same time that policy makers were developing the first environmental laws and regulations, including pesticide laws and regulations, the science of ecology was becoming more sophisticated and more widespread in its use. "Ecology" seems to have played a prominent role in pest control approaches, as well as pesticide policy during the early years of the development of the US environmental regulatory system. For example, the seminal text on crop protection for more than a decade was a 1974 book entitled *Pest Control and its Ecology*. However, by the 1980s, ecology seems to have lost its previous place of honor in the lexicon of pest control and pesticide policy. As Professors van Emden and Service have pointed out, in 1989, the book was renamed by dropping the term "ecology" from the title to be simply *Pest Control*.[3] Although the naming of the later edition of the book is merely one small example, it does reflect the trend of "ecology" falling out of favor during the 1980s. A more substantive example of this shift can be seen by comparing early EPA pesticide regulatory actions to those that occurred later. Perhaps the most substantial role of ecology in an EPA pesticide regulatory decision was in EPA's first major action of cancelling the registration of the now infamous pesticide, DDT. Not only was the basis of the cancellation an ecological risk, such as bioaccumulation rather than human health concerns or "one species" concerns, in the 1972 final cancellation order then-Administrator, William Ruckelshaus, provided a relatively sophisticated ecological analysis of the environmental risks of DDT. After this decision, ecological concerns seemed to take a back burner in EPA's pesticide regulatory decisions.

Only fairly recently have new ecological understandings highlighted the fact that current environmental laws are wholly inadequate to address ecological impacts of pesticide use. Recent studies demonstrate that the actions taken in the 1970s and early 1980s to ban or restrict certain ecologically harmful pesticides, such as DDT and its relatives, only partially protect wildlife, including threatened and endangered species, or ecological systems from the harms of pesticide use. A 2004 Center for Biological Diversity Report concludes that 375 threatened or endangered species are currently at risk of extinction from pesticide use in the

---

3 H.F. van Emden & M.W. Service, Pest and Vector Control ix–xii (2004). *See* H.F. van Emden, Pest Control and its Ecology (1974, rev 1989 as Pest Control).

US.[4] Another study shows that millions of birds are killed by the legal use of pesticides annually.[5] Chemical pesticides also cause or contribute to countless other adverse impacts to many species and degrade ecosystem services.

Moreover, a 2006 study demonstrates that the impacts from pesticides extend to international economy. A recent study concludes that non-pest insects, which are frequently non-target victims of pesticide use, provide ecological services such as pest control, pollination, and grazing land clean up, amounting to more than $57 billion per year in the US alone.[6] A 2006 National Research Council Report concludes that populations of pollinators and other insects providing ecological services are in serious decline, due at least in part to pesticide use.[7]

The interactions of humans with their natural environment have created a seemingly perpetual cycle of the evolution of pests leading to the evolution of pesticides, resulting in ecological harms leading to the need for the evolution of environmental laws. As humans have altered the ecology, they have facilitated the emergence of new pests and the expansion of existing pest problems. These newly created pest problems create a need for new pest controls, which ultimately result in the need for new environmental regulations to address the risks posed by the controls.[8] The irony is that many of the pest controls that have been developed to address pest problems result in new or worsened pest problems, creating a need for new or more aggressive pest controls, which frequently carry with them new, or more insidious, environmental harms.

While the focus of this book is on agricultural pests, it must be recognized that there are significant issues with pests that transmit diseases or otherwise harm humans. Pesticides are used to mitigate against these pests, particularly in the developing world. For example, many insects such as mosquitoes carry serious diseases such as malaria, yellow fever, dengue fever, West Nile virus and equine encephalitis. Lice transmit typhus, tsetse flies carry sleeping sickness, and ticks are vectors of a number of diseases including Lyme disease. Many other examples exist of invertebrate disease vectors of both humans and domestic animals. Control measures including pesticide application are frequently employed to control the invertebrate vectors and thereby prevent the spread of these diseases.[9]

---

4 Brian Litmans & Jeff Miller, Silent Spring Revisited: Pesticide Use and Endangered Species, Center for Biological Diversity Report 19–44 (2004), *available at* http://www.biologicaldiversity.org/publications/papers/Silent_Spring_revisited.pdf (CBD Report).

5 David Pimentel et al., *Assessment of Environmental and Economic Impacts of Pesticide Use*, *in* The Pesticide Question: Environment, Economics and Ethics 47, 68–9 (David Pimentel & Hugh Lehman, eds, 1993).

6 John E. Losey & Mace Vaughan, *The Economic Value of Ecological Services Provided by Insects*, 56 BioScience 311, 312 (2006).

7 Committee on the Status of Pollinators in North America, National Research Council, Status of Pollinators in North America 75 (2006).

8 Van Emden & Service, *supra* note 3, at 52.

9 *Id.* at 1.

Although it is not the focus of this book, it is important to acknowledge that in addition to the ecological risks associated with pesticide use, pesticides pose significant risks to human health. The World Health Organization estimates that approximately three million humans are poisoned by chemical pesticides each year. Of these poisonings, approximately 220,000 result in death and 735,000 result in chronic illness.[10]

When the limited pest control abilities of chemical pesticides are considered in light of the undeniable substantial human health and environmental consequences of chemical pesticide use, it is not clear why society would choose to continue to rely so heavily on chemical pesticides. The reasons farmers continue to use chemical pesticides despite the problems associated with them are complicated; they include the fear of losing one's livelihood, risk aversion, encouragement from the chemical industry, government research and extension services, the need to meet cosmetic standards to accommodate market demand and government regulations, and a flawed agricultural subsidy system that encourages high-intensity, high-yield practices.

New ecological understandings of the complex nature of pest ecology, natural pest controls, and adverse effects of chemical pesticides suggest that there may be better ways to manage pests and protect human health and the environment at the same time. Concerns about the long-term sustainability of industrial agriculture and the environmental harms associated with it are leading to a reevaluation of our agricultural system, including the way we control pests. The new focus on eco-agriculture—i.e., treating the farm as a functioning ecosystem that performs many pest control functions naturally or with non-chemical human assistance—can provide a roadmap for shifting away from a predominately industrialized agriculture system toward a more sustainable system. The related concept of ecologically based pest management can provide the tools needed to reduce our dependence on chemical pesticides, thereby reducing the harms associated with them. However, despite the scientific basis for such a shift, our current environmental laws and agricultural policies are geared toward maintaining the status quo. Changes to our laws and policies will be necessary to move away from a chemical input-dependent agricultural system to an ecologically based one.

This book is structured as an organized exploration of the parallel development of pest control and pesticide law. After an introductory overview of the issues, the book begins with an overview of basic ecological concepts and ecologically related pest management tools, placing them in an agricultural pest management context. A detailed discussion of the ecology of pests is then followed by a discussion of the history of the agricultural revolution and farm policy, and a description of how they have influenced farm practices, including our current heavy reliance on chemical pesticide inputs. Next, the book provides an in-depth treatment of the ecological impacts of current pesticide use, as well

10 Report of the WHO Commission on Health and Environment. World Health Organization, Our Planet, Our Health (1992).

as the legacy of past pesticide uses. This is followed by a detailed discussion of current domestic and international laws that relate to pesticide use, and an analysis of the relationship between these laws and their failure to adequately incorporate ecological concepts. Finally, the book concludes with an analysis of how recent developments in ecological science can be used to inform the law and proposes a number of potential ways in which the law could be changed to respond to increased ecological understanding.

PART I
Ecological Science and Policy as Applied to Pests and Pesticides

Chapter 1
Ecological Science

Basic Ecological Concepts

Fly over almost any farm field today and you will see endless rows of crops aligned in giant monocultures, likes stripes on the landscape. Walk into that farm field and look more closely and you are likely to encounter many more living organisms: insects feeding on crops, weeds growing between crop rows, nematodes and other organisms living in the soils. Look even more closely, and you will see the multifaceted mosaic of living organisms with intricate relationships among them. Birds and predatory insects may be feeding on pest insects, parasitic and parasitoid insects may be laying their eggs in the bodies of pest caterpillars where their larvae may literally eat the caterpillars from the inside out, and a host of other species with complex interactions between the species and their environments may be present. In other words, the agricultural field, even the field that appears to be a mere monoculture, is a living ecosystem, governed by the principles of ecology. Humans trying to improve crop yields by reducing pest damage impact the agricultural ecosystem, as well as the surrounding natural or human-managed ecosystems, in an almost indefinite number of ways depending on the type of system, the pest management chemical or approach used, and a number of other factors. Pest management through chemical or other pest management approaches not only impacts the population of the pest species themselves, but also impacts populations of non-target species and disrupts complex interactions among the entire community of species and their abiotic environment.

An agricultural field, although managed by humans and often consisting of large-scale monoculture, is not a sterile laboratory. It is a complex assemblage of living organisms with relationships to each other and the abiotic community in which they exist. Accordingly, to understand the effects of pesticides and other pest management in an agricultural system, it is first necessary to have a basic understanding of ecology and ecological systems. Ecological principles apply to all living organisms, whether those organisms exist in a pristine natural environment or whether they exist as pests or beneficial organisms in a human-managed agricultural environment.[1]

To understand the role of ecology in environmental law addressing pesticide-related issues, it is necessary to have a basic understanding of some fundamental ecological concepts and how they relate to pests, pest management, and pesticides.

---

1 Robert F. Norris, Edward P. Caswell-Chen & Marcos Kogan, Concepts in Integrated Pest Management 66 (2003).

This chapter provides an overview of well-established ecological concepts that come into play in discussions and analysis in other portions of the book. Because this chapter is primarily descriptive of basic concepts that have appeared extensively in published works by ecology scholars, it draws heavily from a few such works, most significantly, Robert Norris, Edward Caswell-Chen and Marcos Kogan's 2003 book entitled *Concepts in Integrated Pest Management*,[2] and two other books in the *Ecology and Law in Modern Society* series.[3]

A good starting point for an attempt to understand the complexity, uncertainty, and changing nature of natural systems is an understanding of the numerous fields of inquiry and terminology used to describe these systems. "Ecology" is the science of the interactions of living organisms with each other and their biotic and abiotic environment.[4] The term "ecology" has been described in more detail as follows:

> Ecology is the study of the relationship of organisms and their environment. The environment includes other individuals in its population, other populations of plants and animals with which an organism and its population interacts, and the physical and chemical factors that influence life (i.e., the abiotic environment). 'Relationships' include such diverse subjects as the adaptation of an individual's physiology to environmental extremes (e.g., low moisture and high temperatures), the killing and eating of prey by predators, and the flow of carbon through the living and non-living components of the environment. To study such a diversity of natural phenomena, ecology takes an interdisciplinary approach, drawing from biology, chemistry, physics, and the earth sciences to achieve an understanding of how the natural world functions, and in so doing, sustains human life.[5]

In the first book in this series, Brooks et al., outline the evolution of ecology as a scientific discipline starting in the latter half of the nineteenth century. As Brooks et al. explain, by the 1920s, ecology "had become the detailed, empirical study of dynamic systems of interactions between the living components and environmental factors making up a local community."[6] The field of ecosystems ecology developed later and was the subdiscipline of ecology that played an influential role in the development of the major federal environmental laws that grew out of the environmental movement of the 1960s and 1970s.[7]

---

2 *See generally*, Norris et al., *supra* note 1.

3 Richard O. Brooks, Ross Jones and Ross A. Virginia, Law and Ecology: The Rise of the Ecosystem Regime (2002) and Janet E. Milne, Julia LeMense & Ross A. Virginia, eds, Mountain Resorts: Ecology and the Law (2009).

4 *See* Brooks et al., *supra* note 3, at 7.

5 *Id.*

6 *Id.* at 69.

7 *Id.* at 66–72.

Ecosystem Organization

One of the organizing principles of ecology is levels of ecosystem organization. Norris et al. present the levels of organizations within the context of pest management as follows:

> *Individual*: a single organism such as a bacterium, a weed, a nematode, or a lygus bug.
>
> *Population*: all individuals of a species within a defined area usually isolated to some degree from other similar groups, such as weeds or boll weevils on an acre of cotton field.
>
> *Community*: the assemblage of all species occurring together within a limited geographic area. Within an agricultural field, the community would include the crop plants, weeds, pathogens, bacteria, fungi, nematodes, insects, and vertebrates.
>
> *Ecosystem*: a community of organisms together with the abiotic environment in a specified region …. In the pest management context, ecosystems can be thought of as the crop fields and the surrounding ditches, hedges, woodlands, and so forth, within a defined area such as a valley or a watershed. Agricultural ecosystems are termed agro-ecosystems.[8]

An important concept with significance to pest management is "succession," which is "the chronological sequence of successive species in a biotic community …."[9] The classic example of succession is the sequence of life that arises after a forest fire: first the quick-growing weedy species move in, followed by shrubs, and finally by slow-growing trees. Succession is significant in agriculture and pest management because, as Norris et al. point out, "[a]griculture represents human attempts to arrest succession … so that the net increase in biomass (energy) can be harvested for human benefit. Arresting succession requires human inputs to the system, and the greater the input, the earlier in the successional stage the system can be arrested and the greater the return that humans can harvest."[10] In particular, pest management disrupts succession by attempting to eliminate weeds, the primary stage of succession.[11]

---

8 Norris et al., *supra* note 1. *See also* Richard B. Primack, Essentials of Conservation Biology 41–3 (1993).

9 Norris et al., *supra* note 1, at 68.

10 *Id.* at 68–9.

11 *Id.* at 70–71.

Ecological Perspective	Agricultural Perspective
Ecosystem	Agro-ecosystem *e.g., Colorado potato beetles, potato crop, beneficial insects, weeds, other organisms in and around a field of potatoes, and the abiotic environment*
Population	Pest Population *e.g., Population of Colorado potato beetles in a field of potatoes*
Organism	Individual Pest *e.g., Individual Colorado potato beetle*

Figure 1.1 Ecosystem organization from ecological and agricultural perspectives

Another ecological concept important in pest management is that of trophic dynamics. Trophic relationships in an ecosystem are often referred to in common parlance as the "food chain." A simplistic view of trophic dynamics is a linear "bottom-up" system of energy flow coupled with a top-down system of consumption of organisms. As illustrated in Figure 1.2, at the bottom of the food chain are the inputs of energy via sunlight, water, CO_2 and nutrients. The next level up, or the next "trophic level", consists of plant producers who use these inputs to photosynthesize and make plant biomass.[12] At each level some energy is lost and generally there are higher numbers of smaller individuals at the lower levels, and lower numbers of larger individuals at the higher trophic levels.[13] The next trophic layer up consists of the herbivore primary consumers who feed on the plant producer, followed by a top trophic level of primary carnivores, or secondary consumers, who feed on the herbivore below.[14]

In an example of trophic levels in a farm field, the crop plant would be the producer, an insect pest of the plant would be the herbivore/primary consumer, and a predatory insect that consumes the pest insect would be the primary carnivore or secondary consumer.

12 *Id.*; *see also*, Primack, *supra* note 8.
13 Norris et al., *supra* note 1.
14 *Id.*

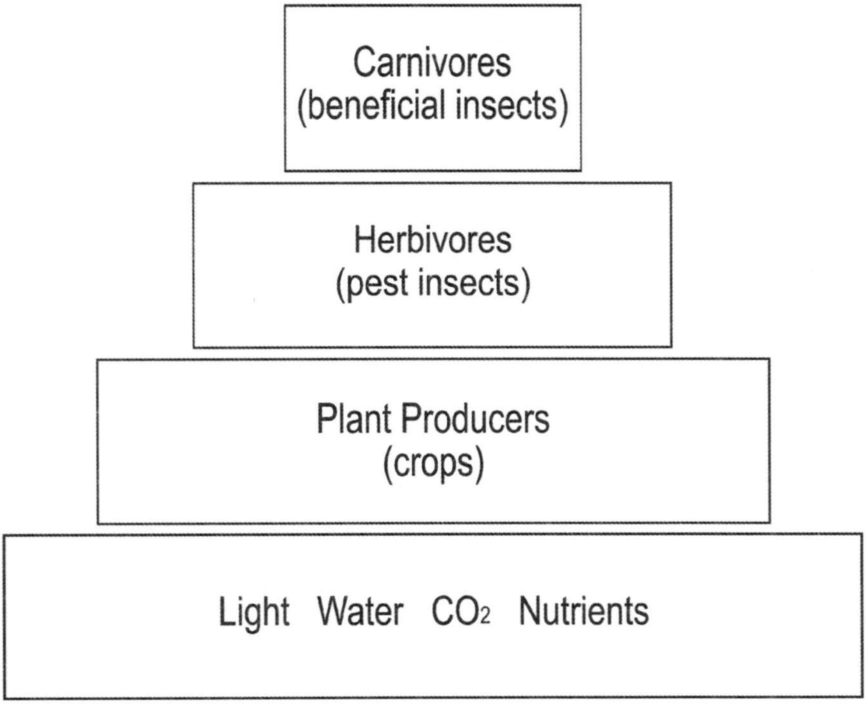

Figure 1.2 Diagram of simple agricultural food chain trophic levels

The organisms at the higher trophic levels consume organisms at lower levels, thereby controlling the population of the organisms in the lower levels. If high-trophic-level organisms, such as species that prey on pest insects, are removed from the system, populations of organisms at the next trophic level down, in this example, the pest insects, will increase. An increase in pest population will in turn result in more crop damage.[15] Thus, from a pest management standpoint, high-trophic-level predators and parasites are extremely important in managing population levels of lower trophic-level pests and therefore keeping crop damage to a minimum.[16] As will be discussed in more detail in later chapters, one significant negative side effect of using chemical pesticides that kill a wide range of organisms (or broad-spectrum pesticides), is that they kill not only the target pest, but also the higher trophic-level carnivores and parasites, thereby potentially resulting in increased pest populations. Figure 1.3 illustrates how predator–pest ratios can be altered by pesticide use.

15 *Id.*
16 *Id.*

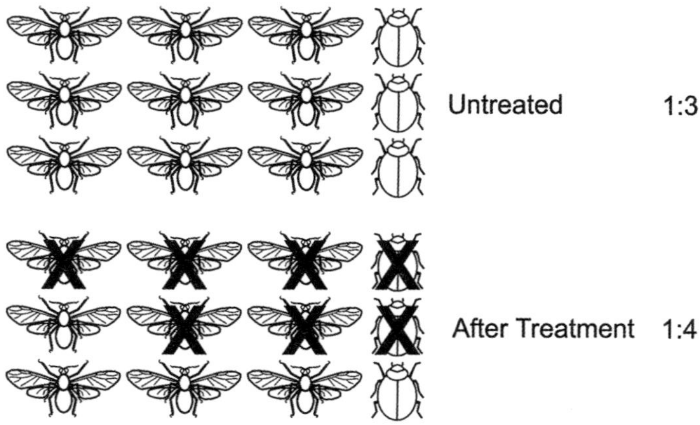

Figure 1.3 Predator–pest ratios after pesticide treatment

When pesticide use kills the predators and parasites that naturally keep pest populations in check, there is often a resurgence of pest populations. As illustrated in Figure 1.4 (below), such resurgence frequently results in a greater pest problem than existed prior to the pesticide treatment.

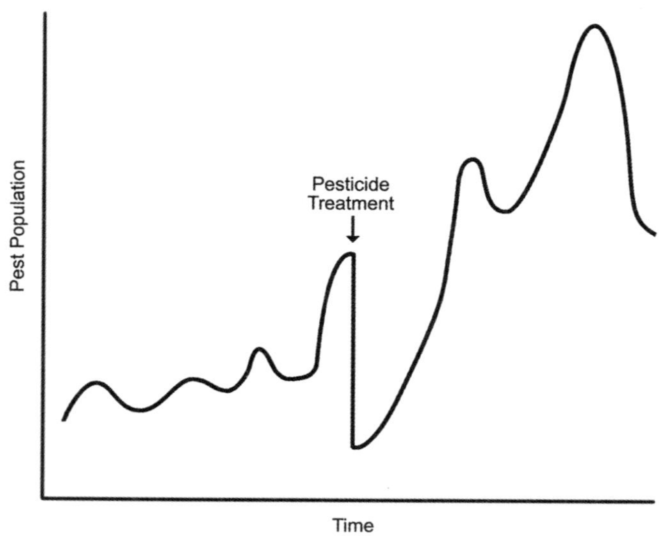

Figure 1.4 Pest resurgence following pesticide treatment that kills the beneficial predators and parasites that previously kept the pest population under control

Another significant concept involving trophic levels and pesticides, illustrated in Figure 1.5, is that of bioaccumulation, wherein certain types of pesticides found in low concentrations in the bodies of organisms at low trophic levels will accumulate increasingly as organisms in each higher trophic level feed on organisms at the next level down, until the bodies of the highest trophic level predators or parasites contain dramatically greater concentrations of the pesticide.[17] These high concentrations of the pesticide may be extremely harmful to the top predators and parasites, whereas the low concentrations in organisms at lower trophic levels may be low enough that they have little or no harmful effect.

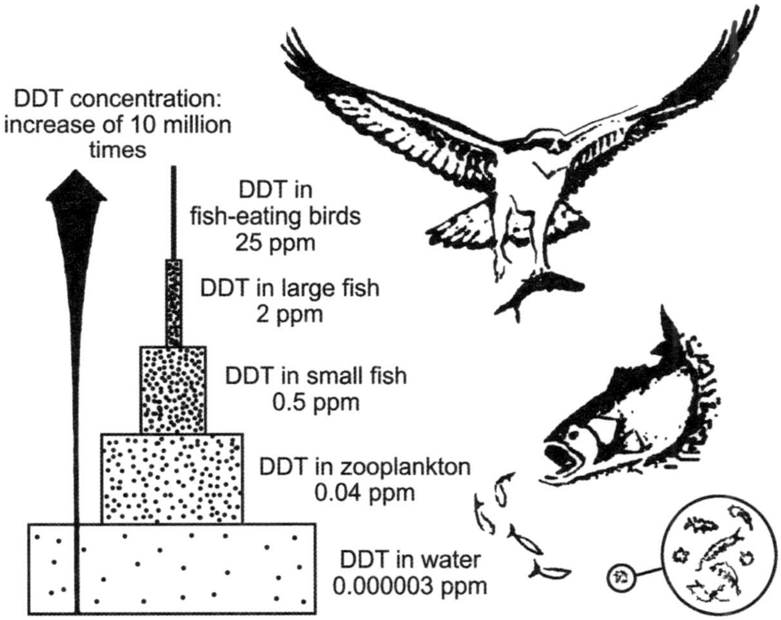

Figure 1.5 Illustration showing phenomenon of bioconcentration
Source: US Fish and Wildlife Service.

As described above, predators at higher trophic levels in a food web contribute to controlling pest population levels. Several other factors constrain population growth. The carrying capacity of an ecosystem is the maximum population size that can be supported by the ecosystem. A number of factors including resource

17 For a comprehensive discussion of the phenomenon of bioaccumulation of DDT in wildlife, *see* Bill Freedman, Environmental Ecology: The Impacts of Pollution and Other Stresses on Ecosystem Structure and Function 191–8 (1989).

limitations, competition—"the simultaneous demand by two or more organisms for a common resource"—for resources, predation and parasitism determine the carrying capacity of any particular system. Resource limitation and competition for resources are substantial determinants of carrying capacity in natural ecosystems. In agricultural systems, however, if the resource is the crop plant, which is intentionally grown at high density and high growth rate to obtain maximum yield, there can seemingly be an unlimited supply of food resource for pest species. A large, dense, highly productive monoculture of corn, for example, can be seen as an "all you can eat" buffet for pests of corn. Typically, herbivorous pest populations are not limited by inadequate food supply. Of course, at some point, the density of the pest species may increase to such a great extent that there is significant crop damage and therefore, there will be greater competition for resources. Weeds, on the other hand, are producer pests, which must compete with crops and other plant species for ecosystem resources such as sunlight, water, and nutrients.

Another important ecological concept related to pest population and pest management is the different life-history strategies employed by different types of organisms. Ecologists categorize species by placing them on a continuum which at one end includes small, rapidly reproducing species referred to as "r-selected," and at the other end "K-selected" species, which are larger, more slowly reproducing organisms. R-selected organisms are small-bodied, produce large numbers of offspring, have short generation times, and are short-lived. This strategy enables these species to rapidly use resources, adapt to changing environments, and quickly colonize new environments. Most agricultural pest species, including many pest insects and weeds, are considered to be r-selected. At the other end of the spectrum are K-selected organisms, which tend to be large-bodied, have longer life spans, have longer generation times, and produce fewer offspring. These species tend to be very competitive in stable environments. Most mammals are K-selected. Whether a pest is r-selected or K-selected is a significant factor in determining effective and appropriate pest management strategies.

Biodiversity

The field of ecology has existed for many decades, but only in the last 30 years or so has ecological study advanced to the point where we are beginning to understand the importance of biodiversity to natural systems. "Biodiversity" has been defined as "the richness, abundance, and variability of plant and animal species and communities and the ecological processes that link them with one another and with soil, air, and water."[18] The measure of biodiversity is not merely an accounting of the number of species present in an area; it also has a functional

18 Malcolm L. Hunter, Jr., Fundamentals of Conservation Biology 20–21 (1996) (providing Wildlife Society 1993 definition of biodiversity).

component.[19] Accordingly, a focus on preserving individual species from extinction is not sufficient to preserve biodiversity unless ecological processes are also maintained.[20] Thus, an overriding goal of biodiversity protection is to maintain all of the elements of biodiversity, including both structural diversity (i.e., forms and levels of organization) and functional biodiversity (i.e., ecological and evolutionary functions).[21]

While biodiversity certainly is an important indicator of ecosystem health, ecosystem health also relies on the integrity of the physical and chemical components of such systems, as well as the relationship between species and the physical world. Of course, one measure of ecological integrity is the presence of a diversity of species. Nevertheless, not all species within an ecosystem provide the same level of value to the other members of the ecosystem. Conservation biologists have identified certain types of species that provide greater value to ecosystems, and accordingly should be afforded greater protection. One such type of species, known as the "controller species," plays a major role in controlling the movement of energy and nutrients within an ecosystem. Controller species typically include decomposer species.[22] Another type of species of special importance is known as the "keystone species." These species provide more value to the ecosystem than would be predicted by their abundance in the ecosystem.[23] One commonly cited example of a keystone species is the red-cockaded woodpecker, which excavates cavities in living trees, providing a habitat for a variety of species who live in tree cavities, but who are not capable of excavating their own cavities.[24] Some conservation biologists believe that the extinction of a keystone species has the potential to result in a domino effect, whereby numerous species go extinct and the entire ecosystem is drastically altered.[25] "Umbrella species" are species that typically have very large home territories, such that protecting a keystones species habitat will have the effect of protecting many other species as well.[26] Finally, species that are very sensitive to pollutants and other environmental stresses, such that the health of these populations serves as a good indicator of overall ecological health, are known as "indicator species."[27] The pelican's sensitivity to DDT made it a good indicator species for organochlorine pesticide contamination. Of course, while certain types of species such as controller species and keystone species may warrant greater protection, due to our extremely limited understanding of the

---

19 *Id.* at 21.
20 *Id.*
21 *Id.* at 21, 28.
22 *Id.* at 52.
23 PRIMACK, *supra* note 8, at 43–8.
24 *Id.*
25 *Id.* at 74.
26 *Id.* at 54.
27 *Id.* at 55.

workings of ecological systems, it would be imprudent not to assume that every component of an ecosystem has some value unless proven otherwise.[28]

Both individual species and ecosystems exhibit a wide range of resilience to man-induced changes.[29] Because of the dramatic differences between species and between ecosystems, and because so little is understood about the ability of a particular species or particular ecosystem to tolerate or bounce back after damage from human activities, it is very difficult to determine whether a particular change will result in significant habitat destruction.[30] While it may be a gross overstatement to say that all ecosystems are so delicate and intricate that removing one part will necessarily cause the demise of the entire system, it is not an overstatement that removing one species from an ecosystem can negatively impact other species and, in the words of Princeton conservation biologist David Wilcove, "there is no way of knowing ahead of time where the chain reaction will end."[31]

Ecological Resilience

Ecological resilience has been described as "a measure of the amount of change or disruption that is required to transform a system from being maintained by one set of mutually reinforcing processes and structures to a different set of processes and structures."[32] The concept of ecological resilience is based on the understanding that ecosystems can exist in multiple stable states.[33] Ecological resilience should not be confused with "engineering resilience," which is a measure of the time it takes for a system to return to a steady state after experiencing a perturbation. In contrast, the concept of ecological resilience focuses on conditions that exist far from a steady state, where perturbations can shift a system into a different state, or in other words, into another "regime of behavior." As Holling and Gunderson explain, the fundamental flaw in using engineering resilience as a metric with regard to natural ecosystems is that it "reinforces the dangerous myth that the variability of natural systems can be effectively controlled, that the consequences are predictable, and that sustained maximum production is an attainable and

---

28 David S. Wilcove, The Condor's Shadow: The Loss and Recovery of Wildlife in America 12 (1999).

29 *Id.* at 6.

30 *Id.* at 9.

31 *Id.* at 11–12.

32 Garry Peterson, Contagious Disturbance and Ecological Resilience 216, PhD dissertation, University of Florida (1999) (contrasting ecological resilience with engineering resilience, which is defined as "the rate at which a system returns to a single steady or cyclic state following a perturbation").

33 *Id.* at 217.

sustainable goal."[34] Ecological resilience, a measure of the magnitude of a perturbation that a system can absorb before the disturbance causes the system to shift into a different regime of behavior with different controlling processes, provides a more realistic view of natural systems.[35] As such, ecological resilience captures the strength of redundancies in the system stemming from reinforcing processes and compensating functions provided by more than one species. These redundancies endow the system with an ability to absorb disturbances and persist despite the disruption.[36] When viewed in the context of environmental management, restoration, or regulatory decision making, ecological resilience is a measure of a system's ability to withstand failed management or regulatory decisions.[37]

> [t]he concept of resilience, in which the different distinct modes of behavior are maintained because of, rather than despite, variability, is suggested as an overall criterion for policy design. The more that variability in partially known systems is retained, the more likely it is that both the natural and management parts of the system will be responsive to the unexpected. The very process and techniques we recommend, while aimed in part at reducing uncertainty, are designed as a changing adaptive process of policy design.[38]

Historically, environmental managers and regulators have operated on the outdated notion that the goal of environmental protection is to achieve or maintain a "balance of nature," in which an ecosystem exists in one static "steady state." New understandings of ecosystems suggest that ecosystems are not static because they are continually subject to a variety of both natural and human perturbations, which cause a range of changes. Most perturbations are within a range of type and magnitude such that, while certain changes may occur within the ecosystem, it does not "flip" into a new and different state. Certain changes, however, are either of a type or of sufficient magnitude that they can cause the ecosystem to change to an entirely different state. In many cases, the ecosystem may be able to revert to the prior state once the perturbation is removed, or if certain previous conditions are reintroduced to the system. If the perturbation is extreme enough, or if it is maintained for sufficiently long duration, however, the new state may become permanent, and it may be virtually impossible for the ecosystem to revert to the

---

34 C.S. Holling & Lance H. Gunderson, *Resilience and Adaptive Cycles*, *in* PANARCHY: UNDERSTANDING TRANSORMATIONS IN HUMAN AND NATURAL SYSTEMS 28 (Lance H. Gunderson & C.S. Holling, eds, 2002).

35 Lance H. Gunderson, C.S. Holling, Lowell Pritchard, Jr. & Garry Peterson, *Resilience of Large-Scale Resource Systems*, *in* RESILIENCE AND THE BEHAVIOR OF LARGE-SCALE SYSTEMS 4 (Lance H. Gunderson & Lowell Pritchard, Jr., eds, 2002).

36 *Id.* at 6.

37 *Id.* at 6–7.

38 C.S. HOLLING et al., ADAPTIVE ENVIRONMENTAL ASSESSMENT AND MANAGEMENT 19–20 (1978).

original state absent some extreme circumstances. This new understanding of natural systems makes it evident that the objective of environmental managers and regulators should not be to achieve and maintain a "fixed" condition, but rather to seek to keep man-made perturbations within the range of types, magnitudes, and durations that will not result in the system flipping to a different state or, at a minimum, such that if a system does flip to a different state, it is not a permanent, irreversible condition. The goal, then, of environmental restoration projects is often to reintroduce the conditions necessary to revert the system back to the previous preferred state. Ensuring that the ecological resilience of an ecosystem is maintained or reintroduced, makes it more likely that the ecosystem will be able to withstand a greater range of perturbations without undergoing a state shift. So how can environmental mangers and regulators assure that ecosystems are resilient? First, it is necessary to understand what factors increase a system's ecological resilience. Research suggests that one of the most significant factors to increasing a system's ecological resilience is to increase its species richness. However, mere numbers of species may not be the complete answer. Scientists have developed a variety of hypotheses that attempt to describe how species diversity affects ecological resilience.

A number of factors can contribute to the stability of an ecosystem. One factor in ensuring stability is biological diversity. Because individual species are only able to perform limited ecosystem functions, the greater the species richness (i.e., the greater the number of species), the greater functional diversity will be obtained in the ecosystem.[39] Thus, the extent to which one species can compensate for the loss of a function previously provided by another species, impacts the ability for the system to be able to dampen the effects of perturbations.[40] However, mere numbers of species may not tell the whole story. Ecologists have developed a number of hypotheses that attempt to discern the influence of species richness on ecological resilience. One such hypothesis analogizes species as "rivets" on an airplane wing. A certain number of rivets can be lost without the wing coming loose from the airplane, however, at some point a threshold is reached and the next rivet removed causes the wing to come loose. Under this hypothesis, a certain number of species can be lost from a system without the system flipping to another state, but at some point, the next species lost will result in the system flip.[41]

The elimination of an important component of an ecosystem, such as the loss of a species that performs important functions, can result in the irreversible loss of a former stable state such that the system is no longer able to shift back into that state.[42] However, once a system is disturbed to the extent that it shifts into a different state, it may be extremely difficult, if not impossible to revert the system back to the prior state, because the new state will exhibit its own ecological resilience.

39 Peterson, *supra* note 32, at 209.

40 Gunderson et al., *supra* note 35, at 9.

41 *Id.*

42 Peterson, *supra* note 32, at 221; Gunderson et al., *supra* note 35, at 7.

Resource managers and regulators must be cognizant of this fact as well as the fact that even a very slow and gradual erosion of an ecosystem's controlling processes can result in a flip into a different state.[43] Thus, an objective of environmental management and regulation should be to maintain a level of ecological resilience, including a sufficient amount of redundancy ecosystem controlling processes, such that unexpected disturbances, whether anthropogenic or natural, can be absorbed without causing the system to shift states.

Another hypothesis, the "driver/passenger" hypothesis, posits that it is not the mere numbers of species that is critical, but rather it is the role that the species play that determines the impact to the system if that species is removed. This hypothesis identifies certain species as drivers, i.e., those playing a critical role in the controlling processes of the ecosystem. Loss of these driver species can have a much greater impact on the ecosystem than the loss of a passenger species.[44]

One of the most significant lessons from our new understanding of the dynamism and resilience of ecological systems is an acknowledgement that, although we can never completely prevent all human disturbances, we do not necessarily need to. Human disturbances that are similar in character, magnitude, and duration to natural disturbances, are not as likely to pose significant risks to species or ecosystems as human disturbances that are substantially different in character, magnitude, or duration from natural disturbances.[45]Accordingly, a goal in developing ecologically sound pest management approaches should be to use natural ecosystems as a model for agricultural systems. This will allow agricultural systems to benefit from natural pest management mechanisms, and will reduce the likelihood of anthropogenic perturbations that exceed the systems resilience.

---

43 Gunderson et al., *supra* note 35, at 7.

44 *Id.* at 9.

45 Stephen R. Carpenter & Kathryn L. Cottingham, *Resilience and the Restoration of Lakes*, *in* RESILIENCE AND THE BEHAVIOR OF LARGE-SCALE SYSTEMS 52 (Lance H. Gunderson & Lowell Pritchard, Jr., eds, 2002).

Chapter 2
Ecologically Related Management Tools

The past 30 years have evidenced some dramatic changes in the study of natural systems. During this period, the field of ecology has come into its own and the new fields of conservation biology and ecosystem management have been born. Moreover, unprecedented developments in our understanding of natural systems have occurred. Our once-accepted view of the "balance of nature" has been challenged, and we now are beginning to see nature as considerably more complex and dynamic than we once believed it to be.[1] Accordingly, environmental decision makers must accept, and incorporate into their decision making, the continually changing nature of ecosystems.[2]

Conservation Biology

A relatively new applied scientific discipline, "conservation biology," has emerged to attempt to address the complex relationship between human activities and the protection of ecological systems. The emergence in the mid-1980s of the field of conservation biology has led to novel ways of viewing environmental regulation and natural resources management. Conservation biology has been defined by one of its founders as "a new stage in the application of science to conservation problems, [which] addresses the biology of species, communities, and ecosystems that are perturbed, either directly or indirectly, by human activities or other agents. Its goal is to provide principles and tools for preserving biological diversity."[3]

---

1 *See generally*, DANIEL B. BOTKIN, DISCORDANT HARMONIES: A NEW ECOLOGY FOR THE TWENTY-FIRST CENTURY (1990) and Judy L. Meyer, *The Dance of Nature: New Concepts in Ecology*, 69 CHI-KENT L. REV. 875 (1994).

2 J.B. Ruhl, *Working Both (Positivist) Ends Toward a New (Pragmatist) Middle in Environmental Law*, 68 GEO. WASH. L. REV. 528 (2000) (book review).

3 Michael E. Soule, *What is Conservation Biology? in* ENVIRONMENTAL POLICY AND BIODIVERSITY (R. Edward Grumbine, ed., 1994). Conservation biology has also been described as a "regulatory science that seeks to develop scientific standards that can be applied to regulatory criteria and then to develop the management strategies to meet these standards." A. Dan Tarlock, *The Nonequilibrium Paradigm in Ecology and the Partial Unraveling of Environmental Law*, 27 LOY. L.A. L. REV. 1130 (1994). The seminal works on conservation biology include: CONSERVATION BIOLOGY: AN EVOLUTIONARY-ECOLOGICAL PERSPECTIVE (Michael E. Soule & Bruce A. Wilcox, eds, 1980) and CONSERVATION BIOLOGY: RESEARCH PRIORITIES FOR THE NEXT DECADE (Michael E. Soule & Cordon H. Orians, eds, 2001).

Conservation biology arose in the late 1970s and early 1980s as a reaction to the unprecedented worldwide extinction crisis that many biologists believe is occurring due to the activities of humans.[4] Accordingly, conservation biology is considered to be an applied "crisis" discipline: it is to biology what surgery is to physiology.[5] The driving force behind conservation biology is the conviction that, without serious efforts to reverse the trend of mass extinctions, millions of species may be at risk of extinction.[6] The loss of species may have devastating consequences for humans, as well as for the planet as a whole.[7] Due to the permanence of extinction, rare species are often the focus of conservation biology.[8]

Although conservation biology is still considered a relatively young and not well-defined science, certain postulates have been proposed. Soule divides these postulates into two categories: (1) functional postulates; and (2) ethical or normative postulates.[9] He describes the functional postulates as a set of fundamental rules gleaned from basic sciences, including ecology, that are geared toward the maintenance of both the form and function of natural biological systems.[10] Perhaps most pertinent for legal analysis are Soule's ethical or normative postulates, which can be utilized as a fundamental ecological baseline for an environmental management or regulatory system: (1) biological diversity is good; and (2) ecological complexity is good.[11]

To date, most of the law and policy initiatives pertaining to biodiversity protection have been focused on deciding which lands to preserve based on their diversity, uniqueness, connectivity to other lands, or other factors. While the importance to biodiversity of preserving significant lands cannot be overstated, such preservation would be in vain if steps are not taken to address the contamination of air, water and land, both inside and outside of preserved areas. Environmental pollution is the most subtle form of ecological degradation, and despite the banning of numerous pesticides over the past 30 years, pesticide pollution remains one the most serious

---

4 Malcolm L. Hunter, Jr., Fundamentals of Conservation Biology 14 (1996).

5 Soule, *supra* note 3, at 727.

6 Richard B. Primack, Essentials of Conservation Biology 1, 5 (1993).

7 *Id.* at 5.

8 Edward O. Wilson, The Diversity of Life 228 (1992).

9 Soule, *supra* note 3, at 729–32.

10 Soule's functional postulates include the following: (1) Many species are products of co-evolutionary processes—i.e., in most communities, species make up a significant part of other species' environment; (2) Species are interdependent—i.e., mutualistic relationships exist between many species and thus, there are always uncertainties about the impact that the extinction of one species will have on another Species; (3) Many species are highly specialized—i.e., the majority of animal species depend on a particular host. Thus, if such a host becomes extinct, many other species may be adversely affected; (4) Extinction of "keystone species" can have long-term consequences; (5) Introduction of generalist species may reduce diversity; and (6) Many ecological processes have thresholds below which they become discontinuous, chaotic or suspended. *Id.* at 729.

11 *Id.* at 730–31.

forms of ecological degradation.[12] No matter how strictly protected an area is from human activity, it will never be truly protected if contaminants such as pesticides enter the area through the water or air, or if species that use the area are exposed to harmful contaminants whenever they leave the confines of the protected area.

Although the primary threat to species and ecosystem diversity is habitat destruction, even when habitats are preserved, many populations have been reduced in size to such an extent that they are extremely vulnerable to other environmental stresses, such as hazardous pollutants.[13] Moreover, the vast majority of land in the in United States is in private ownership, and it would be naive to believe that enough property could be put into perpetual preservation to achieve true ecological integrity on a large scale. Thus, it is necessary to complement land preservation programs with effective regulatory programs that ensure a level of protection of widespread ecological integrity while allowing human activities to proceed. Although some legal scholars have analyzed environmental law in a general sense from a conservation biology perspective,[14] to date, the attempts that have been made to evaluate specific management or regulatory schemes using such an approach have focused primarily on natural resources management or the Endangered Species Act (ESA),[15] rather than pesticide law or pollution control law, more generally.

Ecosystem Management

Another important ecological concept is that of ecosystem management. It has been said that conservation biology leads to ecosystem management. In other words, ecosystem management is a tool to carry out the principles of conservation biology. Although environmental literature is rife with varying descriptions and definitions of ecosystem management, there appears to be general consensus that the overriding goal of ecosystem management is to protect ecological integrity over the long term.[16] Ecological integrity has been defined as the total native

---

12 PRIMACK, *supra* note 6, at 145–6.

13 *Id.* at 145.

14 *See, e.g.*, Walter Kuhlmann, *Making the Law More Ecoccentric: Responding to Leopold and Conservation Biology*, 7 DUKE ENVTL. L. & POL'Y 133 (1996); Robert B. Keiter, *Conservation Biology and the Law: Assessing the Challenges Ahead*, 69 CHI-KENT L. REV 911 (1994); Fred P. Bosselman & A. Dan Tarlock, *The Influence of Ecological Science on American Law: An Introduction*, 69 CHI-KENT L. REV. 847 (1994); and William H. Rodgers, *Adaptation of Environmental Law to the Ecologists' Discovery of Disequilibria*, 69 CHI-KENT L. REV. 887 (1994).

15 *See, e.g.*, Holly Doremus, *Adaptive Management, the Endangered Species Act, and the Institutional Challenges of "New Age" Environmental Protection*, 41 WASHBURN L.J. 50 (2001).

16 R. Edward Grumbine, *Introduction*, *in* ENVIRONMENTAL POLICY AND BIODIVERSITY (R. Edward Grumbine, ed., 1994).

diversity of species, populations, and ecosystems, and the ecological patterns and processes that maintain such diversity.[17]

The protection of ecosystem[18] diversity is crucial to the maintenance of overall ecological integrity. Ecosystems themselves provide numerous services, some of which are easily quantifiable, and others of which are not. As Professor Laura Westra has stated "[h]uman survival depends on many of nature's 'goods and services' that are invisible to markets and the economy; some are no doubt invisible to scientists."[19] One example of an ecosystem service is a wetland (a type of ecosystem) serving as a filter to treat water supplies.[20] Other examples include beach dunes buffering upland properties from the effects of storms, and saltmarshes supporting fisheries, air purification, soil renewal, climate stabilization, and crop pollination.[21] In the pest management arena, natural enemies of pest species, including predators, parasites, and diseases of pests, provide the important ecosystem service of keeping pest populations in check. Natural resource economists have identified numerous ecosystem services that, if humans had to replace them, would be extremely expensive, if not technically infeasible. Of course, as with individual species, our lack of understanding of the workings of these complex systems leaves open the very likely possibility that ecosystems perform many other valuable services that we do not yet fully understand or that we are not yet able to quantify.

Adaptive Management

The adaptive management concept originated from the works of C.S. Holling and Carl Walters in 1978 and 1986 respectively,[22] but it can be traced back to

---

17 *Id.*

18 An ecosystem is defined as a group of interacting organisms, or a community, and the physical environment they inhabit a particular point in time. Hunter, *supra* note 4, at 62.

19 Laura Westra et al., *Ecological Integrity and the Aims of the Global Integrity Project*, *in* Ecological Integrity: Integrating Environment, Conservation and Health 25 (David Pimentel et al. eds, 2000).

20 Hunter, *supra* note 4, at 69.

21 *Id.*; *see also*, Ruhl, *supra* note 2, at 544 (citing Nature's Services: Societal Dependence on Natural Ecosystems (Gretchen C. Daily ed., 1997)).

22 Warren T. Coleman, *Legal Barriers to the Restoration of Aquatic Systems and the Utilization of Adaptive Management*, 23 Vt. L. Rev. 177, 186 (1998). *See also* C.S. Holling et al., Adaptive Environmental Assessment and Management (C.S. Holling ed., 1978); Carl Walters, Adaptive Management of Renewable Resources (1986); and Bradley C. Karkkainen, *Adaptive Ecosystem Management and Regulatory Penalty Defaults: Toward a Bounded Pragmatism*, 87 Minn. L. Rev. 943 (2003) (outlining different varieties of adaptive management, including scientific hypothesis testing, macro-adaptation, and adaptive management as used in federal agencies).

Charles Lindblom's article *The Science of "Muddling Through"* published in 1959.[23] Holling incorporated the concept of resilience into policy design as an alternative to environmental assessment,[24] which he found to be a "reactive approach" that "will inhibit laudable economic enterprises as well as violate critical environmental constraints."[25] Holling described adaptive management as "integrat[ing] environmental with economic and social understanding at the very beginning of the process, in a sequence of steps during the design phase and after implementation."[26]

Walters described adaptive management as a way to deal with scientific uncertainty when managing renewable resources, especially since resource managers had begun relying on quantitative modeling as a tool to predict responses to alternative harvesting policies.[27] According to Walters, renewable resource scientists had failed by not putting greater emphasis on socioeconomic dynamics in their research and management, and in their approach to dealing with scientific uncertainty.[28] Instead of cautiously regulating harvests while seeking better understanding through more and more detailed analyses, Walters suggested using an adaptive management process "where management activities themselves are viewed as the primary tools for experimentation."[29]

The need for an adaptive approach to management became apparent in light of new understanding of ecosystems as dynamic, rather than as having only one equilibrium state.[30] Since then, government agencies have been trying to account for the disparity between science and environmental law and formulate a system that can adjust to confront scientific uncertainty.[31] Environmental regulation that

23 19 Pub. Admin. Rev. 79 (analyzing the decision-making processes of public administrators).

24 Holling et al., *supra* note 22, at 19–20.

25 *Id.* at 1.

26 *Id.*

27 Walters, *supra* note 22, at vii.

28 *Id.* at 2.

29 *Id.* at 2–3.

30 Timothy Profeta, *Managing Without a Balance: Environmental Regulation in Light of Ecological Advances*, 7 Duke Envtl. L. & Pol'y F. 71, 71–4 (1996). *See also* Tarlock, *supra* note 3, at 1121 (discussing the non-equilibrium paradigm).

31 "The law tends to encourage regulatory inaction in the face of uncertainty." Thomas T. Ankersen & Richard Hamann, *Ecosystem Management and the Everglades: A Legal and Institutional Analysis*, 11 J. Land Use & Envtl. L. 473, 493 (1996). Adaptive management is being recognized and adopted in varying degrees by federal government agencies responsible for managing natural resources, including the National Forest Service, the US Fish and Wildlife Service, the US Department of the Interior, the Bureau of Reclamation, the US Army Corps of Engineers, and the Bureau of Land Management. Coleman, *supra* note 22, at 177, 187 (1998). Federal agencies have been using adaptive management for the restoration of critical ecosystems, such as the Pacific Northwest Forests, the Colorado River, and the Everglades. *Id.*

can provide "feedback loops to update regulatory efforts as information increases" is counterintuitive to the American legal system.[32] Thus, adaptive management has not been seriously incorporated into environmental law.[33]

Environmental law often requires that regulation be based upon the "best available science," which is a principle of ecosystem management.[34] According to J.B. Ruhl, "[e]cosystem management is exactly what it sounds like—managing ecosystem-level problems through ecosystem-level approaches—and it almost always calls for creative and adaptive use of policy instruments as varied as inflexible commands at one extreme to generous incentives at the other."[35] Adaptive management, also a principle of ecosystem management,[36] has become increasingly synonymous with ecosystem management.[37]

In 1992, the National Research Council (NRC) conducted a study on the use of adaptive management for the restoration of aquatic ecosystems.[38] The study is cited as an example of how legal academics view adaptive management in terms of how resource management should be conducted.[39] The study suggests using the Adaptive Environmental Assessment (AEA) developed by C.S. Holling as an appropriate "process for involving scientists, resource managers, policy analysts,

---

32 Profeta, *supra* note 30, at 71, 86.

33 *Id.* "The legal challenge is to maintain enough flexibility for institutions to manage systems that are in a constant state of flux, while providing the legal certainty required to satisfy procedural and substantive due process." Coleman, *supra* note 22, at. 177, 178 (1998).

34 Ankersen & Hamann, *supra* note 31, at 473, 492.

35 J.B. Ruhl, *Taking Adaptive Management Seriously: A Case Study of the Endangered Species Act*, 52 U. Kan. L. Rev. 1249, 1250 (2004). Edward Grumbine provided the following working definition of ecosystem management, "Ecosystem management integrates scientific knowledge of ecological relationships within a complex sociopolitical and values framework toward the general goal of protecting native ecosystem integrity over the long term." R. Edward Grumbine, *What is Ecosystem Management?* 8 Conservation Biology 27, 31 (1994).

36 Ankersen & Hamann, *supra* note 31, at 473, 492.

37 J.B. Ruhl, *Thinking of Environmental Law as a Complex Adaptive System: How to Clean Up the Environment by Making a Mess of Environmental Law*, 34 Hous. L. Rev. 933, 999 (1997) (proposing the adaptive management process as one of the three legs of a revolutionized environmental law). "[V]irtually every collection of domestic ecosystem management principles, however varied, explicitly incorporates adaptive management as a guiding principle." Ankersen & Hamann, *supra* note 31, at 473, 494. Grumbine found that adaptive management was one of the dominant themes emerging from a review of ecosystem management articles in peer-reviewed journals up to 1993. Grumbine, *supra* note 35, at 27, 39, 31.

38 Tarlock, *supra* note 3, at 1121, 1140. *See* Committee on Restoration of Aquatic Ecosystems, National Research Council, Restoration of Aquatic Ecosystems: Science, Technology, and Public Policy 357 (1992).

39 Coleman, *supra* note 22, at 177, 186–7.

and decision makers interactively in designing resource management problems."[40] In formulating a National Restoration Strategy, the NRC established adaptive management as a principle for priority setting and decision making in the face of scientific uncertainty.[41] The example used by NRC was Chesapeake Bay's nutrient management strategy, in which the initial goal was set to reduce nutrient loading by 40 percent.[42] The policy makers committed to a continuous study of the goal itself, as well as the cost and effectiveness of the chosen means. As a result, both the goals and approaches of the nutrient management strategy are subject to revision over time.[43]

Both the legal and scientific scholarly literature of the past several years is rife with calls for the increased use of adaptive management in a variety of environmental regulatory, management, and restoration contexts.[44] Unfortunately, although numerous examples exist where resource agencies adopted adaptive management policies, at least in name, as part of a variety of environmental management and/or restoration projects, examples of successful adaptive management are hard to find.

The Relationship between Adaptive Management and Resilience

Adaptive management and ecological resilience are two sides of the same coin. Ecological resilience is necessary for adaptive management approaches to be viable. Adaptive management theory teaches us that environmental management and regulatory decisions are unavoidably based upon incomplete information. This imperfect information, coupled with the complexity and changes inherent in all ecosystems, suggests that ecological resilience is necessary to provide a cushion for humans to learn and adapt without risk of transforming a system into

40 Committee on Restoration of Aquatic Ecosystems, National Research Council, RESTORATION OF AQUATIC ECOSYSTEMS: SCIENCE, TECHNOLOGY, AND PUBLIC POLICY 345 (1992).

41 *Id.* at 357.

42 *Id.* at 358.

43 *Id.*

44 *See, e.g.*, J.B. Ruhl, *Regulation by Adaptive Management—Is It Possible?* 7 MINN. J.L. SCI. & TECH. 21 (2005); Bryan G. Norton, *The Rebirth of Environmentalism as Pragmatic, Adaptive Management*, 24 VA. ENVTL. L.J. 353 (2005); Joy B. Zedler, *Adaptive Management of Coastal Ecosystems Designed to Support Endangered Species*, 24 ECOLOGY L.Q. 735 (1997); Alfred R. Light, *Tales of the Tamiami Trail: Implementing Adaptive Management in Everglades Restoration*, 22 J. LAND USE & ENTL. L. 59 (2006); Julie Thrower, *Adaptive Management and NEPA: How a Nonequilibium View of Ecosystems Mandates Flexible Regulation*, 33 ECOLOGY L.Q. 871 (2006); John H. Davidson & Thomas Earl Geu, *The Missouri River and Adaptive Management: Protecting Ecological Function and Legal Process*, 80 NEB. L. REV. 816 (2001).

a different state.[45] To allow for the experimentation and experiential learning that is required by adaptive management, a system must have sufficient resilience to be able to withstand the inevitable mistakes or unintended consequences that will manifest whenever we act in the absence of perfect knowledge.[46]

On the other side of the coin, adaptive management is an integral part of assuring that our environmental management decisions restore and/or maintain ecological resilience. In the past, many of our environmental management and regulatory decisions were based on the outdated notion that if we chose the "right" technology or the "right" ecological indicators, we could maintain an ecosystem in a steady state.[47] We now know that there is no simple "right" answer and that attempts to maintain ecosystem stasis can cause its own set of environmental degradation. Ecosystems are complex and dynamic. The complexity and dynamic nature of ecosystems contributes to the resilience of ecosystems. Thus, to properly manage or restore an ecosystem, we must be certain that we do not intentionally or inadvertently dilute the complexity, or constrain the dynamic nature, of the system so as to reduce the ecological resilience of the ecosystem.

Both adaptive management and ecological resilience can play an important role in developing an ecologically based approach to pest management. As described above, the current dominant approach of relying on synthetic chemical controls to manage pests is antithetical to the concept of ecological resilience. A monoculture, almost by definition, is not resilient. Monocultures are vastly more susceptible to disease and pest outbreaks than more diverse systems. Moreover, broad-spectrum pesticides, by their very nature, eliminate or reduce the natural resilience of ecosystems. By killing predators and parasites, chemical controls eliminate the natural checks and balances that would otherwise keep pest populations from reaching infestation proportions. An agricultural system that is designed to be more ecologically resilient can eliminate or reduce the need for chemical controls and instead rely on and enhance the naturally occurring resilience features of the agricultural ecosystem.

Adaptive management approaches can help to ensure that pest management approaches are appropriately targeted to the specific agricultural system and the ecosystems surrounding it. Moreover, by employing adaptive management, we can improve pest management approaches and reduce environmental impacts as new or different information becomes available.

45 Lance H. Gunderson, C.S. Holling, Lowell Pritchard, Jr. & Garry Peterson, *Resilience of Large-Scale Resource Systems* 7, *in* RESILIENCE AND THE BEHAVIOR OF LARGE-SCALE SYSTEMS (Lance H. Gunderson & Lowell Pritchard, Jr., eds, 2002).

46 *Id.* at 4.

47 *Id.* at 7.

Ecological Integrity Baseline

An important concept in developing ecologically based management or regulatory systems ecology is the idea of starting with an environmental baseline. Professor Daniel Farber, in his book *Eco-Pragmatism*,[48] argues that there must be some overarching principle to help guide decision making and to serve as a "tie-breaker" in the close cases.[49] Farber argues that the baseline should have a presumption in favor of the environment.[50] However, Farber does not provide a comprehensive explanation of what the baseline should be. Ecological principles can be used to assist in developing an appropriate baseline.

As Farber explains, for any environmental regulatory scheme there are at least three possible baselines that could be employed. The first such baseline would be a neutral baseline in which there is neither a presumption in favor of the environmental pursuits, nor one in favor of industrial/economic pursuits. The second such baseline is one that holds a presumption in favor of industrial/economic pursuits. Under this baseline, regulated interests begin with a presumptive entitlement.[51] The third such baseline is one in which the presumptive entitlement is assigned to those who stand to benefit from the regulatory program.[52] In other words, under the third baseline, there is a presumption in favor of environmental protection. Farber asserts that the environmental baseline is the appropriate baseline to use.

Farber's primary justification for starting with an environmental baseline is the long-term commitment to environmental protection in the US. Farber posits that there is a general recognition that nature has inherent value that goes far beyond its purely utilitarian uses for human purposes. Moreover, to the extent to which nature provides utility, it is a public good that should not be frivolously plundered to benefit a few.[53] Another justification for the environmental baseline is the limited scientific understanding, enormous data gaps, complexity, and uncertainty of environmental issues. Without a full understanding of the complex issues involved, prudence suggests a presumption in favor of environmental protection. Environmental law should thus take a stance in favor of environmental protection instead of attempting to be neutral. Under Farber's eco-pragmatism, there is a presumption in favor of environmental protection, but that presumption can be rebutted where the costs of protecting the baseline are disproportionately large.[54]

---

48 Daniel A. Farber, Eco-Pragmatism: Making Sensible Environmental Decisions in an Uncertain World 35 (1999).

49 *Id.* at 93, 104.

50 *Id.*

51 *Id.*

52 *Id.*

53 *Id.* at 108–9.

54 *Id.* Although beyond the scope of this chapter, Farber also asserts that the ecological baseline should be adopted by the judicial system as canon of statutory interpretation, which he refers to as a "green canon." *Id.* at 124. Farber looks to legislative intent, as well

Deciding to establish an "environmental baseline," raises the issue of what an environmental baseline is, particularly in the area of ecological protection. Is the environmental baseline a zero-risk baseline, or is some other measure of environmental protection more appropriate? In the area of ecological protection, establishing an environmental baseline takes on additional complexity beyond that of human health protection. For example, in human health protection, a "no carcinogenicity" baseline could be established. Alternatively, a *de minimis* risk baseline could be established (e.g., a baseline of one cancer death per million). Other approaches could include establishing a baseline that is no greater than the risk from ambient pollutant levels. In the ecological arena, however, the issues are more complex.

This book suggests the use of the maintenance of ecological integrity as an environmental baseline. One significant component of ecological integrity is biodiversity—a measure of species abundance and richness used by ecologists to assess the health of ecosystems. In addition to having considerable scientific justification for protecting biological diversity, it is important to keep in mind that the protection of biodiversity is mandated under international law. The international obligation to protect biodiversity was first created by the 1992 United Nations Environment Programme Convention on Biological Diversity.[55] However, although the United States is a signatory, the Convention has not been ratified, and therefore, the United States is not formally bound by it. Nevertheless, Article 18 of the Vienna Convention on the Law of Treaties provides that a State that has signed a Treaty must refrain from "acts that would defeat the object and purpose of the Treaty."[56] Accordingly, the United States arguably is prohibited from taking any action that would defeat the purpose of the 1992 Convention to Protect Biological Diversity. Moreover, as Professor Dan Tarlock has pointed out, Chapter 15 of Agenda 21, which was approved by the United Nations Conference on Environment and Development, obligates states to develop strategies to conserve biodiversity and for the sustainable use of biological resources.[57] Accordingly, there is at least some basis in international law for arguing that the United States is obligated to have in effect programs that conserve biological diversity.

Support for an ecological integrity baseline also exists in US domestic law. A number of statutes express goals and policies directed toward maintenance of ecological integrity. For example, the Clean Water Act provides the broad

---

as to the more formalistic view of the statutory mandate of the National Environmental Policy Act, 42 U.S.C. 4321 *et seq.* (2004), for the government to use all practicable means, consistent with other essential considerations of national policy to achieve certain prescribed environmental goals, as bases for his green canon. *Id.* at 124–7.

55 United Nations Convention on Biological Diversity, June 5, 1992, S. TREATY DOC. 20 (1993), reprinted in 31 I.L.M. 818 (entered into force December 29, 1993).

56 May 23, 1969, 1155 U.N.T.S. 331.

57 A. Dan Tarlock, *Biodiversity Conservation in the United States: A Case Study in Incompleteness and Indirection*, 23 ENVTL. L. REP. 10529 (2002).

environmental objective "to restore and maintain the chemical, physical, and biological integrity of the Nation's waters."[58] Perhaps the strongest support for such a baseline, however, can be found in the National Environmental Policy Act.[59] In particular, section 101 of NEPA provides, among other things, that it is the continuing policy of the US to "create and maintain conditions under which man and nature can exist in productive harmony ..."[60]

Because ecology encompasses the relationships between the physical, the chemical, and the biological, ecological integrity necessarily includes physical integrity, chemical integrity, and biological integrity. To fully address ecological integrity, a variety of indicators of ecosystem health must be considered, including not only the number of organisms or species, or the status of threatened or endangered species, but also the state of physical, chemical, and biological processes, which, together with biological considerations, comprise ecological integrity. Ecological integrity can be measured by starting with a baseline condition found "at a site with a biota that is the product of evolutionary and biogeographic processes in the relative absence of the effects of modern human activity."[61] Then existing or predicted conditions can be compared to this baseline to measure the extent of deviation.[62]

---

58 33 U.S.C. § 1251(a).

59 42 U.S.C. § 4321 to 4370(f) (2004).

60 42 U.S.C. § 4331 (2004). This section declares the policy of the United States to use all practicable means to achieve a list of broad environmental goals. These goals include the following: 1. Fulfill the responsibilities of each generation as trustee of the environment for succeeding generations; 2. Assure for all Americans safe, healthful, productive, and aesthetically and culturally pleasing surroundings; 3. Attain the widest range of beneficial uses of the environment without degradation, risk to health or safety, or other undesirable and unintended consequences; 4. Preserve important historic cultural and natural aspects of our national heritage, and maintain, wherever possible, an environment which supports diversity and variety of individual choice; 5. Achieve a balance between population and resource use which will permit high standards of living and a wide sharing of life's amenities; and 6. Enhance the quality of renewable resources and approach the maximum obtainable recycling of depletable resources. *Id.* For a detailed discussion of the possible use of NEPA as an environmental baseline which can inform statutory interpretation, *see generally*, Mary Jane Angelo, *Crouching Textualist, Hidden Intentionalist: Reclaiming Our Stolen "Green Destiny" Out of the Judicial Sparring Over the Interpretation of Environmental Statutes*, *in* Strategies for Environmental Success in an Uncertain Judicial Climate (Michael Wolf, ed., 2005).

61 Westra et al., *supra* note 19, at 23.

62 Numerous attempts have been made by scholars of conservation biology to define qualitatively, if not quantitatively, biological or ecological integrity. For example, James R. Karr has developed an index of biological integrity (IBI) to address the objective articulated in the Clean Water Act, 33 U.S.C. § 1251(a) (2004), "to restore and maintain the chemical, physical, and biological integrity of the Nations Waters." James R. Karr, *Health, Integrity, and Biological Assessment: The Importance of Measuring Whole Things*, *in* Pimentel, *supra* note 19, at 209, 219–23. The IBI has been compared to the index of leading economic

Whatever metrics are employed to evaluate whether an environmental regulatory program preserves ecological integrity, some system of measurement is needed to take into account diversity, health of significant species, and integrity of the physical and chemical world. By clarifying that ecological integrity will serve as the environmental baseline in eco-pragmatism, a clear reference point can be established against which ecological changes of proposed man-induced activities can be measured. Accordingly, whenever an environmental regulation or other risk-reduction measure is proposed, an evaluation of the regulation will start with the presumption that the regulation must be sufficient not to completely preserve the status quo, but to maintain ecological integrity. Then the proposal can be evaluated to determine whether it will accomplish the goal of maintaining ecological integrity.

Ecological Risk Assessment

As discussed in more detail in Chapter 6, pesticides can have significant impacts on species as well as entire ecosystems. To develop regulatory or other policy approaches to minimize these impacts, it is critical to be able to characterize the types of risk, the mode of action, and the extent of the risk. The process for characterizing ecological risks is a type of risk assessment known as known as "ecological risk assessment."

In his seminal text on ecological risk assessment, Glenn W. Suter, defines the term "risk assessment" as "the process of assigning magnitudes and probabilities to the adverse effects of human activities or natural catastrophes."[63] He describes risk assessment as an essential element in environmental decision making. Risk assessment is premised on the facts that it may be extremely costly, if not impossible, to eliminate all risk and that environmental decisions often must be made in the absence of complete information. Risk assessment provides an approach designed to facilitate environmental decision making.

Ecological risk assessment provides "flexible problem-solving approaches capable of linking ecological measurements and data with the decision-making

---

indicators, which considers a number of financial measures to assess the overall state of the economy. Similarly, the IBI takes into consideration a number of metrics to assess the overall condition of an ecological system. Examples of metrics used in determining the IBI include species richness, stress intolerant and tolerant indicator species, relative abundance of trophic guilds, presence of alien species, and the incidence of disease, lesions tumors or other anomalies in the biota. The IBI does not call for the measurement of physical or chemical parameters, but instead assumes that their impacts will be manifest in biological indicators. Because of the regional differences between ecological systems, the IBI must be calibrated on a regional basis. *Id.*

63 Glenn W. Suter, II, Ecological Risk Assessment 3 (1993).

needs of environmental managers."[64] Proponents of ecological risk assessment cite its primary benefits as providing a structure process for gathering and evaluating data, and recognizing and providing transparency where uncertainties exist due to lacking data or conflicting data.[65] Ecological risk assessment is beneficial to decision makers because it provides "quantitative bases for comparing and prioritizing risk."

Ecological risk assessment is used to evaluate many diverse types of environmental risks related to exposure to one or more environmental stressors, including chemical stressors, physical stressors, and biological stressors. All three types of stressor are relevant to evaluating risks posed by pesticides and pest management. In this context, chemical stressors would be the pesticides released into the environment to carry out their pesticide purpose.[66] As described in Chapter 1, chemical pesticides can have a wide range of impacts on individual animals or plants, population of species, and ecosystems. Physical stressors would include the growing of monocultures and clearing of hedgerows and other refugia that would otherwise harbor beneficial predators and parasites. Biological stressors would come into play where biological control agents, whether genetically modified or not, are used to control pest populations.

Ecological risk assessments have been used since the 1980s, albeit in a more limited and less sophisticated manner that its current uses. During this decade, the use of ecological risk assessment expanded and began to be used by regulatory and policy decision makers. For many years, the seminal publication on risk assessment in general was what is commonly referred to as the "Red Book," a report issued by the National Research Council entitled "Risk Assessment in the Federal Government: Managing the Process."[67] This report described risk assessment in terms of probability or likelihood—i.e., an evaluation of the "probability that an adverse effect may occur as a result of some human activity."[68] Among other things, the report emphasized the need for a standardized methodology for conducting assessments. The Red Book led to a number of other governmental efforts to provide guidance for ecological risk assessment, most significantly the 1992 EPA document entitled "Developing a Work Scope for Ecological Assessments."[69] In this document, EPA set forth how ecological risk assessment could apply to a wide array of ecological harms. The guidelines were refined in 1998 in another EPA

64 Bruce K. Hope, *An Examination of Ecological Risk Assessment and Management Practices*, 32 Environment International 983–95 (2006).

65 *Id.* at 983.

66 *Id.*

67 NRC (National Research Council), Risk Assessment in the Federal Government: Managing The Process, National Academy Press, Washington, DC, USA (1983).

68 *Id.*

69 USEPA, Developing a Work Scope for Ecological Assessments. Publication 9345.0-05I, Eco update intermittent bulletin, vol. 1(4) Office of Solid Waste and Emergency Response, US Environmental Protection Agency, Washington, DC, USA (1992).

report.[70] Among other things, the 1998 report articulated the need for a planning process that involves stakeholders.

The specific processes used by various entities in carrying out an ecological risk assessment may vary to some degree, but there are key aspects that form the basis of any sound ecological risk assessment. One of the key ingredients of any ecological risk assessment is the recognition of inherent scientific uncertainty.[71] Due to the difficulty of predicting future environmental outcomes with a strong degree of certainty, environmental risk assessment relies on expressing risk in terms of probabilities.[72] For example, in the pesticide context, an ecological risk assessment could provide a probability that the use of a particular pesticide in a defined manner would result in an inability to reproduce in a particular species of bird. It is important to note that ecological risk assessment is merely a tool to provide a quantitative estimate of the probability and magnitude of a particular harm occurring. Although it is designed to be used to inform environmental policy decision makers, ecological risk assessment in itself does not involve making policy choices.[73]

Ecological Risk Assessment typically follows a fairly standardized process which involves first identifying the problem to be assessed and the specific assessment endpoints of concern. For example, in the pesticide context, one ecological problem may be identified as the impact of the use of a particular pesticide on the population of a particular endangered species. Relevant assessment endpoints may include the probability of a reduction in reproduction, the proportion of the individuals of the species killed within the area of use, and the rate of population decline of the species. Laboratory and field studies designed to determine the toxicity of the pesticide would be considered the "measurement endpoints," which Suter describes as "the toxicologist's or field biologist's input to risk assessment."[74] Typically, measurement endpoints are obtained by laboratory studies or local field tests, whereas the ultimate assessment endpoint addresses the effects on populations of organisms or entire ecosystems.[75] Thus, the toxicity measurement endpoints are only one component of an ecological risks assessment.

While both assessment and measurement endpoints can be defined at any level of organization—i.e., individual, population, community, ecosystem—typically the effect of concern, or the assessment endpoint, is at the population, community or ecosystem level, whereas the measurement endpoint is typically at the individual level due to the difficulties and costs of testing at higher organizational

---

70 USEPA, Guidelines For Ecological Risk Assessment. EPA/630/R-95/002F. Risk assessment forum, US Environmental Protection Agency, Washington, DC, USA (1998).

71 Suter, *supra* note 63, at 3–4.

72 *Id.* at 4.

73 *Id.*

74 *Id.* at 24.

75 *Id.*

levels.[76] Extrapolation from the individual level to higher levels of organization is accomplished through modeling.

Models are assessment methods used to quantitatively describe an effect and its underlying causes. Models enable assessors to predict outcomes outside of the range of available data and to predict future events based on current information. They also enable the integration of complex sets of data including field and laboratory data from a variety of different tests conducted over periods of time and under varying conditions.[77]

One of the most challenging aspects of ecological risk assessment is characterizing and quantifying uncertainty.[78] Ecological risk-assessment models typically employ methods of uncertainty analysis, such as the Monte Carlo method, which enable uncertainties to be quantified and characterized, and properly propagated throughout the assessment.

The end result of an ecological risk assessment provides a "probabilistic analysis that can help to clarify the relationship between decision making and uncertainty. They can be used to justify a particular degree of conservatism in the face of uncertainty or can be used to justify making additional measurements or conduct additional tests to reduce uncertainty."[79]

As can be seen from the discussion above, by doing a better job of incorporating our understanding of the ecology of pests and other ecological concepts into designing a pest management system, we can reduce the environmental impacts associated with pest management while at the same time achieving improved pest management.

76 *Id.*

77 *Id.* at 37–8. A detailed discussion of modeling is beyond the scope of this book. For a comprehensive discussion of modeling in the context of ecological risk assessment, *see generally*, *id.*

78 *Id.* at 41.

79 *Id.* at 47.

Chapter 3
The Ecology of Pests

What Makes a Pest a Pest?

Species are not created as "pests." Whether a species constitutes a pest is a direct result of the preferences of humans. Animals or plants that are in places or at times where humans do not want them to be are considered pests. For example, there is nothing inherent to a particular plant that makes it a weed. It is the fact that a plant is not where humans want it that makes it a weed. Additionally, species that cause some form of harm either to humans or to other organisms, or things that humans value, are also considered to be pests. Pests can directly harm humans. For example, biting flies and mosquitoes cause pain, itching, and annoyance. Some pests are merely nuisance pests. For example, certain ants, which cause no particular harm and may be perfectly acceptable when outside, become pests when they find their way into human kitchens. Some pests, such as termites and carpenter ants, may cause harm by damaging human structures. Other pest species may not harm humans directly, but instead may be vectors that transmit human diseases, such as malaria, dengue fever and other insect-borne diseases. Similarly, pests can transmit diseases to domestic animals, whether pets or livestock. Of course, one of the most significant forms of pest are those of agricultural crops and products.

Virtually every crop and animal species has pests associated with it. Some pests are generalists and will feed on a wide range of species. Other pests are more host-specific and will only be attracted to certain taxa of organisms. For example, with crop pests, the level of attractiveness of specific host taxa to specific pest species is frequently due to specialized secondary chemicals produced by the host species, to which the pest species has evolved an attraction.[1]

There are literally hundreds of species that are considered "major" pests. Major pests exist for humans, animals, and plants. One of the most significant human medical pests includes the *Anopheles* mosquito, which carries and transmits malaria throughout vast regions of the world. Fleas have caused substantial harm to humans by transmitting the deadly plague, which has been responsible for multiple waves of pandemics killing millions of people throughout human history,[2] and lice transmit typhus. Livestock has a number of major pests that transmit disease, including certain ticks and flies.

1 H.F. van Emden & M.W. Service, Pest and Vector Control 10 (2004).

2 *Id.*

It is estimated that there are more than 300 major crop pests.[3] Some of the most serious of these include locusts, which can devastate virtually any crop, and a number of plant and leaf hoppers, and bollworms, which are significant pests of major crops, like grain and cotton. Some pests, such as locusts, are major pests because of the large mass of plant material they consume. Others, such as the hoppers, are pests because they are vectors of plant diseases, which cause substantial crop loss.

Because pests are defined as such by being in a place and time where they are not desired, they can be distributed geographically and temporally. Some pests are localized, whereas others may be ubiquitous and distributed over vast areas of the earth. Some pests are only problems temporarily during certain seasons or under certain conditions, whereas others persist through a range of seasons and conditions.[4] Jason Clay describes this phenomenon as follows: "[b]y providing vast areas of food and by eliminating the diversity that would support competition from a wider array of species, modern agriculture has reduced the number of species that can live in large areas of the landscape while allowing a few to become dominant."[5] It is these dominant "few" that we consider to be pests.

The Role of Pests in Ecological Systems

Because many of the species that we consider to be pests, and thus that we use pesticides to control, are invertebrate species, to understand their role in ecological systems it is first necessary to understand the role of invertebrate species in ecosystems. It is estimated that there are as many as 8 million species of insects, a large category of invertebrates, on earth, making up four-fifths of all metazoan species.[6] In fact, in evolutionary terms, insects are considered to be one of the most successful forms of animal, due to their ability to rapidly adapt to changing conditions. They are able to thrive in virtually all types of ecosystem. Samways has described the insects as an "immensely successful life form that has taken … many diverse evolutionary paths …."[7] Insects' ability to adapt to so many different environments, including man-made ones, has made them one of the most significant categories of pest.

While any living organism can be considered a pest if it is found in a place and at a time it is unwanted, or if it causes economic loss or health problems, invertebrates, particularly insects and arachnids, are second only to weedy plants as the most significant targets of pesticide use. Crop damage from insects typically

3 *Id.* at 10.

4 *Id.* at 9.

5 Jason Clay, World Agriculture and the Environment 49 (2004) (citing Kimbrell 2002).

6 Michael J. Samways, Insect Diversity Conservation 16 (2005).

7 *Id.* at 18.

results from the sheer numbers of insects feeding on a crop. In most cases, small numbers of pest insect species can be tolerated without significant crop damage. However, when insect populations reach a threshold level, cumulative damage from many thousands of insects becomes unacceptable. It is estimated that there could be as many as 25 million insects per hectare of soil and 25,000 in flight over a hectare at any given point in time.[8] Of course, most of these are not significant pests, and in fact many may be beneficial. Nevertheless, even the most insignificant pest can become a major problem if present in high enough numbers. For example, as many as 22 million fruit flies may be found in one hectare of oats and an estimated 200 million black bean aphids in a hectare of sugar beets during infestation.[9]

Long before humans were faced with the challenges of agricultural pests, hunter-gatherers had to deal with nuisance and disease-carrying biting insects.[10] Even today, a major category of what we refer to as pests includes species such as the tsetse fly and numerous species of mosquitoes, which act as vectors transmitting diseases to humans and domestic animals. The other major category is agricultural pests, which include, among many others, the well-known locusts, armyworms, bollworms, and diamond-back moths. These kinds of pest have existed since the beginning of agriculture. Six thousand-year-old Chinese cave paintings and 4,300-year-old Egyptian artifacts depict pests attacking crops.[11] The Bible references the "plague of flies" and the "plague of locusts."[12] Even with regard to agricultural pests, frequently the invertebrate pest does not directly cause the damage to the agricultural crops, but instead is a vector that carries and transmits various viruses, fungi and other plant diseases that result in crop loss.[13]

Not only have pests always been inextricably tied to agriculture, but agricultural pests have also had profound effects on human civilization. For example, van Emdem describes how from 1958 to 1962, more than 30 million people died of starvation in the Great Chinese Famine, which was attributed to uncontrolled pest populations resulting from a bizarre bird eradication program in which huge numbers of people were mobilized to beat metal pans and make noise to keep sparrows from landing until the birds died of exhaustion. Without the sparrows to keep them in check, insect pest populations exploded, resulting in agricultural devastation.[14]

8 Van Emden & Service, *supra* note 1, at 31.

9 *Id.* at 32.

10 *Id.* at 1.

11 *Id.* at 4.

12 *Id.* Citing Exodus 10:13–15.

13 *Id.* at 2.

14 *Id.* at 5.

How Species become Pests

Nature provides a complex system of pest control that has evolved over millions of years. Pest, predator, parasite and parasitoid interactions are extremely complex, with parasitoids and predators often functioning in keystone species roles. Removal of key players in these complex community interactions may mean that natural pest-control systems have been disrupted; and this can result in population explosions of herbivorous pest species. In controlling agricultural pests, the challenge is that the act of growing agricultural crops inherently disrupts these communities' interactions; certain forms of agricultural production, such as monoculture planting and use of synthetic pesticides can further disrupt these systems, to the point where natural control from predators and parasitoids is virtually eliminated. The challenge is to find ways to accomplish needed pest control without eliminating natural controls, or even better, by enhancing natural controls.

Insect populations have the potential to explode to infestation levels under the right circumstances. Species will naturally have an upper limit to continued population growth and competition for resources; overcrowding and disease will eventually decrease population growth. However, these natural controls may not come into play until the population is well beyond the level at which the pest can do significant damage. One of the most critical factors in determining pest population is the density of predators and parasites, which can dramatically reduce pest populations under the right conditions.[15] Thus, removal or reductions in density of predators and parasites can lead to pest population explosions. Another significant factor that influences pest population levels is the availability of food. In diverse ecosystems, a particular pest will have to expend energy to find food sources. Industrialized agricultural practices that rely on large areas of monoculture, however, provide ideal conditions for pest population growth. Ideal food sources are readily available in great abundance, and pests do not have to expend energy or travel distances to find food or other members of the species with which to mate.[16] Moreover, monocultures make it extremely easy for pests to find food, because in a monoculture there are large numbers of plants emitting the volatile chemicals that attract the pest species.[17] Monocultures can transform pest problems to epidemic levels.[18] Other agricultural practices designed to maximize yields can also encourage pest problems. For example, breeding crop varieties to achieve desired characteristics such as high yield, improved flavor, or enhanced shippability, often reduces many of the chemical defenses that can deter pest damages found naturally in plants.[19]

---

15 *Id.*

16 *Id.* at 41.

17 *Id.*

18 *Id.*

19 *Id.*

Photo 3.1 Corn monoculture
Reprinted with permission. Howard F. Schwartz, Colorado State University, Bugwood.org.

Humans developed agriculture more than 10,000 years ago. How did farms survive without the current chemical pesticides that are employed worldwide today? Studies and experience have demonstrated that intensive monoculture agricultural practices are transforming some species previously considered non-pests into pests.[20] Humans' efforts to control other pests with chemical pesticides and to obtain higher and higher yields have created numerous new pests.[21] Pesticides used to control one type of pest can destroy the predators and parasites that previously kept another species population in check, thereby resulting in increased populations of the other species which then becomes a pest in itself.[22]

When chemical pesticides are sprayed in the field, less than 0.1 percent of the toxic chemicals actually reach the pests themselves.[23] The vast majority of pesticides reach non-target species, often including beneficial predatory or parasitic organisms that normally keep pest populations in check. In fact, several important agricultural economic pest species have dramatically increased as a result of the combination of the introduction of chemical pesticides, the use of new crop

20 *Id.*
21 *Id.*
22 *Id.*
23 *Id.*

varieties, and the advent of industrialized farming practices.[24] Such pests include, among others, army worms, hoppers, cotton bollworms, and the diamondback moth.[25] These insects were not major pests prior to 1950, but became some of the most serious crop pests as a result of chemical pesticide use destroying their naturally occurring enemy species.[26] This phenomenon was artfully described by Rachel Carson in *Silent Spring:*

> The whole process of spraying seems caught up in an endless spiral. Since DDT was released for civilian use, a process of escalation has been going on in which ever more toxic materials must be found. This has happened because insects, in a triumphant vindication of Darwin's principle of survival of the fittest, have evolved super races immune to the particular insecticide used, hence a deadlier one always has to be developed—and then a deadlier one than that. It has happened also because … destructive insects often undergo a 'flareback,' or resurgence, after spraying, in numbers greater than before. Thus the chemical war is never won, and all life is caught in its violent crossfire.
>
> …
>
> Under primitive agricultural conditions the farmer had few insect problems. These arose with the intensification of agriculture—the devotion of immense acreages to a single crop. Such a system set the stage for explosive increases in specific insect populations. Single-crop farming does not take advantage of the principles by which nature works; it is agriculture as an engineer might conceive it to be. Nature has introduced great variety into the landscape, but man has displayed a passion for simplifying it. Thus, he undoes the built-in checks and balances by which nature holds species within bounds.
>
> …
>
> If the Bill of Rights contains no guarantee that a citizen shall be secure against lethal poisons distributed either by private individuals or by public officials, it is surely only because our forefathers, despite their considerable wisdom and foresight, could conceive of no such problem.
>
> …
>
> The public must decide whether it wishes to continue on the present road, and it can do so only when in full possession of the facts.[27]

---

24 *Id.* at 5.

25 *Id.*

26 *Id.* at 5–6, 11.

27 Rachel Carson, Silent Spring 8–13 (1962).

Photo 3.2 Colorado potato beetle
Reprinted with permission. USDA APHIS PPQ Archive, USDA APHIS PPQ, Bugwood.org.

In addition to pesticide use itself leading to the creation of new pests or increasing the significance of existing pests, other human activities contribute to the development and spread of pest species. The movement of humans, crops, and insects has led to creation of new pest problems. Travel and trade are two of the largest contributors to the creation and movement of pest species. The dramatic increase in travel and trade during the latter part of the twentieth century has also contributed to the creation of new pests.[28] By moving crop plants and crop pests to geographic areas in which they previously did not exist, new crop/crop pest combinations have been created.[29] Similarly, human travel has been responsible for the spread of insect vectors of human diseases.[30] As humans travel, they can bring with them pest vectors such as fleas and lice that can spread disease to new geographic areas. Not only do humans carry pests with them to new locales, but they also frequently move to new locales where they are exposed to new pests. For example, when farmers moved into the Amazon rainforest to grow crops, they became vulnerable to malaria borne by mosquitoes and later carried the disease with them when they returned to their homes in the highlands.[31]

28 Van Emden & Service, *supra* note 1, at 5–6, 11.
29 *Id.*
30 *Id.*
31 *Id.* at 43.

When certain species are introduced into new geographic regions, whether intentionally or unintentionally, the result is often the formation of a new pest. Herbivorous insects introduced into new areas may find new abundant food sources at the same time they are able to thrive in an environment where there are not any predators, parasites or diseases that have evolved to keep insect populations in check. Examples of pest population explosions when species are introduced into new areas abound.

The movement of goods through trade can either move pests to new areas or lead to the creation of new pests in areas where the pest species previously was not considered a pest. For example, introducing a new crop to a new geographic area can cause an existing sparsely populated herbivorous insect species to become a pest. The classic example of this phenomenon is the creation of the Colorado potato beetle. This beetle existed in the Rocky Mountains of the US as a non-pest species until potatoes started to be grown in the foothills of the mountains. The introduction of potato crops provided a new plentiful food source for the beetle, which quickly spread throughout potato-growing regions and became a major pest in both the US and Europe.[32]

Our current approach to agriculture, with its heavy reliance on monoculture farming and intense pesticide inputs, is a significant factor in the creation of new pests and the exacerbation of existing pest problems. We will never eliminate pests, but by doing a better job of incorporating ecological principles into the design of agricultural systems and pest management approaches, we will be able to take advantage of nature to help keep pest populations at manageable levels.

32 *Id.* at 6.

PART II
Agriculture and Pesticides

Chapter 4
The Agricultural Revolution and Farm Policy

The current US industrialized agriculture system, with chemical and Genetically Modified (GM) pesticides as its centerpiece, is not merely a product of market forces driving increased yields per acre. Rather, a long history of US agricultural policy has supported and encouraged high-yield, monoculture-based, high-synthetic pesticide- and fertilizer-input industrialized agriculture. As described below, for many decades US agricultural policy has provided incentives that are antithetical to building a more ecologically based sustainable agriculture system.

History of Agricultural Policy in the US

The face of farming in the US has changed dramatically from the small family farmer of the early part of the twentieth century to the large corporate industrial producer of today. Between 1950 and 2000, the total number of US farms declined by 60 percent, while the total acreage of all farms remained the same.[1] This trend was a consequence of larger farms buying out smaller farms.[2] From 1900 to 1997, the number of farms over 1,000 acres grew from 340 to 5,887.[3] Currently, there are approximately 2.204 million farms with a total of 922 million acres of farmland in the US. Of the over 2 million farms, approximately 1.3 million have actively harvested cropland.[4] Of these 1.3 million farms, approximately 918,000 use chemical pesticides. The two largest categories of pesticide use are for insects on crops and hay, and for weed control.[5] Some of the greatest contributing factors to the revolution in farming can be found in US agricultural economic incentive programs, particularly those contained in the various versions of what is known as the "Farm Bill."[6]

---

1 William S. Eubanks II, *A Rotten System: Subsidizing Environmental Degradation and Poor Public Health with Our Nation's Tax Dollars*, 28 STAN. ENVTL. L.J. 213, 229 (2009).

2 *Id.*

3 *Id.*

4 EPA Pesticide Industry Sales and Usage 2006/2007, at 20, *available at* http://www.epa.gov/opp00001/pestsales/ (last visited May 26, 2011).

5 *Id.* at 8, 20.

6 Eubanks, *supra* note 1, at 228.

The United States has a long history of government-supported agriculture that dates back to the late eighteenth and nineteenth centuries when the government opened vast areas of public land for agricultural settlement.[7] By the mid-nineteenth century, the federal government began to play a significant role in encouraging and funding agricultural research. The Morrill Act authorized grants of public lands to states to establish "land grant" colleges,[8] and the Hatch Act of 1887 authorized federal funding of agricultural research.[9] In the early twentieth century, the Cooperative Extensive Service was established to provide practical education to those in the agricultural industry.[10] It was not until the 1930s, as a reaction to the Great Depression and the devastating "Dust-bowl" era, that the federal government began to enact the complex maze of agricultural law and policy that forms the backbone of our current agricultural system.[11] The combination of the Great Depression and the Dust-bowl era of the 1930s led to the first significant economic intervention by the government into what had previously been a relatively free agricultural marketplace.[12] Initially, government intervention was aimed at stabilizing prices by limiting production to limit supply.[13] During the Depression, agricultural land values sank, resulting in staggering farm mortgage foreclosure and bankruptcy in the farm credit sector.[14] Moreover, commodity prices declined and there were stockpiles of certain commodities well in excess of market needs.[15] As part of the 1932 presidential campaign, Franklin Delano Roosevelt vowed to solve the agricultural problems. Against this backdrop was the devastating Dust Bowl of the 1930s. Consequently, the 1930s experienced a rash of New Deal agricultural legislation designed to stabilize agricultural markets and to support prices of certain basic commodities.[16]

The most significant of the Depression-era agricultural enactments was the Agricultural Adjustment Act of 1933 and its successor, the Agricultural Adjustment Act of 1938, which continues to serve as the foundation for the current commodity-price and income-support programs.[17] The 1938 Act was designed

---

7 *See, e.g.*, Donald B. Pedersen & Keith G. Meyer, Agricultural Law in a Nutshell 1–5 (1995) (describing the series of laws enacted in the seventeenth, eighteenth and nineteenth centuries to promote agricultural settlement).

8 Morrill Act of 1862, 7 U.S.C. § 301; 26 Stat. 417; Morrill Act of 1890, 7 U.S.C. § 321.

9 Hatch Act, ch. 314, 24 Stat. 440 (1887) (codified as amended at 7 U.S.C. § 361a (2011)).

10 Smith–Lever Act of 1914, 7 U.S.C. § 343.

11 Eubanks, *supra* note 1, at 219.

12 *Id.*

13 *Id.*

14 *Id.*

15 *Id.*

16 *Id.*

17 *See* Howard B. Pickard, *Price and Income Adjustment Programs*, *in* Agricultural Law 1, 1–6 (John H. Davidson, 1981).

primarily to increase farm income and stabilize prices.[18] These goals were echoed in a succession of later legislative acts including the acts that are now commonly referred to as the Farm Bills. During the 1930s, the federal government also created the Commodity Credit Corporation (CCC), which is a federal corporation within the Department of Agriculture that is authorized to act to stabilize and support farm income and prices.[19]

After the problems of the 1930s, farm subsidies continued as a way to keep prices high, limiting the amount produced by taking some land out of production and controlling the amount of crops making it to market. The goal of stabilizing markets by limiting production continued until 1973 when then-Secretary of the US Department of Agriculture, Earl Butz, dramatically changed policy direction by encouraging farmers to grow the maximum amount possible of commodity crops like corn. Butz's policy shift, coupled with the technological advances of the Green Revolution, which centered around heavy inputs of pesticides and fertilizers, resulted in dramatic increases in per acre yields. Rather than limiting production, the new policies tied payment amounts to production levels, thereby incentivizing the maximum production of certain commodity crops for which subsidies were available. Growers could benefit by substituting the heavily subsidized commodity crops for their previous variety of vegetable crops and grazing lands. As more and more land became devoted to ever increasing densities of corn and other grains, farms grew larger, and large farms squeezed out the smaller family farms that once grew a variety of food. Today, a majority of corn farmers have more than 1,000 acres of corn, and farm size continues to grow, as larger farmers buy out smaller farmers. Because more bushels of corn meant more money, techniques were developed to maximize the yield of corn per acre. Currently, it is not uncommon for corn growers to yield 200 bushels, or 5 tons, of corn from a single acre of land. This represents an approximately four-fold increase in per acre corn yield since the early 1990s. To achieve such high yields it is necessary to plant a staggering 30,000 kernels of corn per acre and to rely on high inputs of fertilizer, pesticides, and irrigation water. A typical Iowa corn farm uses 150 pounds of anhydrous ammonia per acre as fertilizer.[20]

The Green Revolution refers to the transformation of agriculture that began in the 1940s and gained traction during the 1960s, leading to dramatic increases in per acre farm yields.[21] One of the primary changes during the Green Revolution was replacing human labor with technological innovations, fossil fuel-derived inputs, and mechanized farm equipment. These changes resulted from a combination of new government policies that for the first time encouraged high-yield farming of commodity crops by linking subsidy payments to production levels, offering more

---

18 *Id.* at 5–6.

19 *Id.* at 4.

20 Iowa State University Department of Agronomy, Iowa Total Corn Fertilizer Use, 2003, *available at* http://extension.agron.iastate.edu/soils/pdfs/Nuse/NBacktb3.PDF.

21 Eubanks, *supra* note 1, at 222.

government money for research and development on high-yield farming, and a creating a vast network of extension service education and training of farmers in high-yield commodity framing. The Green Revolution is credited with increasing farm production by more than 80 percent in the past 30 years.

Environmental Impacts from Industrialized Agriculture

Although the changes brought about by the Green Revolution have served to significantly increase crop yields, they have also brought with them a variety of adverse social, economic, and environmental consequences. And though the focus of this book is pesticides, many of the impacts of agricultural practices are interrelated, and a brief overview of the environmental impacts associated with agriculture is warranted.

From an economic and social standpoint, non-labor-intensive industrial agriculture has led to the vast majority of crops being produced by a smaller total number of farmers, the virtual disappearance of the traditional family farm, high-risk working and living conditions for farm laborers, increased production costs, and a decline of economic and social conditions in rural communities.[22] With regard to the natural environment, high production industrialized agriculture has contributed to topsoil depletion, contamination of surface and groundwater, loss of biodiversity, and harm to protected species.[23] The high fossil fuel energy inputs required in industrialized agriculture exacerbate the daunting challenge of reducing carbon emissions to stem climate change.

A range of industrial agricultural practices contributes to environmental harms. The Union of Concerned Scientists identifies the key features of industrialized agriculture as: monocultures,[24] few crop varieties, reliance on chemical and other inputs, and the separation of animal and plant agriculture.[25] These features alone and in combination carry with them a panoply of potential environmental, human

22 *See* Eubanks, *supra* note 1.

23 *Id.* For additional discussion on the environmental harms caused by farming, *see* Eubanks, *supra* note 1; Mary Jane Angelo, *Embracing Uncertainty, Complexity and Change to Protect Ecological Integrity: An Eco-Pragmatic Reinvention of a First Generation Environmental Law*, 33 Ecology L.Q. 105 (2006); Mary Jane Angelo, *The Killing Fields: Reducing the Casualties in the Battle Between U.S. Endangered Species and Pesticide Law*, 32 Harv. Envtl. L. Rev. 96 (2008); J.B. Ruhl, *Farms, Their Environmental Harms, and Environmental Law*, 27 Ecology L.Q. 263, 337–8 (2000); and Jan Lewandrowski, James Tobey & Zena Cook, *The Interface Between Agricultural Assistance and the Environment: Chemical Fertilizer Consumption and Area Expansion*, 73 Land Econ. 404 (1997).

24 For a discussion of the global reliance on monoculture farming, *see* Helena Norberg-Hodge, *Global Monoculture: The Worldwide Destruction of Diversity* 13, *in* Fatal Harvest: The Tragedy of Industrial Agriculture (Andrew Kimbrell, ed., 2002).

25 http://www.ucsusa.org/food_and_agriculture/science_and_impacts. *See also*, Fatal Harvest: The Tragedy of Industrial Agriculture, *supra* note 24.

health, and socio-economic impacts. Some of these activities include conversion of undeveloped land into agricultural fields, intensive water use for irrigation, fertilizer use, pesticide use, growing crops in monocultures, and tilling soils. These various practices impact human health, wildlife and biodiversity, natural resources, and ecological services, and significantly contribute to climate change.

One of the significant environmental impacts related to agriculture is that on water resources. All agriculture has the potential to cause adverse impacts on the water resources. Agriculture impacts water quantity as well as quality.[26] Intensive industrialized agriculture with its concomitant large fossil-fuel-derived (energy, fertilizer and pesticides) and water inputs, dramatically increases the potential for harm to water resources.[27] Agriculture, and especially highly intensive industrialized agriculture, is a significant user of water.[28] The high-yield goal of industrial agriculture requires water-intensive agricultural practices that depend on large-scale irrigation.[29] Water quantity impacts are a direct result of irrigation.[30] Commercialized commodity-crop production is responsible for significant reductions in both water quality and water quantity.[31] Irrigation for agriculture constitutes more than one third of the freshwater use in the US, making it the largest use in the nation.[32] An exacerbating problem is that many commodity crops, such as corn, are grown in parts of the country that do not have sufficient water resources for this type of intensive agriculture.[33] Accordingly, water must be diverted from water bodies long distances from the fields.[34] In many western states, water consumed for crop irrigation accounts for approximately 75 percent of the total water consumed.[35] Many of the country's, as well as the world's, water disputes involve agriculture as a major player.[36]

In addition to water quantity impacts, agriculture can cause serious adverse impacts to the quality of both groundwater and surface water.[37] When rain or

26 Eubanks, *supra* note 1, at 252.

27 Daniel A. Farber, *Adaptation Planning and Climate Impact Assessments: Learning From NEPA's Flaws*, 39 Envtl. L. Rep. News & Analysis 10605, 10605 (2009).

28 *Id.*

29 Peter Rosset, Inst. for Food & Dev. Pol'y (IFDP), *Lessons from the Green Revolution*, April 8, 2000, *available at* http://www.foodfirst.org/media/opeds/2000/4-greenrev.html (last visited August 27, 2009).

30 Ruhl, *supra* note 23, at 280.

31 *Id.* at 274.

32 Eubanks, *supra* note 1, at 253.

33 Christine A. Klein, *Water Transfers: The Case Against Transbasin Diversions in the Eastern States*, UCLA J. Envtl. L. & Pol'y 249, 269 (2007).

34 *Id.* at 253.

35 B. Delworth Gardner, *Legal Impediments to Transferring Agricultural Water to Other Users*, *in* Agricultural Policy and the Environment 67 (Rodger E. Meiners & Bruce Yandle, eds, 2003).

36 Eubanks, *supra* note 1, at 254.

37 *Id.* at 255.

irrigation water comes into contact with farm fields, certain agricultural chemicals, including water soluble pesticides such as atrazine, and nutrients, such as nitrites found in fertilizers, easily leach into groundwater.[38] This contamination can render groundwater sources of water unacceptable for drinking.[39] Where groundwater naturally flows into surface water, as it does in artesian springs, surface waters become contaminated as well.[40] Rain and irrigation water that exceeds the amount capable of being absorbed into the soil flows off of agricultural fields as stormwater runoff, carrying with it a variety of pollutants that ultimately end up in surface-water bodies.[41] Stormwater runoff from farm fields frequently contains high levels of sediments resulting from soil erosion from tilled fields, pesticides, and fertilizers.[42]

Fertilizers used to maximize yields in industrial agriculture typically contain nutrients such as phosphorus and ammonium nitrate.[43] Scientific studies demonstrate that agricultural intensification via increased chemical fertilizer and other inputs is directly linked to increased environmental damage.[44] Large quantities of these compounds are carried in rain runoff into water-bodies, where they exert their plant-growth enhancing effect and lead to an overgrowth of algae.[45] When algae become overabundant, they deplete oxygen and reduce sunlight penetration resulting in a condition referred to as eutrophication.[46] Eutrophied lakes are characterized by algae dominance, rather than submersed plant dominance, low oxygen, and reduced fish and other aquatic organisms.[47] When nutrient-laden water finds its way to estuarine areas, it can create "dead zones" in areas previously characterized by high fish and aquatic organism productivity.[48] For example, nutrient-heavy water in the Mississippi River is believed to have caused a large dead zone in the Gulf of Mexico.[49] Seventy percent of the nitrogen entering the gulf can be traced back to agricultural activities in the Mississippi River basin.[50] Similarly, pesticides

38 *Id.*

39 Ruhl, *supra* note 23, at 291.

40 Eubanks, *supra* note 1, at 255.

41 *Id.*

42 For a further discussion of industrial agriculture's contribution to soil erosion, *see* Johan Boardman, Jean Poesen & Robert Evans, *Socio-economic Factors in Soil Erosion and Conservation*, 6 ENVTL. SCI. & POL'Y 1 (2003).

43 Ruhl, *supra* note 23, at 284.

44 Jan Lewandrowski, James Tobey & Zena Cook, *The Interface Between Agricultural Assistance and the Environment: Chemical Fertilizer Consumption and Area Expansion*, 73 LAND ECONOMICS 404 (1997).

45 Eubanks, *supra* note 1, at 256.

46 *Id.*

47 Ruhl, *supra* note 23, at 288.

48 *Id.*

49 Eubanks, *supra* note 1, at 256.

50 *Id.*

used on farm fields can be washed away by rainwater and end up in water-bodies, exerting their own harmful effects on fish and aquatic life.[51]

In addition to fertilizers and pesticides in water-bodies, sedimentation from soil erosion is a major concern. Soil erosion results from tilling practices that dislodge soil, making it vulnerable to being carried off by runoff.[52] The Green Revolution's shift from perennial rotation of crops to large single-crop monocultures, such as most corn fields, has led to erosion of topsoil.[53] Not only are large quantities of topsoil critical to future farm productivity lost by erosion, but it is estimated that more than 2 billion tons of sediment enters the nation's waterways each year.[54] This sedimentation can clog streams and fill in shallow areas in water bodies, thereby reducing habitat, and reduce light availability to submersed plants.[55]

In addition to water resource impacts, agriculture also contributes to impacts to wildlife and ecosystem diversity. Harm to wildlife and biodiversity from agriculture occurs in a number of ways.[56] Conversion of natural areas into farmland reduces or eliminates habitat.[57] Sedimentation from erosion harms aquatic organisms.[58] Eutrophication from fertilizer runoff chokes out oxygen, thereby killing submersed plants and aquatic organisms.[59] Pesticides harm wildlife and aquatic organisms.[60] This occurs through direct contact to animals that are in farm fields when they are treated with pesticides, as well as from aerial drift and runoff from farm fields into non-farm areas where wildlife species are present.[61] Further, certain pesticides bio-accumulate in the food chain, exposing predatory species to highly concentrated pesticides in their food sources.[62]

High-intensity agriculture, such as corn production, has not only a large "water footprint," but also a large "carbon footprint." Many of the inputs relied on in industrial agriculture are derived from fossil fuels. Nitrogen fertilizers are derived from natural gas made from fossil fuels.[63] Most synthetic pesticides are

---

51 Ruhl, *supra* note 23, at 283.

52 Eubanks, *supra* note 1, at 257.

53 *Id.* at 258.

54 *Id.* at 257.

55 *Id.*

56 *See generally*, Fatal Harvest: The Tragedy of Industrial Agriculture, *supra* note 24.

57 Alex Avery & Dennis Avery, *High-Yield Conservation: More Food and Environmental Quality through Intensive Agriculture*, *in* Agricultural Policy and the Environment, 135–6 (Rodger E. Meiners & Bruce Yandle, eds, 2003).

58 Ruhl, *supra* note 23, at 278.

59 Eubanks, *supra* note 1, at 255–6.

60 *Id.* at 258–9.

61 Ruhl, *supra* note 23, at 283.

62 James M. Armitage & Frank A.P.C. Gobas, *A Terrestrial Food-Chain Bioaccumulation Model for POPs*, 41 Envtl. Sci. & Tech. 4019, 4020 (2007).

63 William S. Eubanks II, *The Sustainable Farm Bill: A Proposal for Permanent Environmental Change*, 39 Envtl. L. Rep. News & Analysis 10493, 10504 (2009).

made from fossil fuels.[64] Fossil fuels, especially diesel and gasoline, are used for heavy machinery, including tractors and combines, as well as for transportation of agricultural products to processing facilities and ultimately to retail grocery stores.[65] Agriculture accounts for about 20 percent of US fossil fuel consumption as well as 37 percent of US and 15 percent of worldwide GHG emissions.[66] It is estimated that takes 10 calories of petroleum to yield 1 calorie of industrial food and almost 1 gallon of gasoline to produce 1 bushel of corn.

In addition to the environmental impacts of industrial agriculture, many of which indirectly affect human health, certain agricultural practices can cause direct harm to humans. This is particularly true with regard to pesticide use. As discussed above, pesticides can leach into groundwater and runoff into surface waters causing contamination to drinking water sources, contamination of fish that humans consume, and direct skin contact through bathing or swimming in contaminated waters. Humans are also directly exposed to pesticides through aerial drift and pesticide residues in food.[67] By far, the largest group of people to be directly impacted by pesticides is farmworkers and their families, who are put at significant risk from high levels of pesticide exposure.[68] Finally, industrial agriculture can impact human health indirectly, by influencing the foods people eat. For example, the historic glut of corn resulting from US agricultural policy has resulted in cheap high fructose corn syrup that has been incorporated into virtually all processed foods, which according to some has contributed to the current obesity and diabetes epidemic.[69]

The Farm Bill Commodity Subsidy Programs

Despite the many environmental and human health impacts of industrial agriculture, US Farm Policy continues to encourage high-yield monoculture and intense

64 *Id.*

65 *Id.*

66 *Id.*

67 A detailed discussion of the risks to humans from pesticide exposure from food and water is beyond the scope of this chapter. For a more detailed discussion, *see* Edwin D. Ongley, *Pesticides as Water Pollutants*, *in* Control of Water Pollution from Agriculture—FAO Irrigation Papers, *available at* http://www.fao.org/docrep/W2598e/w2598e07.htm.

68 A detailed discussion of the health effects of farmworker exposure to pesticides is beyond the scope of this chapter. For a more detailed discussion, *see* Eubanks, *supra* note 1, at 276.

69 A detailed discussion of the health effects linked to high fructose corn syrup is beyond the scope of this chapter. For a more detailed discussion, *see* George A. Bray, Samara Joy Nielsen & Barry M. Popkin, *Consumption of High-Fructose Corn Syrup in Beverages May Play a Role in the Epidemic of Obesity*, 79 American Journal of Clinical Nutrition 537–43 (April 2004).

chemical input industrial agriculture. The primary policies that encourage this form of agriculture are the commodity subsidy programs in the Farm Bill. The major agricultural subsidy programs are found in what is commonly referred to as the Farm Bill. The roots of the Farm Bill date back to the 1933 Agricultural Adjustment Act, which was considered to be emergency legislation designed to stabilize farm prices and prevent a collapse of the farming system during the Great Depression. The focus of this early legislation was on stabilizing crop prices by reducing the huge commodity surpluses that existed at that time, which depressed the prices farmers could get for their crops. These programs, which provided financial incentives for farmers who took vast acreages of farmland out of agricultural production, coupled with the literal "dumping" of surplus crops, resulted in a 50 percent increase in gross farm income within three years. The legislation thus served its purpose of averting a crisis. Nevertheless, the policies in the original legislation continued to serve as the backbone of US agricultural policy long after the farming crisis of the Great Depression was averted. Every five years or so, Congress passes another Farm Bill, which primarily consists of a set of amendments to the Agricultural Adjustment Act. Subsidies in the early Farm Bills were primarily designed to limit production to keep prices high, which was accomplished through land "set asides." The result of these acreage set asides was that farmers found more intensive ways to achieve higher per acre yields. Beginning in the 1940s, the Green Revolution led to further yield increases through the development of new technologies to increase yields and policies to grow as much as possible.

More recent commodity subsidy provisions under the Farm Bill comprised a large number of complex programs including price and income support programs. Price support programs are geared toward bolstering and stabilizing prices for certain specified commodities, including corn. One of the major price support programs is the non-recourse loan program. Income support programs are direct payments to farmers. Direct payments and counter-cyclical payments comprise the largest categories of subsidies. From 1995 through 2006, $8,807,823,536 of direct payments and $5,381,622,107 in counter-cyclical payments went to corn producers.[70]

Over the past 70+ years, although the basic structure of the Farm Bill has remained intact, numerous programs have been added and several significant changes have been made. One significant change was the policy shift that took place in the early 1970s, which encouraged farmers to produce as much as possible, rather than limit production. Another change was the addition of certain conservation programs in the 1985 Farm Bill. These programs, while similar

70 Environmental Working Group, Farm Subsidy Database, Total Direct Payments in United States, *available at* http://farm.ewg.org/farm/progdetail.php?fips=00000&progcode=total_dp (last visited June 5, 2011); Environmental Working Group, Farm Subsidy Database, Total Counter Cyclical Payments in United States, *available at* http://farm.ewg.org/farm/progdetail.php?fips=00000&progcode=total_cc (last visited June 5, 2011).

to the early Farm Bill acreage set aside programs, had a different purpose than the early set aside programs. The new conservation programs were not targeted at limiting production, but instead were aimed at conserving certain areas such as highly erodible lands, wetlands, and other environmentally sensitive lands.[71] Ultimately, these conservation programs resulted in farmers attempting to increase yield even further on the lands that were not identified for conservation. Another significant policy shift was adopted as part of the failed attempts in the 1996 Farm Bill to begin to wean commodity producers off federal subsidies. This change "decoupled" certain commodity subsidy programs from productions levels. In other words, for certain specified subsidy programs, the amount of financial incentive received would no longer be tied to the amount of crop produced by the farmer. Although the primary goals of decoupling were to address inefficiency through market distortion, to address variability in federal budget exposure, and to respond to international trade distortion concerns, one of its benefits was that it would no longer provide incentives for farmers to produce large surpluses of certain commodity crops.[72] However, as discussed in more detail below, not all commodity subsidy programs were decoupled, and history has shown that the decoupling that did occur does not appear to have resulted in surplus reductions.

The most recent Farm Bill is the Food, Conservation, and Energy Act of 2008.[73] This Act mostly reauthorizes and makes small changes to programs in previous Farm Bills. Some new programs were added, but the basic structure of the previous Farm Bills is maintained. The primary types of subsidies in the latest Farm Bill include income supports, price supports, and acreage set-asides.[74] The goal of income supports is to keep farmers' income high regardless of the prices they get for crops. The goal of price supports is to keep prices that farmers get for crops at a higher level and more stable. The goal of acreage set-asides is to conserve certain lands for conservation purposes and to limit supply.

The largest category of commodity subsidy programs is the income support programs. Included in this category are direct payments, counter-cyclical payments, and the Average Crop Revenue Election program. Direct payments are made to producers with "eligible historical production of commodity crops."[75] The covered commodity crops are wheat, feed grains (including corn), cotton, rice, and

71 16 U.S.C. §§ 3811, 3821, 3831 & 3837 (2011).

72 Food and Commodity Policy: What is Meant by Decoupling? *available at* http://www.ers.usda.gov/Briefing/FarmPolicy/decoupling.htm (last visited May 27, 2011).

73 Food, Conservation, and Energy Act of 2008, Pub. L. No. 110-246, §§ 1102–1104, 122 Stat. 923–1525 (2008).

74 For a detailed summary of the commodity programs in the 2008 Farm Bill, *see* Jim Monke, *Farm Commodity Programs in the 2008 Farm Bill*, CRS Report for Congress, RL34594, September 30, 2008, *available at* http://www.nationalaglawcenter.org/assets/crs/RL34594.pdf.

75 United States Department of Agriculture Economic Research Service, 2008 Farm Bill Side-By-Side, Title I: Commodity Programs, *available at* http://www.ers.usda.gov/FarmBill/2008/Titles/TitleIcommodities.htm (last visited May 26, 2011).

oilseeds. The payment amount is the product of the payment rate (statutory), the historical payment acres (83.3 percent in crop years 2009–2011 and 85 percent in crop years 2008 and 2012) of the covered commodity, and the historical payment yield for the commodity. Direct payments are limited to $40,000 per person per crop year. Payments must be linked to a person or legal entity, either directly or indirectly. One of the most interesting aspects of the direct payment program is that the producer does not need to actually grow the crops to get the payments. For example, if the producer historically grew a covered crop, such as corn, she can now grow soy and still get the subsidy. It is even possible to get the subsidy for leaving the farmland fallow—i.e., for growing nothing at all.

The 2008 Farm Bill allows the 2002 Farm Bill Direct Payment Program to remain intact, with a few minor changes. In particular, the 2008 Bill provides that payment acres for crop years 2009–2011 are 83.3 percent and eliminates the three-entity rule (which limited the number of farms from which a person could receive program payments—full payment directly and up to half a payment for two additional entities) for payment limits.

Under another income support program, the Counter-cyclical Payment Program, payments are made to producers with "eligible historical production of commodity crops."[76] As with other programs, the covered crops are limited to wheat, feed grains (including corn), cotton, rice, legumes, and oilseeds. Payments are made if the effective price for a commodity is less than the target price, regardless of the market price of the commodity.[77] The limit of counter-cyclical payments is $65,000 per person per year. The effective price for covered commodities, except rice, equals the sum of the higher of: (1) the national average market price received by producers; or (2) the national average loan rate for marketing assistance loan and the payment rate for the covered commodity.[78]

Under the Average Crop Revenue Election (ACRE) Program, which appears for the first time in the 2008 Farm Bill and covers wheat, feed grains (including corn), cotton, rice, legumes, and oilseeds, producers may make an irrevocable election (once enrolled, a farm is enrolled in ACRE until end of the 2012 crop year) to receive a state-based revenue guarantee equal to 90 percent of benchmark state yield multiplied by the ACRE program guarantee price for the crop year.[79] If the producer elects to receive ACRE payments, she is not eligible to receive counter-cyclical payments, direct payments are reduced by 20 percent and marketing assistance loan rates are reduced by 30 percent. This program provides a revenue guarantee each year based on state market prices and average state

76 *Id.*
77 *Id.*
78 *Id.*
79 *Id.*

yields. Payments are limited to $65,000 per farm in addition to the 20 percent reduction in direct payments and 30 percent reduction in marketing loan rates.[80]

The ACRE Program guarantee may not increase or decrease more than 10 percent from the previous year. Two requirements that must be met before ACRE payments may be made are: the Actual State Revenue for the crop year for the covered commodity is less than the ACRE program guarantee; and the Actual Farm Revenue for the crop year for the covered commodity or peanuts is less than the farm ACRE benchmark revenue for the crop year. The ACRE payment is equal to the product of the lesser of: (1) ACRE program guarantee minus Actual State Revenue; or (2) 25 percent of the ACRE program guarantee, 83.3 percent of planted acres in 2009–2011, or 85 percent of planted acres in 2012, and the ratio of the farmer's five-year average yield per planted acre divided by the state's five-year average yield per planted acre. Under the ACRE program, the total number of acres for which producers may receive payments may not exceed total base acreage for all covered commodities and peanuts on the farm. In addition to the income support programs described above, the Farm Bill has specific subsidy programs that apply only to sugar and dairy.[81]

The price support subsidy programs include the Marketing Assistance Loan (MAL) program, which provides a type of "non-recourse loan." Under this program, producers of wheat, corn, grain sorghum, barley, oats, upland cotton, extra-long staple cotton, long and medium grain rice, soybeans, other oilseeds, peanuts, wool, mohair, honey, dry peas, lentils, and small and large chickpeas are allowed to receive a non-recourse loan at a commodity-specific loan rate per unit of production.[82] Producers pledge their production as collateral for the loan. Producers may then store their crops and sell their production when market conditions may be more favorable. If market prices are higher than the loan rate, the producer will sell her crop, pay off the loan and keep the profits. If, on the other hand, market prices are lower than the amount of the loan, the producer may pay off the loan or forfeit the pledged crop to the Commodity Credit Corporation at loan maturity. In years when market prices are lower than the loan rate this program means, in essence, that the federal government purchases crops at above market value.[83] Thus, growers have the security that they can either take advantage of market prices in good years or simply "sell" their crops to the government at above market value in bad years. Because of the glut of many commodity crops most years, the latter occurs frequently. Thus, the federal government is left with vast quantities of surplus corn and other commodities, for which it must either find a market or some method of disposal. These large surpluses are what have created the flood of cheap corn into the marketplace where it can serve as inexpensive industrial feedstock for processed foods.

80 *Id.*
81 *Id.*
82 *Id.*
83 *Id.*

Other commodity subsidy programs include Crop Insurance programs, under which producers of specific crops, primarily commodity crops, are able to purchase insurance policies at a subsidized rate.[84] One of the most significant differences between the existing income support programs and the non-recourse loan program is that while direct payments are limited to $40,000 per person per year and counter-cyclical and ACRE payments are limited to $65,000 per person per year, non-recourse loan payment amounts are not capped under the 2008 Farm Bill.[85] The 2002 Farm Bill placed a $75,000 per person per year limit on these loans, limiting the maximum total amount of commodity payment that could be received to $360,000 per farm per year. Under the 2008 Farm Bill, the limit for direct and counter-cyclical payments continues to be $210,000 per farm couple per year, however there is no longer a limit on non-recourse marketing loans.[86] Thus, at least in theory, farming couples could receive considerably more than the previous maximum of $360,000 per farm per year.

Although not designed to be conservation programs, the Direct Payments, Counter-Cyclical Payments, and Average Crop Revenue Election programs require producers to agree to carry out certain conservation-related practices to qualify for the payments. For example, during the crop year in which they receive payments the producers must agree to the following: not to produce an agricultural commodity on, designate for conservation uses, or set aside land where a majority of it is highly erodible;[87] not to produce an agricultural commodity on a wetland that has been drained, dredged, filled, leveled, or converted in any way for production of an agricultural commodity;[88] to plant any crop on base acres except fruits, vegetables (excluding mung beans and pulse crops), or wild rice;[89] to use the farmland for agricultural or conservation uses; and to control noxious weeds and follow sound agricultural practices.[90] A limited number of exceptions allow the planting of the prohibited crops, however direct payments and counter-cyclical payments are reduced by an acre for each acre planted with a prohibited crop.

---

84 *Id.*

85 *Id.*

86 Renée Johnson, Geoffrey S. Becker, Tom Capehart, Ralph M. Chite, Tadlock Cowan, Ross W. Gorte, Charles E. Hanrahan, Remy Jurenas, Jim Monke, Jean M. Rawson, Randy Schnepf, Joe Richardson, Donald J. Marples, Mark Jickling & N. Eric Weiss, The 2008 Farm Bill: Major Provisions and Legislative Action, CRS Report for Congress, RL34696, October 3, 2008, *available at* http://www.nationalaglawcenter.org/assets/crs/RL34696.pdf.

87 The conservation requirements are found in subtitle B of title XII of the Food Security Act of 1985, 16 U.S.C. § 3811.

88 The wetland protection requirements are found in subtitle C of title XII of Food Security Act of 1985, 16 U.S.C. § 3821.

89 7 U.S.C. § 8717(b) (2011).

90 Farm Service Agency, USDA, FSA Handbook: Direct and Counter-Cyclical Program and Average Crop Revenue Election for 2009 and Subsequent Crop Years, 2009, *available at* http://www.fsa.usda.gov/Internet/FSA_File/1-dcp_r03_a03.pdf.

Cosmetic Standards

Another government policy that can encourage pesticide use is the cosmetic standards for produce and other foods found in both the FDA's action defect levels and USDA's voluntary grading standards. FDA establishes "defect action levels" (DALs) that specify the allowable level of certain defects in specified foods. While many of these are at least arguably health related, for example, the number of rodent hairs or insect parts allowable in a specified volume of food, some of the DALs are for purely cosmetic purposes. For example, DALs specify allowable levels of "damage" in specified produce. "Damage" is defined to include "evidence of pest habitation or feeding," which includes evidence of tunneling and gnawing. Although perhaps not cosmetically appealing, tunneled or gnawed plant parts in themselves do not pose health problems or affect nutrition such that they should not be permitted. In its handbook, FDA states that its position is that "pesticides are not the alternative to preventing food defects," and the "primary use of pesticides in the field is to protect food plants from being ravaged by destructive plant pests ...."[91] Nevertheless, FDA's handbook goes on to explain that "a secondary use of pesticides is for cosmetic purposes—to prevent some food products from becoming so severely damaged by pests that it becomes unfit to eat."[92] However, it is not clear why mere plant damage would make a plant unfit to eat. These action levels should be reevaluated to determine whether they are necessary to protect against health threats or nutritional diminution or whether they are merely designed to make foods meet certain cosmetic standards that American culture has begun to expect without understanding the implications of its visually "perfect" food.

In addition to FDA standards, the grading standards established by USDA also encourage pesticide use to achieve purely cosmetic desires. USDA has established specific grading standards for a wide range of fruits, vegetables and other foods. For example, USDA standards and grades for Florida oranges and tangelos provide that to be graded as "fancy," the fruit may not have more than one-tenth of its surface discolored.[93] To be graded as "bright," the fruit can have up to one-fifth discoloration, and to be graded "US No. 1," the fruit may not have more than one-third of its surface discolored.[94] Discoloration is defined as "russeting of a light shade of golden brown caused by rust mites or other means."[95] Citrus rust mites only cause superficial discoloration to the outside of the fruit, which

91 US Food and Drug Administration, Food, Guidance, Compliance, & Regulatory Information, Defect Levels Handbook, *available at* http://www.fda.gov/food/guidancecompliance regulatoryinformation/guidancedocuments/sanitation/ucm056174.htm (last visited May 26, 2011).

92 *Id.*

93 *Id.*

94 *Id.*

95 *Id.*

does not affect the taste or nutritional content of the fruit. However, to limit the extent of rust mite damage to achieve the desired grading, chemical pesticide use is needed. Although these grading standards are voluntary, they have in essence become a requirement to be able to have a marketable product. Foods that do not meet the higher grades receive significantly lower prices. Moreover, many contracts between buyers and growers require specified grading standards to be met. Markets may not be available at all for foods not meeting the high standards. Thus, the voluntary grading standards have "evolved into mandatory industry requirements."[96] Given the significant human health and ecological impacts of chemical pesticide use, using these chemicals for mere cosmetic purposes does not appear to be justified. Accordingly, the grading standard system should be re-evaluated. Further, however, consumers must be educated so that they understand that cosmetic imperfection does not necessarily affect the taste or nutritional value of food, and that in fact achieving cosmetic perfection may require the use of harmful chemicals.

The evolution of American agriculture from the small family farm with a diversity of crops and high levels of manual labor to one of large industrial monocultures with high levels of chemical impacts, is a result of a variety of social, economic, and government-policy drive factors. While not the sole factor, US agricultural policy has certainly played a significant role in facilitating this transformation. Current agricultural practices contribute substantially to environment and human health harms, including, but not limited to, water quantity and quality impacts, greenhouse gas emissions, air pollution, and impacts to biodiversity. Dependence on chemical pesticide inputs is both a cause and a consequence of current agricultural practices and is one of the most significant contributors to environment and human health impacts associated with modern agriculture.

96 David Pimentel & Hugh Lehman, The Pesticide Question: Environment, Economics, and Ethics 90 (1993).

Chapter 5
The History of the Control of Pests

History

Although we often think that the word "pesticide" is synonymous with the term "chemical pesticide," the truth is that there are many different ways to control pests and many types of pesticides that are not derived from synthetic chemicals. From a legal perspective, the term "pesticide" is defined as "any substance or mixture of substances intended for preventing, destroying, repelling, or mitigating any pest …."[1] Under this very broad definition of pesticide, not only are the toxic chemicals we normally associate with the term covered, but in addition, many naturally occurring substances and biological organisms themselves can be considered to be pesticides. Moreover, while not fitting within the definition of the term pesticide, a range of agricultural practices can be used to control pest populations.

Scientists believe that chemical pesticides have been used by humans for over 2,000 years. Homer described how Odysseus used burning sulfur as a fumigant to control pests.[2] Additional documentation of pesticide use, such as the use of arsenic to kill insects, can be found in literature dating back to the time of Christ. There are records of arsenic being used to control insects dating back to the year 70 AD, when Pliny the Elder recommended its use to kill insects.[3] Arsenic was also used as a pesticide as early as sixteenth-century China. Paris Green, a form of arsenic, was employed in the nineteenth century to combat a major outbreak of the Colorado potato beetle.[4] The remainder of the twentieth century saw a dramatic increase in the types and quantities of chemical pesticide use.

Early chemical pesticides primarily consisted of metals, such as arsenic, lead and copper, and oils, which rather than exerting a toxic effect, tend to act as physical barriers to pest feeding or actually suffocate the pest. After World War II, vast quantities of synthetic organic compounds, such as DDT, chlordane, parathion, malathion, and a plethora of others were used extensively for agricultural pest control and for public health purposes to control vector-borne diseases such as malaria and yellow fever. During this same period, a number of biological and naturally occurring plant-derived pesticides were also used, although to a

1 7 U.S.C. § 136(u) (2011).

2 Homer, The Odyssey (Edward McCrorie, transl. 2004).

3 Clive A. Edwards, *The Impact of Pesticides on the Environment*, *in* The Pesticide Question: Environment, Economics and Ethics 13 (David Pimentel & Hugh Lehman, eds, 1993).

4 H.F. van Emden & M.W. Service, Pest and Vector Control 52 (2004).

much lesser degree. By the end of the twentieth century, genetically modified microorganisms and plants became ubiquitous forms of pest control.

Cultural Controls

Although metals such as arsenic have been used as pesticides for centuries, most pest control up until the early twentieth century was accomplished through non-chemical controls, primarily by means of cultural controls, such as cultivation, sanitation, crop rotation, and sowing and harvesting practices.[5] Many pest insects live out at least part of their life cycles in soil, weeds, or accumulated crop debris in farm fields. Plowing the top layer of soil kills many of these pest insects.[6] Accordingly, soil tillage historically was a critical component of agricultural pest management. It was not until relatively recently that, as a way to minimize soil erosion, tillage was abandoned in favor of zero or minimum tillage systems, which rely on herbicide usage to control weeds.[7] The demise of tillage as a core component of modern agricultural systems has resulted in a dramatic increase in certain soil-dwelling pests.[8] Other cultivation pest control techniques used historically include mulching,[9] compacting,[10] and manuring.[11] Sanitation practices are among the most effective pest control practices used in both ancient and modern agriculture.[12] By destroying residues of crops left in fields after harvesting, many pest populations that live in such residues are destroyed.[13] Related practices such as destruction of weed hosts and selective pruning also serve as effective pest control tools.

Crop rotation, one of the oldest forms of pest control, is very effective for minimizing soil-dwelling pests.[14] By alternating the planting of different crops in a particular field, populations of soil-dwelling insects that feed on a particular crop are prevented from building up during periods when their food crop is not present. Thus, when the crop eventually is planted, populations of the pest species generally will not be high enough to cause serious problems. Timing sowing and planting dates to avoid pest outbreaks or to ensure the crop plant is in a resistant growth stage when pest outbreaks are likely to occur, as well as carefully tailoring seed and planting rates and early harvesting also can be effective tools for avoiding

5 *Id.* at 123.
6 *Id.* at 125.
7 *Id.* at 128.
8 *Id.*
9 *Id.* at 131.
10 *Id.* at 125.
11 *Id.* at 131.
12 *Id.* at 132, 134.
13 *Id.* at 134.
14 *Id.* at 141–2.

pest damage to crops.[15] Mechanical compaction of soils and tailored irrigation can be used to manipulate pest habitat by eliminating wet breeding areas or inundating breeding areas.

Perhaps the most important of the cultural controls are sanitary measures to reduce pest habitat.[16] For example, an effective way to control disease-carrying mosquitoes is by properly disposing of cans, tires, puddles and any other open container or feature that can collect rainfall and become breeding habitat for mosquito larvae.[17] An ambitious effort to eliminate standing water that serves as breeding habitat is credited with eliminating malaria, yellow fever and other insect-borne diseases from regions of Panama, thereby permitting the construction of the Panama Canal. On farms, sanitary methods primarily involve removing crop residues, such as stalks remaining after harvesting, which otherwise become habitat for pests and disease.[18]

Other non-chemical mechanical means of control include covering crops with mesh material or covering the ground with natural or synthetic mulch to keep out pest insects and weeds. One of the most effective mechanical controls is the use of window screening and mosquito netting to prevent mosquitoes from feeding on humans and thus prevent the transmission of mosquito-borne diseases.[19] Cultural controls, including crop rotation, intercropping and sanitation are important components of both integrated pest management programs and organic farming.

Conventional Chemical Pesticides

EPA considers conventional pesticides to be "[a]ny man-made chemical which can be used to kill pests."[20] It was not until the latter half of the twentieth century that the development of synthetic chemical pesticides led to an explosion of global pesticide use. Because these new synthetic chemical pesticides were spectacularly effective at controlling a wide variety of pests, they quickly gained favor and, before long, were ubiquitous. Estimates of global pesticide use are staggering. More than 1,600 types of pesticides are currently available.[21] More than 5 billion pounds of pesticides, with a value of almost 40 billion dollars, are used annually around the world.[22] Pesticide use in the United States accounts for approximately

15 *Id.* at 144.

16 *Id.* at 132.

17 *Id.*

18 *Id.* at 134.

19 *Id.* at 243–4.

20 http://www.epa.gov/pesticides/glossary/index.html#c (last visited June 21, 2011).

21 Edwards, *supra* note 3, at 13.

22 EPA Pesticide Industry Sales and Usage 2006/2007, *available at* http://www.epa.gov/opp00001/pestsales/ (last visited April 13, 2011).

22 percent of global pesticide usage, with more than 1 billion pounds of pesticides with a value of approximately 12 billion dollars used annually in the US.[23] US exports to other countries exceed 450 million pounds of pesticides per year.[24] The use of chemical pesticides has more than doubled since the time large-scale environmental regulation was instituted in the early 1970s. Pesticide use in the US has almost tripled since the time of the publication of *Silent Spring* in the early 1960s.[25] The Pesticide Action Network asserts that, starting in 2004, there was a significant growth in worldwide pesticides sales.[26] The current global pesticide market is approximately 40 billion dollars per year.[27] This number is expected to grow by approximately 3 percent per year, reaching approximately 52 billion dollars per year by 2014.[28] Agricultural pesticides account for approximately 80 percent of the global pesticide market.

The impetus for the dramatic increase in chemical pesticide usage during the very time period that we started to become aware of the serious ecological and health impacts of chemical pesticides is the structure of our farming systems, which rely in large part on vast monocultures. Monocultures eliminate the diversity, and thus the natural forces that can keep pest populations in check, that was formerly derived from intercropping, crop sequencing, or crop rotation. Consequently, large monocultures depend on chemical pesticide inputs to control pests.[29]

The rapid worldwide adoption of synthetic chemical pesticides began during World War II, with the development of two primary categories of chemical insecticides, the organochlorines and the organophosphates. The first category, the organochlorines, includes the notorious pesticide DDT (the abbreviation for 1, 1, 1-trichloro-2, 2-bis (*p*-chlorophenyl) ethane),[30] developed by the Nobel Prize-winning Swiss scientist Paul Mueller,[31] as well as the lesser known chlordane, toxaphene, aldrin, and dieldrin. The organochlorine pesticides were first considered to be highly desirable because, while they are very toxic to a broad range of invertebrates, they are not highly acutely toxic to humans or other mammals. Organochlorine pesticides, such as DDT, are credited with saving thousands of lives from insect-borne diseases, such as typhus and malaria, during World War II.[32]

---

23 *Id.*

24 Edwards, *supra* note 3, at 13.

25 Jason Clay, World Agriculture and the Environment 49 (2004).

26 Pesticide Action Network, Myths about Pesticides, *available at* http://www.panna.org/myths (last visited May 27, 2011).

27 *Id.*

28 *Id.*

29 Van Emden & Service, *supra* note 4, at 41–2.

30 Robert E. Pfadt, Fundamentals of Applied Entomology 755 (3rd ed. 1978).

31 Bill Freedman, Environmental Ecology: The Impacts of pollution and Other Stresses on Ecosystem Structure and Function 191 (1989).

32 Andrew P. Morriss & Roger E. Meiners, *Property Rights, Pesticides, and Public Health: Explaining the Paradox of Modern Pesticide Policy*, 14 Fordham Envtl. L.J. 1 (2002).

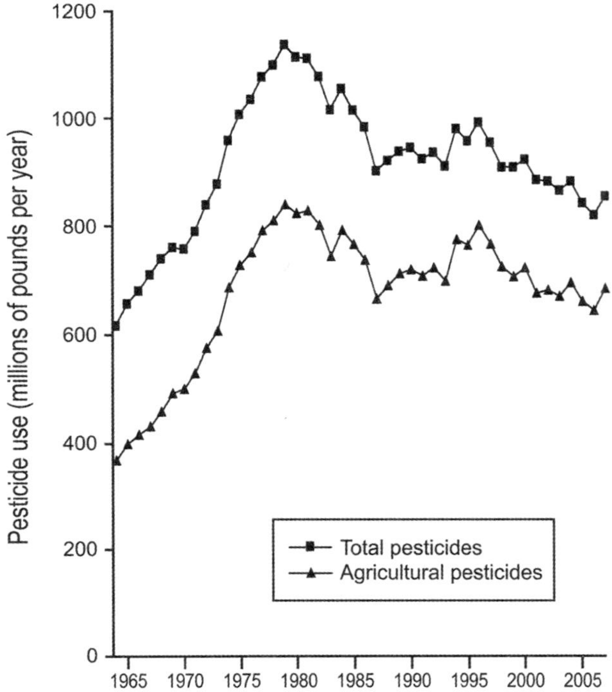

Figure 5.1 Total conventional pesticides and agricultural conventional pesticides used in the United States from 1964 to 2007

Source: Timothy Kiely, David Donaldson & Arthur Grube, *Pesticide Industry Sales and Usage: 2000 and 2001 Market Estimates* US EPA 2004) *available at* http://www.epa.gov/opp00001/pestsales/01pestsales/market_estimates2001.pdf (last visited October 4, 2012); and Arthur Grube, David Donaldson, Timothy Kiely & La Wu, *Pesticide Industry Sales and Usage: 2006 and 2007 Market Estimates* US EPA 2004) *available at* http://www.epa.gov/opp00001/pestsales/07pestsales/market_estimates2007.pdf (last visited October 4, 2012).

The dramatic pest control results initially achieved by the use of DDT and its organochlorine relatives led to such heavy use of these pesticides that large concentrations of the pesticides ended up accumulating in soils and running off into water bodies where non-target fish and wildlife became exposed. Massive amounts of DDT were used in the US in the 1940s, 1950s and 1960s. Worldwide approximately 4 billion pounds of DDT have been used since it was first introduced.[33] Moreover, DDT initially was used as a coating of seeds, which were attractive food for birds, further exacerbating the impacts on wildlife. Although

33 EPA, Persistent Organic Pollutants A Global Issue, A Global Response, *available at* http://www.epa.gov/oia/toxics/pop.html (last accessed May 27, 2011).

Photo 5.1 Aerial application of pesticides
Reprinted with permission. Larry R. Barber, USDA Forest Service, Bugwood.org.

DDT did not typically kill birds and other non-target organisms outright, it nevertheless resulted in substantial avian and wildlife impacts. Many of the results were sub-lethal effects on reproduction and behavior, which had long-term impacts on populations.[34] Moreover, organochlorine pesticides are extremely persistent in the environment, which makes them highly effective for long-term pest control. However, their persistence in the environment became their downfall when the long-term ecological consequences of these pesticides became apparent. It did not take long for it to become evident that these pesticides accumulated in living tissues and bioconcentrated as they moved through the food chain.[35] This resulted in serious impacts to predators at the top of the food chain, including the Bald Eagle and the Osprey. As a consequence, most organochlorine pesticides were either banned or severely restricted, at least in the developed countries of the world.

The other major category of pesticides that was developed during World War II, the organophosphate pesticides, includes the pesticides parathion, malathion, and chlorpyrifos. Chemicals in this category affect the nervous systems of vertebrates by inhibiting the ability of an important enzyme, called cholinesterase, to regulate

34 Van Emden & Service, *supra* note 4, at 112.

35 *See* Carl F. Cranor, Legally Poisoned: How the Law Puts Us at Risk from Toxicants 114–20 (2011).

properly a neurotransmitter known as acetylcholine, which is critical to proper nerve function.[36] Organophosphate pesticides were initially developed as wartime nerve gases.[37] Although these pesticides have the environmental advantage of being far less persistent in the environment than organochlorine pesticides, the organophosphates tend to be acutely toxic to humans, other mammals, and birds.[38] These pesticides became the pesticides of choice in the United States after most organochlorine pesticides were banned or severely restricted. During the 1990s and early 2000s, EPA brought a number of regulatory actions, in response to actions initiated by environmental organizations or as a result of the stringent registration and reregistration review mandates imposed by the 1996 Food Quality Protection Act, which resulted in the banning or severe restriction of some of the most harmful organophosphate pesticides. In 1980, approximately 228 million pounds of organophosphate pesticides were used in the US.[39] By 2007, that number had dropped to approximately 93 million pounds per year.[40] Nevertheless, this category of pesticides remains one of the largest categories of chemical insecticide in use in the United States today.[41] In addition to posing risks of acute poisoning to farmworkers, these pesticides have been implicated in a large number of avian and wildlife poisonings.

Other categories of chemical pesticides include the synthetic pyrethroids and carbamates. Pyrethrum is a naturally occurring pesticide derived from chrysanthemum flowers.[42] Synthetic pyrethroids are synthetically produced versions of pyrethrum. These pesticides have the environmental benefit of having very low mammalian toxicity and low environmental persistence.[43] Nevertheless, they are highly toxic to a broad range of invertebrates, including many beneficial insects.[44] They are also highly toxic to fish and other aquatic organisms. Carbamates are more persistent than organophosphates in the environment and are generally broad-spectrum, having adverse impacts on many different groups of organisms.[45] Other commonly used pesticides are: nematicides, which are not only of high mammalian toxicity and broad-spectrum, but are also very transient in soil; herbicides, which generally are not highly toxic to mammals, but travel easily in

---

36 *Id.*; *see also*, http://www.epa.gov/pesticides/about/types.htm (last visited June 13, 2011).

37 Cranor, *supra* note 35, at 114–20.

38 *Id.*

39 *See* EPA Pesticide Industry Sales and Usage 2006/2007, *available at* http://www.epa.gov/opp00001/pestsales/ (last visited April 13, 2011).

40 *Id.*

41 *Id.*

42 Edwards, *supra* note 3, at 15.

43 *Id.*

44 *Id.*

45 *Id.*

water, where they may be toxic to fish and aquatic organisms; and fungicides, which vary greatly in their toxicity.[46]

Biopesticides

The biopesticides are a category of pesticides derived from animals, plants, and microorganisms. This category includes biochemicals derived from living organisms, macroorganisms biocontrol agents, and naturally occurring microbes. Biopesticides typically are significantly less harmful than most synthetic chemical pesticides primarily because they tend to act through non-toxic mechanisms. In addition, biopesticides tend to narrowly target limited groups of organisms and thus do not affect many non-target organisms. Biopesticides also tend to be used only in very small doses and decompose very quickly. Because biopesticides generally do not act through toxic mechanisms and thereby are of low toxicity, EPA generally requires considerably less data to support a registration for a biopesticide than it would for a conventional synthetic chemical pesticide.

In 1994, EPA created the Biopesticide and Pollution Prevention Division to encourage and facilitate the registration of biopesticides and to promote the use of reduced-risk pesticides as part of Integrated Pest Management (IPM) programs. Finally, when used as part of an IPM program, biopesticides can help to eliminate or reduce the amount of synthetic chemical pesticides needed to maintain yield.[47] Although considered biopesticides by EPA, genetically modified organisms pose unique risks, and thus are treated as a separate category of pesticides for the purposes of this book.

Biochemical Pesticides

Biochemical pesticides have been used in agriculture for many years. Biochemicals are a diverse class of substances that can be derived from natural or synthesized sources. EPA defines "biochemical pesticides" as "naturally occurring substances that control pests by non-toxic mechanisms."[48] Biochemical pesticides include semiochemicals, plant regulators, and insect growth regulators. Semiochemicals, including pheromones,[49] are chemicals that are emitted by an organism that modifies the behavior of the receptor organisms of similar or different species. For example, certain insect mating attractants are used to disrupt mating behavior of insects in agricultural fields. Plant regulators are used, among other things, to

46 *Id.*

47 EPA, Types of Pesticides, *available at* http://www.epa.gov/pesticides/about/types.htm#biopesticides (last visited May 27, 2011).

48 EPA, Pesticides: Glossary, *available at* http://www.epa.gov/pesticides/glossary/#b (last visited May 27, 2011).

49 *Id.*

accelerate or retard the rate of growth of plants, and include such agricultural chemicals as auxins and gibberellins. Insect growth regulators include juvenile hormones, which act to prevent larval stages of insects from metamorphosing into adult forms, thereby preventing reproduction. Biochemical pesticides tend to be less ecologically harmful than many chemical pesticides, in part because they tend to be of low toxicity, to narrowly target certain groups of organisms, and to breakdown quickly in the environment; they are also useful in very small doses.

Biocontrol Organisms

Biological organisms, including predators, parasites, and pathogens of pest species, can effectively control pest populations. According to van Emden and Service, one of the earliest examples of using a living organism to combat a pest is the use of the ladybird beetle to control a citrus scale insect in California in the late 1880s.[50]

As discussed above, modern agricultural practices such as planting monocultures, not rotating crops, and using chemical pesticides can reduce populations of natural enemies of pests, thereby resulting in serious pest problems. However, a number of techniques can be used to employ natural pest enemies to improve pest control. One approach is to encourage population growth of natural enemies, by creating favorable conditions on the farm. This can be accomplished by reintroducing biodiversity on the farm by undersowing, intercropping and rotating crops, and providing refugia for natural enemies. Conserving and enhancing areas that provide refuge for these enemies ensures that adequate natural enemy populations are present when needed to combat pest outbreaks. For example, leaving islands of uncultivated land within the farm field can support populations of natural enemies.

Where increasing diversity and providing refugia is not sufficient, natural enemy populations can be enhanced by introduction predators, parasites, or pathogens of pest species into the system. There are two primary methods for introducing natural enemies. The first is inoculation, which involves introducing a small number of the desired species with the intent that the introduced organisms will reproduce and multiply to the point where sufficient numbers exist to provide some level of pest control. Another approach is to simply inundate the area with large numbers of the natural enemy species to achieve immediate control without waiting for the organisms to reproduce and build their population.

Biological controls have many benefits over most chemical controls. First, they tend to be pest-specific. In other words, a parasitic wasp that lays its eggs on a particular pest larva will not typically lay its eggs on a broad range of organisms, and thus non-target organisms will not be impacted by the parasite. Biological controls also do not typically have side effects and are not polluting. Finally, in general, pests do not develop resistance to biological organisms such

50 Van Emden & Service, *supra* note 4, at 148.

Photo 5.2 Parasitic braconid wasp pupae on caterpillar
Reprinted with permission. David Cappaert, Michigan State University, Bugwood.org.

as predators and parasites.[51] Biological controls, however, are not without their shortcomings. Specifically, biological controls often act very slowly and typically do not completely exterminate the pest.[52]

A subcategory of biological pesticides is that of microbial pesticides—i.e., living microorganisms that kill pests and can be applied to crops or other pest habitats. One of the best-known examples of a microbial pesticide is *Bacillus thuriengensis* (*B.t.*), a naturally occurring bacterium that produces a substance that is toxic to certain groups of insects. EPA has regulated naturally occurring microbial pesticides such as *B.t.* for many years. Although *B.t.* was first registered by EPA for use as a pesticide in 1959, it was not the first microbe to be used in this way. Between 1939 and 1951, another bacterium, *Bacillus popilliae*, an obligate bacterial pathogen that causes a milky disease in the larvae of the Japanese beetle and other scarab beetles, was used in 14 eastern states and the District of Columbia.[53] Biological pesticides, including living predators, parasites, and microorganisms, are important components of integrated pest management and organic farming.

51 *Id.* at 149–50.
52 *Id.* at 151–2.
53 Pfadt, *supra* note 30, at 239.

Pesticidal Genetically Modified Organisms

In the past 20 years, the fastest growing sector of the pesticide industry has been the biotechnology sector. Naturally existing microbes have been genetically modified (GM) to make them toxic to insects and other pests. In addition, agricultural crop plants themselves have been genetically modified to produce substances that have pesticidal effects. GM plant crops have been planted commercially in the US since 1995. By 2002, more than 88 million acres of genetic-engineering derived crops were being planted annually in the US, and more than 130 million acres were being planted annually worldwide.[54] The Office of Science and Technology Policy states that while the increases in GM crops are most dramatic in the US, other nations, such as Canada, Argentina, and China, are also experiencing significant growth in the development and use of GM crops.[55] In 2004, 40 percent of the corn, 81 percent of the soybeans, and 73 percent of the cotton grown in the United States were genetically engineered.[56] While not all of these GM crops have been genetically modified to impart pesticidal effects, one of the largest categories of GM crops is crops that have been genetically modified to produce the *B.t.* toxin that kills certain groups of insects. The other large category of GM crops is crops that have been genetically altered to be resistant to certain herbicides. These pesticidal living organisms pose novel ecological risks, by virtue of their ability to reproduce and spread in the environment.

The terms "genetic engineering" and "genetic modification" are frequently used interchangeably. The USDA defines "genetic engineering" to mean the "[m]anipulation of an organism's genes by introducing, eliminating or rearranging specific genes using the methods of modern molecular biology, particularly those techniques referred to as recombinant DNA techniques."[57] The term genetically modified organism or GMO is commonly used to refer to organisms that are the product of genetic engineering. This genetic manipulation has, in effect, been a directed change in the evolutionary process for the purpose of developing specific commercially desirable traits such as pesticidal characteristics.[58]

This deliberate alteration of the path of evolution has brought with it a panoply of novel environmental, human health, and economic risks that could not have

54 67 Fed. Reg. 50578 (August 2, 2002).

55 *Id.*

56 *Id.*; Permitting Progress into Tomorrow, Biotechnology Fact Sheet, United States Department of Agriculture, Animal and Plant Health Inspection Service (February 2006), *available at* http://www.aphis.usda.gov/publications/biotechnology/index.shtml (last visited, May 27, 2011). For a complete list of all GMOs that have been approved to be released in the United States, *see* United States Regulatory Agencies Unified Biotechnology, *available at* http://usbiotechreg.nbii.gov/ (last visited May 27, 2011).

57 United States Department of Agriculture, Glossary of Agricultural Biotechnology Terms, *available at* http://www.usda.gov/wps/portal/usda/usdahome?contentid=Biotechno logyGlosary.xml&navid=AGRICULTURE (last visited May 27, 2011).

58 67 Fed. Reg. 50578 (August 2, 2002).

been foreseen when US environmental and health protection laws were drafted. What once took evolution centuries to accomplish can now be done in what seems an instant, prompting one commentator to refer to genetic engineering as "Darwin in hyperspeed."[59] Not only has genetic engineering dramatically accelerated evolution, but it also has accomplished things that probably never could have been accomplished through natural evolution, regardless of the passage of time. Through genetic engineering, DNA can be moved across all biological barriers even at the kingdom level (i.e., between microorganisms, plants, and animals). This dramatic biological-barrier jumping does not occur in nature or through conventional breeding practices, which typically involve cross-fertilization or crossbreeding between varieties or breeds of the same species and, rarely, between species in the same genus.[60]

GMOs are typically portrayed as either a panacea, at one end of the spectrum, or the stuff of science fiction horror movies, at the other extreme. The truth, as with most truths, however, probably falls somewhere in the middle. While GMOs hold the promise of important advances in agriculture, medicine, and industry, they are not without risk. Further, the elements of risk associated with GMOs are frequently different in kind and degree than the risks typically addressed by environmental regulatory programs.

Nanotechnology Pesticides

One of the newest forms of pesticides to be developed is nanotechnology pesticides. Nanotechnology refers to the use of "nano"-sized[61] materials to carry out certain tasks, such as killing harmful microbes. EPA does not yet have any regulations specific to the regulation of nanotechnology pesticides, however, it has indicated that it intends to regulate nanotechnology pesticides under the Federal Insecticide, Fungicide, and Rodenticide Act (FIFRA). Because nanoscale materials are so small, they can pass through living membranes, and thus, could present unique hazards to humans and the environment. Although no formal FIFRA registration

59 Charles A. Deacon & Emilie K. Paterson, *Emerging Trends in Biotechnology Litigation*, 20 Rev. Litig. 589, 590 (2001).

60 John Charles Kunich, *Mother Frankenstein, Doctor Nature, and the Environmental Law of Genetic Engineering*, 74 S. Cal. L. Rev. 807, 812 (2001).

61 EPA defines nanotechnology as "the science of the very small. A nanometer is 100,000 times thinner than a strand of hair. Nano scale particles (nanoparticles) are so tiny that they cannot be detected by microscopes. Nanotechnology encompasses nanoscale science, engineering and technology, and involves imaging, measuring, modeling, and manipulating matter at this very small scale." *Available at* http://www.epa.gov/nanoscience/questions.htm#q1.http://www.epa.gov/pesticides/about/intheworks/nanotechnology.htm (last visited April 13, 2011).

applications have been submitted to EPA, several companies are considering commercial production of nanoscale pesticides.[62]

Integrated Pest Management

"Integrated Pest Management" (IPM) is just what the name suggests, an integration of a variety of pest management approaches. The goal is to integrate a variety of pest management techniques to manage pests, maximizing non-chemical techniques to reduce the problems associated with chemical pesticides.[63] IPM typically relies on some chemical pesticide use, some biological controls, and some cultural controls.[64] One of the most significant features of IPM is that, rather than striving to achieve total "control" over pests, IPM seeks the more modest and more realistic goal of pest "management."[65] IPM was developed in part to combat the problems that were created by previous over reliance on toxic chemical pest control, which resulted in the development of pest resistance, loss of biological controls, loss of biodiversity, and the abandonment of historically used cultural controls.[66]

A fundamental principle of IPM is to "prevent or at least delay counter-adaptation by pests to control measures by diversifying the latter."[67] To accomplish this diversification, users of IPM attempt to maximize the numbers and types of non-chemical pesticide pest management tools.[68] By limiting the use of chemical pesticides, non-chemical approaches such as biological controls, whether naturally occurring or introduced by humans, are not rendered ineffective by toxic chemicals. Chemical pesticides are viewed more in terms of "gap fillers," to be employed only where necessary. Some of the key components of IPM are the use of economic thresholds, the use of pesticides in a manner that minimizes the damage to bio-control, the use of host-plant resistance to pests, and the use of cultural controls.

IPM minimizes the need for chemical pesticides by combining fairly sophisticated scientific knowledge of a site's particular pest problems and the available pest management strategies, with an understanding of the economic thresholds at which a site's pest problems cause unacceptable harm. This final prong, economic thresholds, plays a critical part in IPM. The premise of economic thresholds is that chemical pesticide intervention should only be used

62 EPA, Pesticide Issues in the Works: Nanotechnology, the Science of Small, *available at* http://www.epa.gov/pesticides/about/intheworks/nanotechnology.htm (last visited April 13, 2011).

63 Helmut F. van Emden & David Peakall, Beyond Silent Spring 70 (1996).

64 Van Emden & Service, *supra* note 4, at 304.

65 National Research Council, Ecologically Based Pest Management vii (1996).

66 Van Emden & Peakall, *supra* note 63, at 70.

67 *Id.*

68 *Id.*

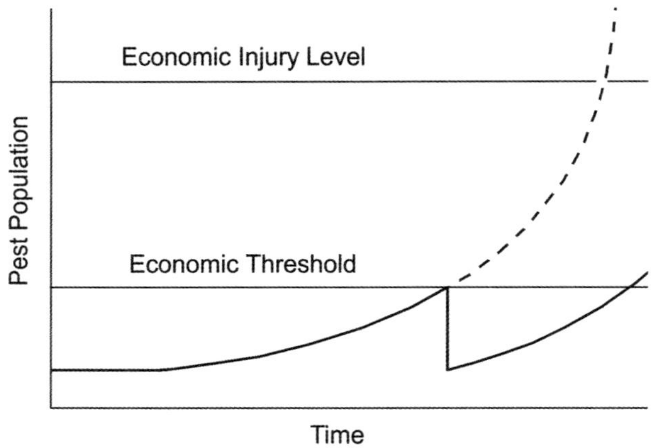

Figure 5.2 Diagram showing IPM economic threshold and economic injury level

when necessary to avoid unacceptable economic harm.[69] Rather than a farmer spraying on a regular schedule, regardless of what is actually happening with pest populations in the field, a farmer using IPM will only apply chemical pesticides at times, in places, and at levels that are determined to be appropriate based on what is actually happening in the field.

It is estimated that simply moving from regular spraying schedules to only spraying when dictated by an economic threshold can reduce chemical pesticide use by 20 to 30 percent.[70] The term "economic threshold" is defined as the pest density at which "control measures should be determined to prevent an increasing pest population from reaching an economic injury level."[71] Accordingly, the "economic threshold" is necessarily lower than the "economic injury level."[72] By applying chemical pesticides when the economic threshold is met, the grower allows time for the pesticide to take effect and reduce pest populations before economic injury occurs[73] (see Figure 5.2).

To determine when economic thresholds are triggered it is necessary to monitor numbers of pests in the field.[74] The basic approach to monitoring is simply to count the number of individuals of a pest species within a given area of the planted crop. A more ecologically sophisticated version of monitoring is to count not only the

69 *Id.* at 71.

70 *Id.*

71 *Id.* (citing V.M. Stern et al., *The Integrated Control Concept*, 29 Hilgardia 81–101 (1959).

72 *Id.* at 71.

73 *Id.*

74 *Id.* at 73.

numbers of pests but also the numbers of predators and parasites of those pests. The ratio of certain pests to certain of their predators and parasites can provide a more valuable indicator of when pest levels are likely to become economically injurious.[75] Of course, the latter approach, while more scientifically sound, is also more complex and time-consuming for farmers.[76] Monitoring data can be entered into computer models that take into account a number of variables including soil conditions and climatic conditions to predict likely pest injury and inform farmers of the best times to apply chemical pesticides.[77]

When chemical pesticides are used in IPM, the goal is to use them very selectively to minimize non-target exposure. This can be accomplished in a variety of ways. First, there are some selective pesticides that are more toxic to the target pest species than they are to beneficial or other non-target species. Although most chemical pesticides are broad-spectrum, affecting a broad range of organisms, some chemical pesticides and many biological pesticides exert their toxic effects primarily on a limited group of organisms. For example, certain varieties of the biological pesticide, *B.t.*, primarily affect insects in the order Diptera (flies and mosquitoes) rather than broadly affecting all insects.[78] Another way to do selective spraying, even where the pesticide itself is not selective, is by using formulations of the pesticide that limit exposure to certain types of organisms.[79] For example, some insecticides can be encapsulated so that small amounts of pesticides are contained in a sticky polymer that adheres to plants, thereby limiting exposure to organisms that actually feed on the plant and avoiding organisms that may simply be present in the area of spraying.[80]

Additionally, reduced doses of certain pesticides will sometimes be toxic to the pest species, while sparing the predators of the pests.[81] Further, selectively choosing the spacial and temporal application of pesticides can focus exposure on pest species and seek to avoid exposure to beneficial species. For example, because natural predators and parasites of pests typically do not arrive or emerge until after pest populations have reached a certain level, early application of short-lived pesticides can result in high exposure to target pest species and low exposure to beneficial predators and parasites.[82] Similarly, in some circumstances, short-lived pesticides can be applied during times when most beneficial insects are protected from exposure, such as during a time when most parasitic wasps are in the pupal stage.[83] Finally, harm to beneficial insects can also be limited by being selective in the application

75 *Id.*
76 *Id.*
77 *Id.* at 74.
78 *Id.* at 76.
79 *Id.*
80 *Id.*
81 *Id.* at 78.
82 *Id.* at 79.
83 *Id.*

of the pesticide in space.[84] For example, insecticides can be contained in traps that are baited with specific chemical attractants, such as pheromones, that are only attractive to specific pest species.[85] Another approach to special selective application is "band spraying," which involves leaving certain crop areas as untreated refuges for predators and parasites, which are generally more mobile than many pests, and thus can move into untreated areas to avoid exposure.[86]

One of the most important tools of IPM is the integration of a variety of biological controls into a pest management program. A key component of most IPM programs is managing to encourage the presence of predators and parasites of the pest species.[87] In addition to encouraging naturally occurring populations of natural enemy species, additional and/or different predators and parasites can be introduced into the farm field.[88] For example, predatory lady beetles and parasitic wasps are sometimes introduced into farm fields as natural bio-control agents. Similarly, pathogens of pest species can be introduced where appropriate. The best-known example of this is the spraying of the naturally occurring microorganism, *B.t.*[89]

In addition to using biological organisms as part of IPM programs, certain biochemical pesticides derived from plants, such as rotenone and pyrethrum, can be part of an IPM program. While these natural plant-derived substances may be toxic, their toxicity may be more selective to certain taxa of organisms than many synthetic chemical pesticides. Other biochemicals, such as semiochemicals and insect growth regulators are also frequently part of IPM programs.[90] Another component of IPM is using pest-resistant crop varieties to reduce the need for pesticide inputs. Pest resistant varieties can be developed with through traditional plant breeding or through genetic modification techniques.[91]

IPM also relies on integrating cultural controls with other pest management techniques. As discussed above, prior to the advent of pesticides, cultural controls were the primary methods used to control pests. Cultural controls are manipulations of the agricultural practices designed to minimize pest problems.[92] Cultural controls that are often incorporated into IPM programs include cultivation, mulching, sanitation, destroying standing crop material, pruning, intercropping (see Photo 5.3), trap crops, and crop rotation.[93] Although very labor intensive, cultural controls are an important part of IPM.[94]

---

84 *Id.* at 80.
85 *Id.*
86 *Id.*
87 *Id.* at 85.
88 *Id.*
89 *Id.* at 86.
90 *Id.* at 123, 153.
91 *Id.* at 134.
92 *Id.* at 113.
93 *Id.* at 115–21.
94 *Id.* at 123.

Photo 5.3 Intercropping
Reprinted with permission. Howard F. Schwartz, Colorado State University, Bugwood.org.

A final technique that has been used successfully, if infrequently, in IPM programs is a form of genetic control that involves rearing and sterilizing large numbers of males of the pest species and introducing them into the environment.[95] Due to the large numbers of sterile males, very few non-sterile males are able to find a female with which they can breed.[96] Thus, most of the breeding is unsuccessful and the next generation of the pest population is greatly reduced.[97] The best-known example of successful use of sterile males is with the screw worm pest of livestock.[98]

Due to the wide range of different types of pesticides and pest management approaches, it is difficult to generalize about the risks associated with pesticide use or how best to regulate pesticides. The ecological risks posed by different pesticides or categories of pesticides must be evaluated independently before any conclusions can be drawn.

95 *Id.* at 131.
96 *Id.*
97 *Id.*
98 *Id.* at 130.

PART III
The Ecology Impacts of Pesticides

Chapter 6
The Ecological Impacts of Pesticide Use[1]

The Risks of Conventional Chemical Pesticides

Overall pesticide usage in the US generally has increased over time. In 1970, the total annual conventional pesticide active ingredient use in the US was approximately 760 million pounds. This number rose to a high of 1,144 million pounds in 1979.[2] By 2007, the annual pesticides usage in the US had declined to 857 million pounds, but was still significantly higher than the 1970 number. Pesticide use in the US does not appear to be declining any further. In 1988, the total annual user expenditure on pesticides in the agricultural section was approximately 4.9 million dollars. By 2007, the number had risen to approximately 7.9 million dollars.[3] On an amount basis, the agricultural pesticide use was approximately 867 million pounds of active ingredient in 1988 and approximately 877 million pounds in 2007.[4]

Despite the significant increases in pesticide use in the US during the latter half of the twentieth century, crop damage from pests has actually increased. Between 1945 and 1989, there was a 10-fold increase in insect pesticide use, but crop loss from insect damaged doubled from 7 to 13 percent.[5] The development of resistance to pesticides contributes to what appears to be a trend in which, as more pesticides are used, pests actually cause more rather than less crop damage. Studies show that between 1948 and 2001, when US pesticide use increased from approximately 50 million pounds per year to almost 1 billion pounds per year, crop loss due to insect pests actually almost doubled from approximately 7 percent

---

1 Portions of this chapter are adapted from Mary Jane Angelo, *The Killing Fields: Reducing the Casualties in the Battle Between U.S. Endangered Species and Pesticide Law*, 32 Harvard Environmental Law Review 96 (2008); Mary Jane Angelo, *Regulating Evolution for Sale: An Evolutionary Biology Model for Regulating the Risks Posed by Genetically Modified Organisms*, 42 Wake Forest Law Review 93 (2007); and Mary Jane Angelo, *Embracing Uncertainty, Complexity and Change to Protect Ecological Integrity: An Eco-Pragmatic Reinvention of a First Generation Environmental Law*, 33 Ecology Law Quarterly 105 (2006).

2 EPA Pesticide Industry Sales and Usage 2006/2007, at 31, *available at* http://www.epa.gov/opp00001/pestsales/ (last visited April 13, 2011).

3 *Id.* at 23.

4 *Id.* at 28.

5 David Pimentel et al., *Assessment of Environmental and Economic Impacts of Pesticide Use*, *in* The Pesticide Question: Environment, Economics and Ethics 47 (David Pimentel & Hugh Lehman, eds, 1993).

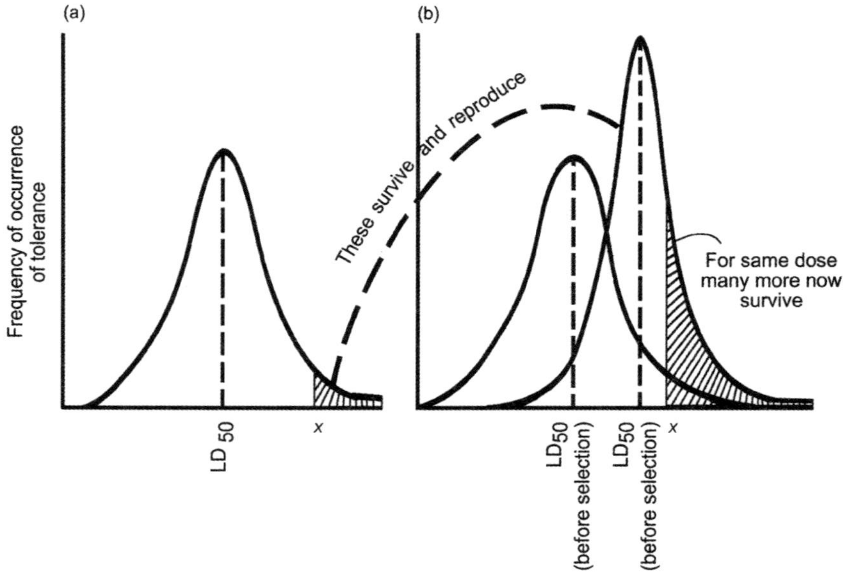

Figure 6.1 Unnatural resistance and the development of resistance to insecticides

Note: (a) = Distribution of tolerance before selection; (b) = Distribution of tolerance after selection compared with distribution before selection; x = The dose exerting selection pressure, kills fewer organisms in the population of progeny than in the population of parents; LD50 = Dose lethal to 50 percent of the population. Reprinted with permission. H.F. van Emden & M.W. Service, Pest and Vector Control 75 (2004).

to 13 percent.[6] In nature, approximately 1–7 percent of plant biomass is lost to pest damage. When plants are grown in agriculture, the loss from pest damage rises to 30 percent. The dramatically increased loss in the agricultural setting is due to a number of factors including growing crops in monocultures and the effects of breeding for taste, cosmetics, and shipability.[7] But, perhaps surprisingly, the large crop loss from pests is in part attributable to the use of pesticides to combat those very pests. A major drawback of broad-spectrum pesticides is that, while the pesticide kills the target pest, it also kills non-target organisms, including predatory and parasitic natural enemies of the pest. In fact, these natural enemies frequently are more sensitive to the pesticides than are the pests themselves. This is due in part to the fact that unlike herbivores, predators have not evolved to

6 Jason Clay, World Agriculture and the Environment 53 (2004) (citing Hawken 1999).

7 *Id.*

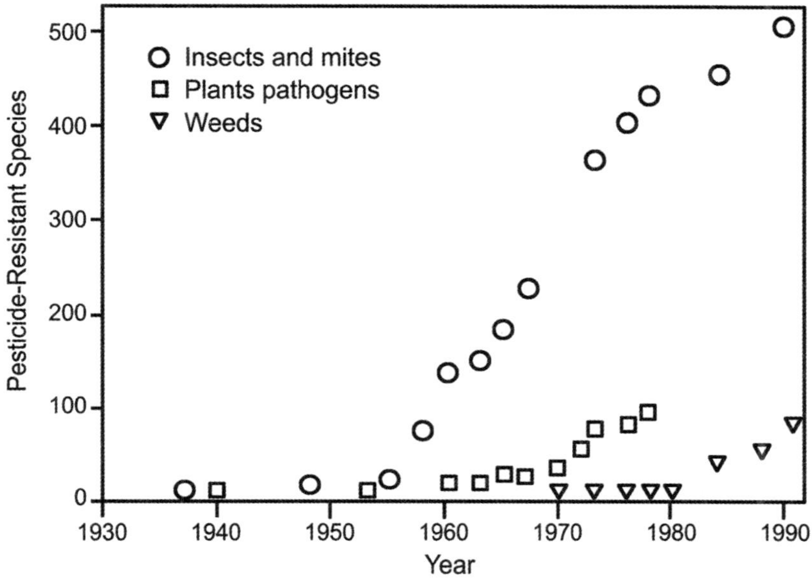

Figure 6.2 Chronological increase in the number of arthropods (insects and mites), plant pathogen, and weed species resistant to synthetic chemical pesticides since their development in the late 1930s

Reprinted with permission. Gould, F. *The Evolutionary Potential of Crop Pests*, 79 AM. SCI. 496–507 (1991).

be resistant to chemicals found in plants.[8] Thus, broad-spectrum pesticides tend to kill higher proportions of predators than pests, resulting in a higher ratio of pest to predator populations.[9] Although from an ecological standpoint, narrow-spectrum pesticides are preferable, broad-spectrum synthetic pesticides continue to dominate US pesticide usage.[10]

One of the major drawbacks of pesticide use is the tendency of pest populations to develop resistance to the very substances that are intended to control the populations. When a population of pests is exposed to a given pesticide, the pesticide will kill individuals in the population that are susceptible to the particular pesticide, leaving resistant individuals to survive and reproduce. Thus, the next generation of individuals will have a higher proportion of resistant individuals and a lower proportion of susceptible individuals (see Figure 6.1). Another round of pesticide use will have a similar effect, resulting in increasing

8 H.F. VAN EMDEN & M.W. SERVICE, PEST AND VECTOR CONTROL 52 (2004).

9 *Id.* at 113.

10 WILLIAM H. RODGERS, ENVIRONMENTAL LAW 407 (West, 2nd ed., 1994).

numbers of susceptible individuals being destroyed before they can reproduce and future populations becoming ever more resistant. This phenomenon creates a treadmill effect, where more and more pesticides are needed to keep up with the ever increasingly resistant pest populations.[11] In other words, pesticide efficacy declines as pesticide use increases. When DDT was first employed to combat malaria by killing the disease-carrying mosquito vector, its ultimate failure, despite the initial dramatic success, was due at least in part to the development of resistance by many populations of mosquitoes.[12] One of the best-known examples of a pest developing resistance to pesticides is that of the diamondback moth, which has become a much more serious pest post-pesticide use than it ever was pre-pesticide use.[13] As of 1990, more than 250 insects were resistant to DDT and organophosphate pesticides.[14] Although perhaps counterintuitive, the widespread use of chemical pesticides has not been particularly effective in controlling pests, and in some ways has exacerbated pest problems. Noted pesticide expert David Pimentel estimates that if the US banned all chemical pesticides, we would experience only a 10 percent increase in crop loss.[15] Nevertheless, the use of chemical pesticides continues to dominate US agriculture and continues to impose significant ecological risks.

Scientists estimate that as many as 10 million species, or 99 percent of the earth's wild biodiversity, not including cultivated and weedy species, are in a "precarious condition."[16] Causes and contributors to the decline of this large number of species include indirect habitat destruction through clearing for agriculture and development, the spread of non-native invasive species, pollution, over harvesting of species, and disease. Although there is no doubt that direct habitat destruction is the leading contributor of species loss (estimated as being implicated in 85 of US species decline), pollution, including pesticide pollution, is implicated in 24 percent of US species decline. While the environmental regulation of the past 40 years has resulted in many industries decreasing pollution, agricultural pollution remains high and is considered to be the major source of pollution in many parts of the globe.[17] Of course, agricultural pollution is not limited to pesticide pollution and includes many other forms of contamination such as fertilizer runoff, animal waste, and siltation from erosion. Nevertheless, pesticide contamination remains a substantial component of agricultural pollution; and due to the toxic nature of

11 Van Emden & Service, *supra* note 8, at 115–16.

12 *Id.* at 109.

13 *Id.* at 5.

14 Robert F. Norris, Edward P. Caswell-Chen & Marcos Kogan, Concepts in Integrated Pest Management 66 (2003).

15 Pimentel et al., *supra* note 5.

16 Edward O. Wilson, *Foreword* to David S. Wilcove, The Condor's Shadow: The Loss and Recovery of Wildlife in America xiv (1999).

17 Clay, *supra* note 6, at 27.

many synthetic pesticides, wildlife, biodiversity, and other environmental impacts from pesticide pollution can be substantial.

Pesticide poisoning of fish and wildlife is a significant factor in species decline. In one study of the decline of fish species in the United States, Canada, and Mexico, it was determined that the destruction of physical habitat was implicated in 73 percent of the declines, the displacement by introduced species was implicated in 68 percent of the declines, the alterations of habitat by chemical pollutants was implicated in 38 percent of the declines, hybridization with other species and subspecies was implicated in 38 percent of the declines, and over harvesting was implicated in 15 percent of the declines. The numbers add up to more than 100 because more than one factor is implicated in many of the fish population declines.[18] Thus, while pesticide usage in itself may not directly destroy habitat (although clearing for agriculture certainly does), chemical pesticides may be a significant contributor to species decline; and pesticidal GMOs, which pose risks of spread in the environment similar to non-indigenous species release, may also be important contributors. Habitat destruction, spread of non-natives, pollution, overkill, and disease have been referred to as the five horsemen of the environmental Apocalypse.[19]

Because pesticides are by definition intended to kill or disrupt living organisms, and because they are intentionally released into the environment, often in large quantities over large areas, it is not surprising that pesticides pose a wide array of risks to individual species as well as to overall ecosystem function. Many pesticides are broad-spectrum, affecting diverse species, including many non-target organisms. Others are more narrowly targeted to pest species. However, even these may have significant impacts on non-target species that are closely related to the intended targets. Some pesticides persist in the environment for weeks, months and even years, while others breakdown relatively quickly. Moreover, living organisms vary significantly in their susceptibility to pesticides. The potential ecological risks of pesticide use depend on a number of factors including toxicity or other hazards of the pesticide, method of application, persistence in the environment, amount used, and susceptibility of non-target organisms. Moreover, there are not many data available on the environmental effects of pesticide usage on many species. Accordingly, the ecological risks of pesticides cannot be easily described or quantified. Nevertheless, some generalizations can be made.

Many pesticides in current use in the US, as well as in other parts of the world, are acutely toxic and are known to cause adverse effects on non-target mammals, birds, reptiles, amphibians, fish, and invertebrates. Birds and other wildlife may be exposed through direct spraying, ingesting pesticide granules, drinking water that has been contaminated by pesticides, or eating prey organisms that have been contaminated by pesticides. While the banning and severe restriction of certain

18 E.O. Wilson, The Diversity of Life 253–4 (1992).

19 David S. Wilcove, The Condor's Shadow: The Loss and Recovery of Wildlife in America 8 (1999) (quoting E.O. Wilson).

pesticides, such as DDT, over the past 30 years has dramatically reduced certain risks to wildlife, many risks remain. One startling example is that when roughly 10,000 dead birds were tested for the presence of West Nile Virus in 2000, the New York State Department of Environmental Conservation determined that pesticides and other chemicals were responsible for more bird kills than the virus.[20] As further evidence of the effects on bird populations, studies have shown substantially higher nesting rates of birds, as well as significantly higher bird abundance and avian species richness, on organic farms as compared to conventional farms that use synthetic pesticides.[21] In addition to effects from direct exposure to pesticides, birds and other wildlife may also be exposed to pesticides by ingesting prey animals that have been contaminated. For example, the New York State Department of Environmental Conservation has found a number of different avian species, such as screech owls, red-tailed hawks, American kestrels, and other raptors that have died as a result of eating small rodents that had consumed rat poison.

Other less visible species are also at considerable risk from exposure to pesticides. For example, for the past decade, there has been considerable concern and debate in the scientific community over the worldwide decline of amphibians. There are now significant data to support a conclusion that certain pesticides, such as the herbicide atrazine, may be contributing to the global decline in amphibian populations.[22] Until very recently, the results of many studies on the effects of pesticides on amphibians have been puzzling because pesticide levels in nature tend to be much lower than levels found to be lethal in the laboratory setting.[23] A recent study sheds new light on this dilemma. Scientists have determined that the combination of the pesticide carbaryl and stress from the presence of predators was more lethal in certain amphibian species than the pesticide by itself. In other words, there appears to be a synergistic effect at work between pesticides and predators, making the combination of the two more lethal than the sum of the parts,

---

20 Laura A. Haight, *Local Control of Pesticides in New York: Perspectives and Policy Recommendations*, 9 Alb. L. Envtl. Outlook J. 37, 51 (2004).

21 Nancy A. Beecher, Ron J. Johnson, James R. Brandle, Ronald M. Case & Linda J. Young, *Agroecology of Birds in Organic and Nonorganic Farmland*, 16 Cons. Biol. 1620 (2002).

22 Hayes, T.B., A. Collins, M. Lee, M. Mendoza, N. Noriega, A.A. Stuart & A. Vonk. 2002. Hermaphroditic, demasculinized frogs after exposure to the herbicide, atrazine, at low ecologically relevant doses. Proceedings of the National Academy of Sciences (US) 99:5476–5480. *See also*, Pesticide Atrazine Can Turn Male Frogs Into Females, Science Daily (March 1, 2010), *available at* http://www.sciencedaily.com/releases/2010/03/100301151927.htm (last visited June 14, 2011).

23 Rick Relyea, *Predators Make Pesticides More Lethal*, *in* Conservation in Practice, vol. 5, no. 2, 2004, at 5, 5 (excerpting Rick Relyea, *Predator Cues and Pesticides: A Double Dose of Danger for Amphibians*, 13 Ecological Applications 1515–21 (2003)). *See also* Carlos Davidson, H. Bradley Shaffer & Mark R. Jennings, *Spacial Tests of the Pesticide Drift, Habitat Destruction, UV-B, and Climate-Change Hypotheses for California Amphibian Declines*, 16 Cons. Biol. 1588 (2002).

and resulting in low concentrations of pesticides in nature being highly lethal to amphibians. Of course, amphibians in nature must cope with other stress, such as the presence of predators, in addition to the stress of pesticides. Accordingly, this study suggests that amphibians in nature may be significantly more sensitive to pesticides than they are in the sterile isolated confines of the research laboratory.[24]

Although the most obvious adverse effects of pesticide use are those to humans and large animals such as mammals and birds, it is likely that the most significant adverse effects of pesticides are those to invertebrates that are closely related to target pest species. Casualties from this "friendly fire" are widespread in the invertebrate world.[25] One example is that the insect Order Lepidoptera contains not only many pest moth species, but also contains many non-pest butterfly species. These butterfly species may be beneficial pollinators and may be aesthetically pleasing, colorful, and interesting species, such as the monarch butterfly. Also, the Order Lepidoptera contains a number of butterfly species that have been listed as threatened or endangered under the federal Endangered Species Act. Pesticides that are used to kill pest moth species generally do not discriminate within the Lepidoptera Order, and will kill non-pest, beneficial butterflies, including endangered species. Mosquito control pesticides have been indicted as one of the threats to the continued survival of the endangered Miami Blue Butterfly over the past few decades. Perhaps equally if not more important than direct acute effects on non-target organisms are the chronic effects upon growth, physiology, reproduction, and behavior. Much less is known about these effects.

Even where a pesticide is not toxic enough to kill an organism, it can have very significant sub-lethal effects on the organism by affecting life span, growth, physiology, behavior, and reproduction. For example, extremely low doses of some pesticides have been determined to disrupt honeybees' homing flight behavior, thereby adversely affecting pollination.[26] Pesticides have been documented to have significant indirect effects on non-target organisms by reducing the populations of animals or plants that serve as food or cover for other species. Moreover, recent studies demonstrate a reduction in the abundance of non-target butterflies on conventional farms as compared to butterflies on organic farms.[27] Pesticides

24 *Id.*

25 May Berenbaum, *Friendly Fire*, *in* WINGS: ESSAYS ON INVERTEBRATE CONSERVATION, Spring 2004, at 8–12 (a publication of the Xerces Society (2004). Jaret C. Daniels & Thomas C. Emmel, *Florida's Precious Miami Blues*, *in* WINGS: ESSAYS ON INVERTEBRATE CONSERVATION, Spring 2004, at 3–7.

26 H.M Thompson, *Behavioural Effects of Pesticides in Bees—Their Potential for Use in Risk Assessment*, 12 ECOTOXICOLOGY 317–30 (2003), *available at* http://www.ingentaconnect.com/content/klu/ectx/2003/00000012/F0040001/05112975 (last visited, June 3, 2011).

27 D.J. Hole et al., *Does Organic Farming Benefit Biodiversity?* 122 BIOLOGICAL CONSERVATION 113–30, *available at* http://dx.doi.org/10.1016/j.biocon.2004.07.018 (last visited, June 3, 2011).

have also been implicated in the decline of global pollinator populations.[28] Many insects, in addition to bees, play a critical role in pollination in natural ecosystems as well as in agricultural systems.[29] These pollinators, which are attracted to flowering crop plants and thus are abundant in certain farm fields at certain times of the year, are vulnerable to insecticide use on these farms. Commercial bees that are brought in to pollinate crops are also vulnerable. Further, pesticides may play a role in "colony collapse disorder," which is disseminating commercial bees' colonies.[30] Although it is not yet known what the cause of this disorder is, pesticide poisoning, alone or in combine with other environmental pressures, is one of the major suspected causes.[31]

More generally, insects are critical to the function of virtually all natural ecosystems. As Samways has stated "[t]he ecological grandeur of insects is in their ability as a group to transfer vast amounts of energy. As such, they are determinants of community structure and shapers of habitat."[32] Insects play a "keystone role" in many ecosystems and frequently act as ecosystem engineers and soil modifiers.[33] As such, insects can influence the structure and composition of an entire ecosystem. For example, some termites modify hydrology and organic matter in a way that results in increased diversity of tree species, and some unique plants can only exist with the presence of certain ants.[34]

Given the critical role that insects and other invertebrates play in the overall functioning of ecosystems, it goes without saying that if insect or other invertebrate populations are impacted by chemical pesticides, the ecosystems that depend on them will also be impacted. Thus, a significant decline in insect keystone species populations can lead to a cascading effect where other species populations decline in response and overall biodiversity is reduced.[35]

One of the most insidious risks posed by pesticides is the tendency of certain synthetic pesticides to mimic hormones, such as estrogen, in humans and wildlife. Only recently has science begun to understand these complex effects.[36] Although

---

28 Pesticide Action Network, Pesticides & pollinator decline, *available at* http://www.panna.org/current-campaign/bees (last visited June 3, 2011).

29 Michael J. Samways, Insect Diversity Conservation 40 (2005).

30 EPA, Pesticide issues in the works: Honeybee colony collapse disorder, *available at* http://www.epa.gov/opp00001/about/intheworks/honeybee.htm (last visited June 3, 2011).

31 *Id.*

32 Samways, *supra* note 29, at 39.

33 *Id.* at 40–41.

34 *Id.*

35 *Id.* at 41.

36 Noah Sachs, *Blocked Pathways: Potential Legal Responses to Endocrine Disrupting Chemicals*, 24 Colum. J. Envtl. L. 289 (1999); *see also* Theo Colburn et al., Our Stolen Future: Are We Threatening Our Fertility, Intelligence And Survival? A Scientific Detective Story (1996); Matthew P. Longnecker, Walter J. Rogan & George Lucier, *Effects of DDT (Dichlorodiphylytrichloroethane) and PCBs (Polychlorinated Biphenyls) and an Overview of Organochlorines in Public Health*, 18 Rev. Of Pub. Health 211 (1997); Louis

the term "environmental estrogen" was coined in the 1970s, it was not until more recent years that scientific studies were conducted to lend support to the hypothesis that environmental exposure to certain synthetic chemicals could cause estrogenic effects. Estrogen-mimicking substances include a number of pesticides as well as a wide variety of other products in common use, such as toiletries, spermicides, and plastics. Exposure to these compounds, particularly when the exposure occurs in the fetus or young children, has been correlated with a large number of effects in humans, including decreased sperm counts, breast and testicular cancer, endometriosis, deformed or stinted reproductive organs, neurological defects, and low birth weights.[37] In addition to these human health effects, these substances have also been implicated in numerous wildlife impacts including deformed alligators and turtles, and reproductive difficulty in birds, fish, and mammals. These estrogenic effects can be extremely complex, unpredictable, and difficult to understand. For example, DDT exposure has been demonstrated to cause gulls to begin sharing nests with other females rather than males, and young gulls had grossly feminized reproductive tracts.[38] Moreover, a large number of studies on various species of fish exposed to estrogenic compounds have shown effects such as increased time to maturity, smaller gonads, and reduced fertility. Similarly, declines in the reproductive rates of mammals, such as minks, have been linked to ingesting fish contaminated with estrogenic substances. The US Fish and Wildlife Service has reported that between 1985 and 1990, 67 percent of male Florida panthers were born with one or more undescended testes, compared with only 14 percent 10 years earlier. Although not fully understood, scientists suspect a link with exposure to estrogenic substances in the environment.[39] Perhaps the most widely cited wildlife examples of endocrine dysfunction in wildlife are the feminization of alligators and the occurrence of masculinized female fish in Florida. Interestingly, although we are only beginning to understand the estrogenic effects of certain pesticides, in her 1962 book, *Silent Spring*, Rachel Carson predicted such effects: "A substance that is not a carcinogen in the ordinary sense may disturb the normal functioning of some part of the body in such a way that malignancy results. Important examples are the cancers, especially of the reproductive system, that appear to be linked with disturbances of the balance of

---

J. Guillette, Jr., Timothy S. Gross, Greg R. Masson, John M. Matter, H. Franklin Percivil & Allan R. Woodward, *Developmental Abnormalities of the Gonad and Abnormal Sex Hormone Concentration in Juvenile Alligators from Contaminated and Control Lakes in Florida*, 8 ENVTL. HEALTH PERSPEC. 680 (1994); D. Crews, E. Willingham & J.K. Skipper, *Endocrine Disruptors: Present Issues, Future Directions*, 75 Q. REV. OF BIOL. 243 (2000).

37 Sachs, *supra* note 36, at 293–8.

38 *See* Susan M. Salvatore, *Estrogens in the Environment*, 69 FLA. B.J. 39, n. 35 (1995).

39 *See* CARL F. CRANOR, LEGALLY POISONED: HOW THE LAW PUTS US AT RISK FROM TOXICANTS 114–20 (2011).

sex hormones … [t]he chlorinated hydrocarbons are precisely the kind of agent that can bring about this kind of indirect carcinogenesis."[40]

Another concern is the uncertainty regarding the potential effects of pesticides on ecologically significant microorganisms. Very little is known about the complex ecology of microorganisms.[41] Although there are not many data to suggest that most types of pesticides pose significant risks to microorganisms, soil fumigants, which are designed to destroy soil microorganisms and are applied at very high doses, may pose substantial risks to beneficial microorganisms. For example, the killing of soil microbes and invertebrates resulting from pesticide use may actually cause crops to become more susceptible to disease, and may thereby reduce crop growth. In addition, populations of nitrogen-fixing organisms may be reduced, thereby requiring higher levels of fertilizer application. Critical ecological services provided by microorganisms, including decomposition, may also be impacted by certain pesticides.[42]

The problem of ecological risk from pesticide exposure is exacerbated by the tendency of certain pesticides to undergo a phenomenon known as bioaccumulation. Bioaccumulation of pesticides became widely recognized during the 1960s as a result of the publication of *Silent Spring*. In her book, Carson explained how DDT and other organochlorine pesticides have the ability not only to persist in the environment for years, if not decades, but also to accumulate in the tissue of animals and humans. These pesticides accumulate in animals on the bottom of the food chain and then are passed from prey to predator until they can be found in very high concentrations in top predators.[43] Pesticides that persist, accumulate, and biomagnify are especially insidious in that they can adversely affect organisms far removed in both time and place from the original release of the pesticide into the environment.[44] Other risks posed by pesticides have only recently begun to be studied. For example, in recent years, the extent of atmospheric transport of pesticides has come to light. Moreover, the pesticide methyl bromide has been determined to be a significant contributor to the thinning of the stratospheric ozone layer.[45]

Moreover, although agricultural systems in themselves are not natural systems *per se*, they are generally located in close proximity to natural ecosystems and often contain within their borders sizable natural and semi-natural ecosystems. Thus, adverse effects from pesticide usage in agricultural systems may negatively impact

---

40 Rachel Carson, Silent Spring 235 (1962).

41 Clive A. Edwards, *The Impact of Pesticides on the Environment*, *in* The Pesticide Question: Environment, Economics and Ethics 18 (David Pimentel & Hugh Lehman, eds, 1993).

42 Pimentel et al., *supra* note 5, at 68–9.

43 Bill Freedman, Environmental Ecology: The Impacts of pollution and Other Stresses on Ecosystem Structure and Function 191–8 (1989).

44 Carson, *supra* note 40, at 22–3

45 69 Fed. Reg. 76982 (December 23, 2004).

ecosystems within the farm boundaries, as well as nearby ecosystems that may be contaminated by pesticide runoff in water, drift through the air, or movement of contaminated organisms. As described above, invertebrate non-target species may be at greatest risk from pesticide use. Loss of invertebrate biodiversity, however, is not the only concern. Equally concerning are the ecological and economic disruptions that frequently occur as a result of non-target predator and parasites being killed by pesticides. Many pest populations are kept in check in nature by the existence of organisms that feed on pest species. Consequently, if these predators or parasites are eliminated or greatly reduced in number, the population of pest species will experience a population explosion. In addition to increases in existing pest species population resulting from pesticide usage, new pest species may also be created as a result of this phenomenon. An example of new pest creation resulting from pesticide use is the bollworm, which is now a major economic pest of cotton. Although the bollworm existed previously, it was not considered a pest until pesticides used to control the boll weevil, another pest of cotton, killed the natural enemies of the bollworm, allowing its population to explode.

An interesting indirect ecological impact from synthetic pesticide use is that unintentional contamination by chemical pesticides can discourage farmers from seeking or maintaining their organic certification. Organic farms that become contaminated by pesticides used on nearby lands can lose the ability to sell their produce under the organic label and in some cases may lose their organic certification. Thus, farmers who wish to "go organic" may be discouraged from doing so based on historic pesticide use on their land or historic or current pesticide use on neighboring lands.

The Risks of Pesticidal GMOs

In the past 20 years, a completely new suite of concerns has emerged regarding the use of pesticidal genetically modified organisms (GMOs). Although many of the risk considerations for biotechnology pesticides are similar, if not the same, as those for traditional chemical pesticides, these new pesticides pose a number of novel risks not presented by chemical pesticides. One of the most significant new risk considerations for pesticidal GMOs is the potential to spread the living organism or its genetic material. For example, plants can reproduce sexually and/or asexually and, as a result, the genetic material that was introduced into the plant, and which enables the plant to produce pesticidal substances, could spread through agricultural or natural ecosystems. Thus, if a plant that produces a pesticide has the capacity to spread in the environment, or to spread its genetic material to other plants, there will be a greater potential for increased exposure to non-target organisms than there would be for a pesticide produced in a plant that can only grow in a limited geographic area, or does not have the ability to cross-fertilize with other plants in the environment.[46] This is a particular concern for pesticides produced in plants that have wild relatives in the United States. If these wild relatives acquire the ability to produce the pesticide through cross-fertilization, many additional non-target organisms could potentially be exposed to the pesticide. The potential for a GMO or its genetic material to spread from one plant to another raises additional risk issues beyond those of exposure to humans and non-target organisms. One potential risk of biotechnology products parallels the risk of the introduction of any non-native species into a new environment.[47] Among the most significant concerns regarding pesticidal GMOs are over the potential for the development of "superweeds" through the out-crossing of pesticidal GMOs to wild relatives. For example, the ability to produce a pesticide that makes a plant resistant to insect or viral pests can be spread to a wild relative, and subsequently passed on to future generations of that relative. Consequently, the wild relative, by virtue of its newly acquired ability to resist insects or viruses, has the potential to become a hardy weed, or superweed.

The concept of risk includes elements of hazard and exposure. GMOs present hazards and exposures that are different in both type and degree from the hazards and exposures presented by traditional environmental chemicals. Toxicity is the typical hazard presented by most traditional environmental chemicals. With regard to GMOs, hazards are expanded well beyond toxicity to include hazards such as the creation of superweeds or pest resistance. In addition, pathways for exposure to hazards may be much greater and more widespread because GMOs are living organisms spreading and reproducing in the environment. Accordingly, there are a number of different types of risks associated with GMOs. Some are very similar to the risks associated with traditional pesticides and chemicals, some are similar to the risks associated with the introduction of non-indigenous organisms into new environments, and some are novel and result from the fact that that many GMOs are intentionally genetically modified to give them an evolutionary selective advantage.

Even very small genetic manipulations can result in significant changes in an organism's ability to survive and flourish in a particular ecosystem. Examples abound regarding the disastrous, but unpredicted, effects of the introduction of non-native species into the environment displacing native species. Genetically modified organisms introduced into the environment could have similar impacts.

46 For a detailed discussion of the potential risks and benefits of pesticidal GMOs, *see* Mary Jane Angelo, *Regulating Evolution for Sale: An Evolutionary Biology Model for Regulating the Risks Posed by Genetically Modified Organisms*, 42 Wake Forest Law Review 93 (2007) and Mary Jane Angelo, *Genetically Engineered Plant Pesticides: Recent Developments in the EPA's Regulation of Biotechnology*, 7 U. Fla. J.L. & Pub. Pol'y 257 (1996).

47 David J. Earp, *The Regulation of Genetically Engineered Plants: Is Peter Rabbit Safe in Mr. McGregor's Transgenic Vegetable Patch?* 24 Envtl. L. 1633, 1666–9 (1994). *See* J.J. Kim, *Out of the Lab and Into the Field: Harmonization of Deliberate Release Regulations for Genetically Modified Organisms*, 16 Fordham Intl. L.J. 1160 (1993).

One of the most significant risks is that of a genetically engineered plant becoming a weed or pest itself or out-crossing to related species to create new weeds or pests. Once released into the environment, the spread of a GMO may be extremely difficult, if not impossible, to control. Development of such a superweed has the potential to result in substantial disruption of agricultural and natural ecosystems.[48] For a GMO plant to transfer its genes to related existing weed species, wild relatives of the GMO plant must grow in the geographic areas where the GMO plant is introduced. Many crops grown in the US are of foreign origin. Moreover, most of the major crops in the US, including soybeans, corn and wheat, have been bred to the point that they have lost their ability to compete with wild species in the environment. Thus, the risk of hybridization between transgenic crops and wild relatives is limited in the US. Nevertheless, crops with weedy relatives in the US may be vulnerable, and even where crops do not have domestic weedy relatives, once these GMOs are exported (intentionally or otherwise) to other parts of the world that have wild relatives of the GMOs, the risks become more profound.

Probably the most well-known example of problems that can arise from the release of pesticidal GMOs into the environment is the recent case involving StarLink corn. StarLink corn contained a protein that was similar to a protein found in peanuts, and peanuts contain many proteins that can cause severe allergies in humans. Thus, the StarLink corn was approved only for animal feed and was not approved for human food use. Despite this limited approval, testing by an environmental organization revealed the presence of the StarLink gene in large batches of taco shells and other human food-corn products.[49] Farmers throughout the US were forced to destroy their crops. The farmers filed class action lawsuits.[50] The defendants, Aventis CropScience USA Holdings and Garst Seed Company, moved for dismissal, arguing that the farmers' claims were preempted by FIFRA and that the economic loss rule barred recovery.[51] The trial court held that FIFRA did not preempt the farmers' claims, that the contamination of crops by neighboring GM-crops provided a claim to which the economic loss rule did not apply, and that the plaintiffs had properly alleged claims based on negligence, public nuisance, and private nuisance. In addition, consumers alleging fraud,

---

48 *See* Earp, *supra* note 47, at 1666–9; *see generally*, Judith E. Beach, *No "Killer Tomatoes": Easing Federal Regulation of Genetically Engineered Plants*, 53 Food Drug Law 181 (1998) and Celeste Marie Steen, *FIFRA's Preemption of Common Law Tort Actions Involving Genetically Engineered Pesticides*, 38 Ariz. L. Rev. 763 (1996).

49 *StarLink Corn: How it Reached the Food Supply*, Associated Press, December 4, 2002, *available at* http://archive.showmenews.com/2000/dec/20001204busi011.asp (last accessed June 3, 2011); *see also* Rebecca M. Bratspies, *Myths of Voluntary Compliance: Lessons from the Starlink Corn Fiasco*, 27 Wm. & Mary Envtl. L. & Pol'y Rev. 593, 628–33 (2003) and William Lin, Gregory K. Price & Edward Allen, *Starlink: Impacts on the U.S. Corn Market & World Trade*, *available at* http://www.ers.usda.gov/Briefing/Corn/pdfs/StarLinkFDS2001.pdf (last visited June 3, 2011)

50 *In re* StarLink Corn Prods. Liab. Litig., 212 F.Supp. 2d 828 (N.D. Ill. 2002).

51 *Id.* at 833, 852.

negligence, and breach of warranty for recklessly exposing millions of consumers to these unapproved and potentially dangerous products filed class action lawsuits. Moreover, corn products with the gene showed up as far away as Japan and Korea, leading to a dramatic decline in imports of US corn products in these countries.[52]

Other documented instances of the risks include a case where the British government ordered the destruction of experimental fields of herbicide-tolerant oilseed rape plants because the genetically engineered forms had successfully pollinated nearby "natural" plants. Such spread might have created a new breed of superweeds resistant to herbicides and capable of displacing other plant life.[53] In Canada, GM corn cross-pollinated a neighboring field of "organic" corn, which could no longer be sold as "organic."[54] Another interesting twist on the economic consequences of genetic contamination occurred in Canada when Percy Schmeiser, a farmer in Saskatchewan, was sued by Monsanto for patent infringement when his canola crop was found to contain the company's patented modified genes, which made the crops herbicide-resistant.[55] Schmeiser did not have a license for use and did not purchase the company's crop, yet 95 to 98 percent of his 1998 crop consisted of the genetically modified canola. The Supreme Court of Canada upheld the patent infringement claim, but struck down the damages award as Schmeiser used the canola solely for feed purposes and did not gain any particular advantage from his usage of herbicide-resistant crops.[56]

GM corn pollen has also been found to cause significant risk to Monarch butterfly larvae in laboratory conditions.[57] Moreover, a number of insect pests have demonstrated resistance to *Bacillus thuringensis*, an important low-risk

---

52 *See* Lin et al., *supra* note 49; Bratspies, *supra* note 49; Rebecca M. Bratspies, *Consuming Fear of Corn: Public Health and Biopharming*, 30 Am. J.L. and Med. 371, 387 (2004).

53 "Farmers advised to destroy GM crops," BBC News Online: UK Politics (May 27, 2000), *available at* http://news.bbc.co.uk/1/low/uf_politics/766539.stm.

54 Thomas Hayden, *Bad Seeds in Court: When Genetically Modified Plants Contaminate Their Crops, Organic Farmers Fight Big Biotech*, US News & World Report, Science & Tech., January 28, 2002.

55 Monsanto Canada Inc. v. Schmeiser, 2004 CarswellNat 1391 (Supreme Court of Canada 2004).

56 *Id.*

57 Wendy Thai, *Transgenic Crops: The Good, the Bad, and the Laws*, 6 Min. J.L. Sci, & Tech. 877 (2005); J.J.E. Losey, *Transgenic Pollen Harms Monarch Larvae*, 399 Nature 214 (1999). At least one other study has found negligible harm to Monarch butterflies. Marc Kaufman, *2nd Study Links Gene-Altered Corn, Butterfly Deaths*, Wash. Post, August 22, 2000, at A2; *Gene-Spliced Corn No Big Threat to Butterflies, Studies Say*, S.F. Chron. (East Bay Edition), November 3, 1999, at A11. EPA has prepared a document evaluating risks to Lepidoptera in general from *B.t.* PIPs, which can be viewed at http://www.epa.gov/pesticides/biopesticides/pips/executive_summary_and_preface.pdf (last visited June 3, 2011).

non-engineered biological pesticide that is a naturally occurring soil bacterium,[58] which has been genetically engineered into a wide variety of crops including corn, potatoes, and cotton.

Perhaps the most serious concern with pesticidal GMOs stems from the fact that the risks of GMOs are uncertain. Moreover, although the risk of a GMO organism released into the environment creating a new superweed or disrupting the balance of natural ecosystems may be small, the consequences could be disastrous and potentially irreversible.[59] The precise nature and magnitude of the risk is difficult to predict because of the almost infinite variety of potential genetically modified organisms, the ability of GMOs to reproduce and spread, the complexity inherent in natural ecosystems, and the dearth of long-term data on the effects of GMOs.[60]

This discussion of the risks of GMOs is not intended to suggest that these pesticides have no benefits. In fact, many scientists believe that GMO pesticides may provide a less risky alternative to chemical pesticides. Many GMOs are less toxic than chemical pesticides, more narrowly targeted toward the intended pest, and released into the environment in smaller quantities. Nevertheless, the purpose of this discussion on the unique ecological risks posed by GMO pesticides is merely to highlight the complex ecological risks at issue and the large amount of uncertainty regarding such risks. It is worth noting that although there has not been any catastrophic damage caused by the GMOs that have been in the marketplace for the past several years, the pesticidal GMOs that have been commercialized to date are largely relatively innocuous from an ecological standpoint. All of the pesticidal GM products in commercialization to date include genes from microorganisms that are non-toxic to humans, which are naturally ubiquitous in the environment and have been applied widely to food crops in their microorganism form for decades. For example, the *B.t.* delta-endotoxin produced by numerous plant-incorporated protectant in wide use, is essentially the same *B.t.* toxin that occurs naturally in soil and is ubiquitous in the environment. However, there are literally thousands of GM products in the research and development stage that are not so innocuous or well understood. Research is being done on just about every GM product that human ingenuity can conceive including crop plants that produce a scorpion toxin. These new products must be approached with much greater caution.

---

58 *See* Steve Hill, *Science Hopes to Keep One Step Ahead of Adaptive Bugs*, September 6, 1996, *available at* http://agnews.tamu.edu/stories/ENTO/adbugs.htm (last visited June 3, 2011).

59 *See* John Charles Kunich, *Mother Frankenstein, Doctor Nature, and the Environmental Law of Genetic Engineering*, 74 S. Cal. L.R. 807, 819 (2000–2001).

60 *See* Steen, *supra* note 48, at 764.

Chapter 7
The Legacy of Past Pesticide Use: One Example[1]

As described in previous chapters, the ecological risks from pesticides remain high despite decades of environmental regulation and the banning of some of the most harmful pesticides. Despite the bans, however, many pesticides that pose substantial risks to wildlife and the environment in general remain on the market and are released into the environment in large quantities. Even where the US has banned or severely restricted a pesticide deemed to pose unreasonable adverse effects, however, environmental risks remain. As discussed in Chapter 13, a pesticide banned or severely restricted in the US may still be sold and used in other parts of the world. Equally disturbing, however, is the legacy of pesticide risk that remains in the US from pesticides banned decades ago. Despite EPA's ban of most organochlorine pesticides in the 1970s and 1980s, the enormous historic use of these pesticides, coupled with their tendency to persist in the environment has resulted in continued environmental harm. Pesticides that were banned years ago can still be found on farms, in warehouses, and in garages throughout the US. Although these pesticides cannot be used legally, their continued existence creates the possibility that they will be illegally used, accidentally released, or improperly disposed of.

Another long-term concern with persistent organochlorine pesticides is that pesticides that are bound in soils may become biologically available when the soil is disturbed or where an area becomes intentionally or unintentionally flooded with water. This is of particular concern for large environmental restoration projects, such as those carried out under the federal Wetland Reserve Program (WRP), which typically involve restoring farmlands to historic wetland conditions. The most dramatic example of this occurred in 1998 when a massive bird die-off occurred on farmlands that had been rehydrated to their prior wetland state on lands adjacent to Lake Apopka in Florida.

---

1 Portions of this chapter are adapted from Mary Jane Angelo, *The Killing Fields: Reducing the Casualties in the Battle Between U.S. Endangered Species and Pesticide Law*, 32 Harvard Environmental Law Review 96 (2008); Mary Jane Angelo, *Regulating Evolution for Sale: An Evolutionary Biology Model for Regulating the Risks Posed by Genetically Modified Organisms*, 42 Wake Forest Law Review 93 (2007); and Mary Jane Angelo, *Embracing Uncertainty, Complexity and Change to Protect Ecological Integrity: An Eco-Pragmatic Reinvention of a First Generation Environmental Law*, 33 Ecology Law Quarterly 105 (2006).

Lake Apopka, Florida is located in central Florida near the city of Orlando. It is a large, approximately 31,000-acre lake, which served as a major sports fishing venue dating back to the 1800s.[2] At one time, more than 20 fish camps were located on the shoreline, and sports fishermen traveled for long distances to take advantage of the excellent fishing and recreational opportunities on the lake.[3] During the 1940s, the State of Florida gave away thousands of acres of wetlands along the north shore of the lake to encourage row crop, or "muck" farming operations on the nutrient-rich peat soils.[4] To farm these wetlands, it was necessary to build a large levee between the wetlands to be farmed and the open-water area of the lake and then to pump water out of the farmlands into the lake proper.[5] Consequently, approximately 20,000 acres of sawgrass marsh was isolated from the remainder of the lake by levees.[6] The water pumped from the farms into the lake was laden with high levels of phosphorus.[7] Water continued to be pumped on a regular basis from the farms into the lake from the 1940s until the 1990s.[8] By the mid-1960s Lake Apopka was Florida's most polluted large lake.[9]

After many years of efforts to control nutrient loadings into the lake through regulatory and non-regulatory mechanisms, the state of Florida decided that the only way to achieve the dramatic nutrient reductions that were necessary to restore the lake was for the state to purchase the lands surrounding the lake and take them out of agricultural production.[10] The buy-outs involved 26 different property owners. The funding for the buy-outs came from both Florida's Preservation 2000 land acquisition program and from the federal WRP.[11] Because the Lake Apopka project was one of the first major WRP purchases, the WRP program did not yet have established rules or procedures. What turned out to be a major shortcoming in the WRP at that time was that it was assumed that individual Endangered Species Act (ESA) consultations for each restoration project would not be necessary because WRP staff believed that all of the projects

2 St. Johns River Water Management District (SJRWMD), The Lake Apopka Marsh Flow-Way Fact Sheet (2008) and SJRWMD, Lake Apopka Fact Sheet 1 (2008) (on file with author).

3 SJRWMD, The Ocklawaha Chain of Lakes Fact Sheet 1 (2003) (on file with author).

4 Victoria R. Hoge, Roxanne Conrow & David L. Stites, *SWIM Plan for Lake Apopka* (15 July 2003), *available at* http://www.sjrwmd.com/lakeapopka/pdfs/2003_Lake_Apopka_SWIM_Plan.pdf.

5 SJRWMD, Lake Apopka, *available at* http://www.sjrwmd.com/lakeapopka/index.html (last visited May 26, 2011).

6 *Id.*

7 Lake Apopka Fact Sheet 1, *supra* note 2, at 1. This document no longer exists.

8 *Id.*

9 *Id.* at 25 (citing US Environmental Protection Agency, Environmental Impact Statement, Lake Apopka Restoration Project, Lake and Orange counties, Florida (1979), Region 4, Atlanta).

10 Mary Jane Angelo, *Stumbling Toward Success: A Story of Adaptive Law and Ecological Resilience*, 87 Nebraska Law Review 950 (2009).

11 *Id.*

would be covered by a programmatic consultation that had been conducted on the WRP as a whole.[12]

Prior to purchasing any of the farmlands, the St. Johns River Water Management District (SJRWMD), which is one of Florida's five water management regulatory agencies and has jurisdiction over the Lake Apopka area, conducted environmental site assessments to determine if soils on the land previously used for farming were contaminated with pesticides.[13] If significant contamination was discovered, SJRWMD required the seller to clean up the property prior to the purchase.[14] In response to the site assessments, approximately 33,000 tons of soil contaminated with organochlorine pesticides were removed by the landowners, for a total remediation cost of more than $1.4 million.[15] The environmental assessments revealed low levels of contamination over most of the farm fields tested, with higher concentrations in areas of the fields that had been used for the mixing and loading of pesticides. The federal partners in the WRP agreed to go forward with the purchases provided that SJRWMD conducted a risk assessment to determine what level of risk would be acceptable.[16] SJRWMD conducted soil sampling on portions of the land to be flooded and used the data to conduct the risk assessment, and the WRP partners signed off on the risk assessment.[17] In addition to the federal partners' agreement with the risk assessment, the Florida Department of Environmental Protection also agreed to the conclusions of the risk assessment. The conclusion of the risk assessment was that the risk posed by organochlorine pesticides (OCPs) could be managed through wetland creation and natural attention.[18]Although the scientists had concerns that the lower concentrations of pesticide residues in the field could pose a risk, based on best scientific judgment, founded in part on information from the scientific literature regarding how pesticides act in the environment and are metabolized by fish and wildlife, they concluded that the risks were not significant and the project could proceed.[19]

---

12 Edgar Lowe, St. Johns River Water Mgmt. Dist., Address at the University of Florida (April 3, 2007).

13 Hoge et al., *supra* note 4, at 89 (citing Parsons Engineering Science, Inc. and GeoSyntec Consultants, Inc. Review of documentation for the location and remediation of possible pesticide/chemical sites at the North Shore Restoration Area, Lake Apopka, Florida (2000), Winter Park, Fla.).

14 SJRWMD, The Lake Apopka Agreement: Questions and Answers (2007).

15 Gianfranco Basili, Edgar Lowe, St. Johns River Water Mgmt. Dist., Address at the University of Florida (January 30, 2007).

16 Hoge et al., *supra* note 4, at 89.

17 ATRA, Inc., Environmental risk assessment of a Lake Apopka muck farm wetlands restoration, Special Publication SJ98-SP7 (1997), St. Johns River Water Management District, Palatka, Florida; Personal correspondence, Dr. Edgar Lowe, January 9, 2009.

18 *Id.*

19 *Id.*

By August 1998, the SJRWMD, in partnership with the WRP of the USDA, had purchased most of the farms on the north shore of Lake Apopka.[20] SJRWMD's plan was to reflood the farmlands to mimic their pre-agricultural wetland state. Ultimately, SJRWMD planned to treat the fields with alum prior to reflooding to trap excess phosphorus before it entered the water column.[21] The farmers in Unit 2, a 6,000-acre area on the northeast side of the lake, were asked to leave their fields shallowly flooded following their final crop harvest in the summer of 1998.[22] Historically, during late summer and early fall, some farmers flooded their fields to minimize soil subsidence and erosion and to control nematodes.[23] Short-term shallow flooding before pumping wastewater back into the lake was thus standard farming practice at the end of each year's growing season.[24] This pumping covered the area with approximately 18 inches of water for up to six weeks.[25] The shallow-water habitats created by the flooding attracted large numbers of shorebirds, wading birds and other aquatic species.[26] Even though the decades of farm flooding in late summer and early fall coincided with shorebird migration, and many birds had been visiting the flooded fields for decades, there had never been any reports of bird die-offs during previous flooding.[27] In 1998, however, because the lands would no longer be farmed, the water was allowed to remain on the farms through the fall and into the winter.[28] By eliminating the post-season pumping, the influx of phosphorous and pesticides to the lake would be reduced, and the growth of terrestrial vegetation on the farm fields would end.[29]

The fields were to be drained during the winter and treated to prevent phosphorus release.[30] The late summer weather and farming conditions of 1998 were similar to previous years.[31] However, as water levels began to rise with seepage and rainfall, and as fields remained flooded into late fall and early winter, more and more birds arrived.[32]

---

20 *Id.* at 146.

21 Lowe, *supra* note 12.

22 *Id.*

23 Hoge et al., *supra* note 4, at 146 (citing W.E. Johnson and J.E. Crumpton, Lake Apopka fish population assessment (Final report for Contract #97W121 for St. Johns River Water Management District, Palatka, Fla.) (1998), Florida Game and Fresh Water Fish Commission, Eustis, Fla.).

24 *Id.*

25 *Id.*

26 *Id.* (citing P.W. Sykes, Jr. and G.S. Hunter, *Bird Use of Flooded Agricultural Fields During Summer and Early Fall and Some Recommendations for Management*, 6 Florida Field Naturalist 36–43 (1978)).

27 *Id.*

28 *Id.*

29 *Id.*

30 *Id.*

31 *Id.*

32 *Id.*

Lake Apopka is one of the most diverse areas for birds of any place in the southeast US. For example, the total species list for Lake Apopka is 336, as compared to the only slightly longer list of 343 species for the Florida Everglades.[33] Because Lake Apopka is located in a migratory flyway, it is on the path of many thousands of migratory birds. In December of 1998, when the fields around the lake were flooded, the Audubon Christmas Bird Count (CBC) documented more than 46,000 birds, consisting of 174 species, present at the site, the highest recorded species diversity at an inland site in North America in the 100-year history of the CBC.[34] More than 3,500 American white pelicans were seen on a single day in December 1998 in the former farming area—an unprecedented number.[35] Approximately 10,000 to 12,000 white pelicans winter in Florida each year. Thus, almost one half of all of the Florida pelicans were observed at Apopka on a single day. Birders and scientists alike were excited about what promised to become one of the premiere birding sites in the world.

Excitement over bird diversity was tempered by the first reported mortalities of American white pelicans in November 1998.[36] While the first signs of the die-off appeared on November and December of 1998, the major die-offs occurred in January and February of 1999. Over those four months, 676 birds, including 441 American white pelicans, 58 great blue herons, 43 wood storks, 34 great egrets, and smaller numbers of 20 other bird species died on-site.[37] Although fewer than 700 birds were found dead on the site, the actual number of deaths is very likely higher, in part because it is likely that at least some birds were able to fly away from the site before they succumbed to the poisoning, and in part because research consistently demonstrates that the number of bird carcasses recovered from any mortality event is much lower than the actual number of deaths, since many carcasses are eaten or carried off site by predators and scavengers, or simply are missed by humans collecting carcasses because sick and dying birds often find secluded places to hide. In addition to actual deaths from pesticide exposure, many birds suffered from non-lethal reproductive impairment due to exposure to sub-lethal concentrations of the chemicals. The total number of birds that died and birds that were unable to reproduce will never be known,[38] but total losses sustained by the birds in the Lake Apopka area are estimated to be 5,213 "bird-

33 SJRWMD, Lake Apopka Restoration Area Land Management Plan 16 (2006), *available at* http://www.sjrwmd.com/landmanagementplans/pdfs/2006_Lake_Apopka_RA.pdf.

34 Hoge et al., *supra* note 4, at 146.

35 *Id.*

36 *Id.*

37 *Id.*

38 Gianfranco Basili, St. Johns River Water Mgmt. Dist., Address at the University of Florida (January 30, 2007).

years," a measure that incorporates the losses sustained by both adult birds and their progeny over time.[39]

Analysis of soils and birds showed that the lake resources and specifically the parts of the wetland restoration area were mainly contaminated by a group of compounds called organochlorine pesticides (OCPs).[40] Commonly applied to agricultural lands for decades, OCPs include chemicals such as DDT (dichlorodiphenyltrichloroethane), toxaphene, dieldrin, and chlordane. Flooding the fields caused these contaminants to become available for accelerated bioaccumulation to the birds which, according to the US Fish and Wildlife Service (USFWS), constituted a release that was not exempt under the pesticide provisions of Comprehensive Environmental Response, Compensation, and Liability Act (CERCLA).[41]

The massive number of bird deaths was attributed to organochlorine pesticide poisoning.[42] Most of the birds that were poisoned were fish eating birds.[43] It is believed that as the farm fields were flooded, fish left the canals adjacent to the fields and swam into the flooded fields where they were contaminated by organochlorine pesticides, such as DDT and its breakdown product DDE, toxaphene, dieldrin, and chlordane.[44] Although farm fields had been flooded regularly for years without negative effects, in the past, the fields were not flooded during winter months and white pelicans are only in Florida during the winter months.[45] It is interesting to note that most of these organochlorine pesticides were banned by EPA in the 1970s or 1980s. EPA cancelled the registration of DDT in 1972,[46] the registration for most uses of dieldrin in 1969, the registration for most uses of chlordane in 1978, and the registration for most uses of toxaphene in 1982.[47] However, when toxaphene was banned for most uses in 1986, EPA allowed its continued use on corn at Lake Apopka.[48] The registrations for all remaining uses of dieldron were cancelled by

---

39 Industrial Economics Incorporated, Final Lake Apopka Natural Resource Damage Assessment and Restoration Plan 6 (June 2004) (on file with author).

40 Hoge et al., *supra* note 4, at 146–9.

41 Industrial Economics Incorporated, *supra* note 39, at 8; *see also*, Memorandum of Understanding Between the St. John's River Water Management District and the United States of America, October 3, 2003 (on file with author). CERCLA is codified at 42 U.S.C. § 9603.

42 Hoge et al., *supra* note 4, at 146.

43 *Id.* at 309.

44 *Id.* at 43–4.

45 *Id.*

46 US Environmental Protection Agency, DDT—A Brief History and Status, *available at* http://www.epa.gov/pesticides/factsheets/chemicals/ddt-brief-history-status.htm (last updated on May 27, 2011).

47 US Environmental Protection Agency, International Programs, Toxics, Persistent Organic Pollutants: A Global Issue, A Global Response, *available at* http://www.epa.gov/international/toxics/pop.html (last visited May 26, 2011).

48 *Id.*

1982, the registrations for all remaining uses of chlordane were cancelled by 1988, and the registrations for all remaining food crop uses of toxaphene were cancelled by 1993.

At the onset of the bird deaths in the fall of 1998, SJRWMD began draining the site.[49] Pumping accelerated in January 1999, and by mid-February 1999, the entire north shore farming area had been drained.[50] Since then, the fields have been kept dry and have become vegetated with upland grasses, herbs, and shrubs.[51]

Documented effects of organochlorine pesticides on avian species include behavioral changes, reproductive impacts, and death.[52] Although other compounds such as PCBs (polychlorinated biphenyls) and heavy metals (e.g., cadmium, copper, and lead) were also detected, concentrations were low and any adverse effects sustained by birds due to exposure to these other chemicals were considered small compared to the effects of OCPs.[53] Therefore, the USFWS considered the bird deaths to be a result of exposure to OCPs in the area.[54]

Shortly after discovering the bird deaths, scientists from SJRWMD, the US Fish and Wildlife Service, and the Florida Audubon Society began working together to try to determine what was causing the bird deaths and what should be done to address the problem. However, once the US Justice Department initiated its criminal investigation into the bird deaths, cooperation ceased. The federal government went into criminal investigation mode and the scientists were no longer able to work together or to share information. The Justice Department seized all of the bird carcasses as evidence so the scientists no longer had access to them to conduct testing in order to determine the cause of death. The criminal investigation hindered the ability of scientists and environmental managers to be able to cooperate to determine what steps should be taken to address the problem and prevent further harm. After years of investigation and more than two years of legal negotiations, the US Justice Department and SJRWMD reached a settlement agreement. The settlement is a global agreement that addresses the criminal issues under the Endangered Species Act (ESA), the Migratory Bird Treaty Act (MBTA) and the Bald and Golden Eagle Protection Act (BGEPA), as well as civil natural resources damages under CERCLA.

On October 8, 2003, SJRWMD and the United States entered into a Memorandum of Understanding (MOU)[55] to resolve the criminal and civil issues related to the bird kill. The MOU addresses both the civil and criminal issues involving CERLCA, the ESA, the MBTA, and the BGEPA, and resolves the

---

49 Hoge et al., *supra* note 4, at 146.

50 *Id.*

51 *Id.*

52 Industrial Economics Incorporated, *supra* note 39, at 8.

53 *Id.*

54 *Id.*

55 Memorandum of Understanding Between the St. Johns River Water Management District and the United States of America, October 3, 2003 (on file with author).

federal investigation as to both the SJRWMD and its employees. The MOU sets forth a number of processes governing future action by the SJRWMD. First, the MOU sets up a clear framework for conducting ESA consultations for actions that may affect listed species. Second, the MOU contains an Avian Protection Plan (APP), which will guide SJRWMD's land management activities to avoid or minimize impacts to migratory birds. Pursuant to the APP, SJRWMD is required to address 38 management areas related to MBTA and BGEPA. The APP covers projects within the 38 management areas that will "more than minimally" alter the condition of land or vegetation, and therefore the habitat value and bird use of those areas. Each project has been assigned a potential avian impact level—low, medium, or high—to determine where risk-minimizing efforts should be concentrated. The impact levels are based primarily on the contaminants found at the site. Each project has then been given a project description, a definition of the habitat prior to start of the project, a predicted definition of the habitat upon completion of the project, a description of the avian use of the area, a prediction of the potential effect of the project on avian composition, a risk assessment and, finally, a plan of action for minimizing risks and impacts on habitat. Finally, the APP includes a Post-Project Monitoring Plan and a Contingency Plan. The Post-Project Monitoring Plan covers projects with a high potential impact level that is not significantly offset by the project design. In those cases, the birds on site will be monitored on a regular basis. SJRWMD will report any sick, injured, or dead birds to the USFWS within 48 hours, in accordance with the MOU. Further, SJRWMD will report any large die-offs of amphibians, as indicative of poor system health. The Contingency Plan will be used if monitoring identifies early warning signs of a bird die-off (birds acting strangely, not responding to loud noises or human activity, wobbling, unable to fly, or not flying well). In such cases, SJRWMD may undertake activities to discourage avian use of the site. These activities include altering water levels and hazing to keep animals away from the area.[56]

Under the MOU, SJRWMD will compensate for damages to natural resources by purchasing, preserving, and managing land containing a major wood-stork rookery. The rookery site purchased by the SJRWMD is an 8,465-acre property that contains approximately 150 wood-stork nests, making it the second largest breeding colony of wood storks in northeast Florida. The site includes five miles of frontage along the Matanzas River and its acquisition creates a nearly 16,000-acre conservation area. SJRWMD contributed $10 million to the purchase price of the site. This property provides multiple restoration benefits, including protection for a breeding colony of the endangered wood stork, and habitat for a myriad of other bird species such as American white pelican and great blue heron. The MOU also requires SJRWMD to do employee training on federal laws that protect avian wildlife and to participate in a number of cooperative efforts with wildlife agencies.

Subsequent to the extensive soil testing and research that occurred in response to the bird kills, SJRWMD has experimented with a variety of remediation plans for

56 *Id.*

the portions of the farmlands that have unacceptable levels of pesticide residue.[57] Based on the new research, SJRWMD's plan is to restore farmlands to vegetated marshes rather than open water to reduce the risk of future poisonings.[58] To reduce pesticide levels in areas that exceed acceptable levels, SJRWMD has tested a variety of remediation methodologies including removing soil and bringing in clean soil to replace it, a "soil blending" approach in which the top several feet of soil are mixed to dilute the pesticide levels in the higher strata of the soil, and using a special plow which takes the top 2–4 feet of soil and flips it over so that the contaminated portion of the soil is buried.[59]

The Lake Apopka example illustrates the long-term ecological legacy of past pesticide use. Decades after a pesticide has been banned, severe ecological impacts can continue to occur. This is particularly problematic with attempts to restore farmlands back to their prior wetland conditions. Disturbing farmland soils that contain pesticides and rehydrating agricultural lands can result in previously tied-up pesticides becoming physically and biologically available. Thus, programs such as the WRP that seek to revert and conserve farmland in wetlands can result in unanticipated ecological harm if care is not taken. Current large-scale projects involving restoring agricultural lands back to their previous wetland condition, including the Comprehensive Everglades Restoration Program in south Florida, may also have to address the legacy of past pesticide use. The Lake Apopka story serves as a stark reminder that we must take into consideration residues from past pesticide use when undertaking environmental restoration efforts.

57 Hoge et al., *supra* note 4, at 148.

58 *Id.* at 149.

59 SJRWMD, Lake Apopka Restoration Area Land Management Plan 56 (2006), *available at* http://www.sjrwmd.com/landmanagementplans/pdfs/2006_Lake_Apopka_RA.pdf. The total project is 5,050 acres, including 4,150 acres of wetlands.

PART IV
The Legal Response to Pesticides

Chapter 8
The Evolution of Domestic Law: FIFRA

The Federal Insecticide Fungicide and Rodenticide Act (FIFRA)

A number of different federal statutes address the ecological risks associated with pesticides in the US. The primary statute directly governing the regulation of pesticides is FIFRA. Several other statutes such as the Clean Water Act, the Endangered Species Act, the Migratory Bird Treaty Act, and the Organic Foods Production Act, indirectly address the ecological risks of pesticides and are discussed in the next chapter.

The most comprehensive federal law addressing pesticides is FIFRA.[1] The primary federal law addressing toxic substances in general, the Toxics Substances Control Act (TSCA), explicitly excludes from its regulatory jurisdiction "any pesticide [as defined by FIFRA] when manufactured, processed, or distributed in commerce for use as a pesticide."[2] Although a version of FIFRA has been in place since 1910, the environmental movement of the 1960s directly led to substantial amendments to FIFRA that, for the first time, brought environmental considerations into the purview of the law. Brooks et al., provide a detailed exploration of the role that the ecological science of pesticides played in the development of environmental law in general and pesticide law in particular. According to Brooks et al., the 1968 Wisconsin DDT hearings, in which an environmental organization filed a petition with the Wisconsin Department of Natural Resources seeking a declaratory ruling that DDT was a pollutant, was one of the first attempts to use ecology "in a legal forum to promote environmental protection."[3] According to these authors, the Wisconsin DDT petition "rode on the fame of Rachel Carson's *Silent Spring* written in 1962 and presented to a lay audience, a view of pollution as an unwelcome intruder in the "house of nature," a complex ecosystem which magnified the effects of the pollutant."[4] Ecological concerns, including concerns with the manner in which DDT moved through the ecosystem, the way DDT

---

1 7 U.S.C. §§ 136–136(y).

2 15 U.S.C. § 2602(2)(B)(ii).

3 Richard O. Brooks, Ross Jones & Ross A. Virginia, Law and Ecology: The Rise of the Ecosystem Regime 86 (2002).

4 *Id.* at 86. Carson's work not only set the foundation for the use of ecological science in pesticide law, but also played an important role in introducing ecological considerations into other legal fora, including the US Supreme Court. In his 2012 book, The Supreme Court and the Environment: The Reluctant Protector, Michael Allan Wolf, describes the significant influence that Rachael Caron's work had on Supreme Court Justice William

bioaccummulated in the food chain, and the impact of DDT on individual bird species as well as on the community as a whole, dominated the testimony in the Wisconsin DDT hearings.[5] These same ecological considerations later played a predominant role in the adoption of the 1972 amendments to FIFRA and in the subsequent cancellation of DDT by EPA.

In the US, EPA has the primary responsibility for regulating pesticides under the Federal Insecticide Fungicide and Rodenticide Act (FIFRA). Under FIFRA, EPA has the authority to regulate the sale, distribution, and use of pesticides in the US. FIFRA has its roots in the 1910 Insecticide Act, which marked Congress' first attempt to regulate the sale and distribution of pesticides. The 1972 FIFRA amendments, which form the backbone of the current FIFRA, were not enacted as a new freestanding environmental protection initiative. Instead, the amendments were an attempt to integrate an environmental component into a then 60-year-old statute, the Insecticide Act of 1910.[6] The Insecticide Act was designed to protect consumers from ineffective insecticide products and fraudulent claims about such products, which could cause crop losses. Environmental concerns did not play any role whatsoever in either the Insecticide Act of 1910 or its subsequent amendments in 1947. In fact, when President Truman signed the 1947 legislation[7] amending the 1910 Act, the *New York Times* printed a small blurb in the "News on Food" section on page 26, describing it as a law to "color poisons."[8] At the time of its passage, the primary groups concerned about pesticides were farmers, whose interests in government were advocated by the United States Department of Agriculture (USDA). DDT was seen as a magic bullet against the pests and crop disease that cost farmers $360 million in 1945 alone.[9] Passage of the Act and the 1947 amendments was non-controversial in large respect because there were few opponents to the concept of widespread pesticide application, much less well-organized opponents.[10]

Pesticide regulatory reform moved slowly, partially as result of the makeup of the primary regulating Congressional committees.[11] Of particular importance was James Whitten, who chaired the subcommittee on agricultural appropriations for the House Appropriations Committee.[12] As subcommittee chair, Whitten was called the "Permanent Secretary of Agriculture," and held this post from 1947

O. Douglas. *See* Michael Allan Wolf, The Supreme Court and the Environment: The Reluctant Protector (2012).

5 *Id.* at 87.

6 Act of April 26, 1910, ch. 191, 36 Stat. 331, repealed 61 Stat. 163, 172 (1947).

7 61 Stat. 163 (1947).

8 Christopher J. Bosso, Pesticides & Politics 21 (1987).

9 *Id.* at 22.

10 *Id.* at 34 (noting that few groups understood the potential effects of widespread use).

11 *Id.* at 60 (describing how the seniority of Southern Democrats, who were generally against pesticide regulation, allowed them to head various committees).

12 *Id.* at 67.

until 1992.[13] He encouraged the USDA to pursue the means necessary to eradicate pests and advocated widespread pesticide application to accomplish this goal.[14]

The first backlash against unremitting pesticide application was seen in the late 1950s with federal government campaigns against the gypsy moth and fire ants. Several groups complained about the gypsy moth program's use of DDT suspended in oil, which led to high fish kills in northern states and, similarly, groups were concerned about the application of dieldrin in high concentrations to fight the so-called fire ant "threat."[15] The most dramatic public backlash began with the *New Yorker* magazine's publication in 1962 of three articles by Rachel Carson arguing that pesticides were over used, despite the fact that their effects were poorly understood.[16] The resulting public debate pitted scientist against scientist, arguing over the benefits of pesticide usage and the respective hazards.[17] In the following years, pesticide regulation and reform came to the forefront and resulted in the 1964 FIFRA amendments, which required registration numbers for pesticides and eliminated the "protest registration," which had allowed chemical makers to keep a product on the market while protesting cancellation.[18] Of additional importance was that environmental values were discussed in the 1964 bill whereas in 1947 such issues had no influence in the resulting legislation.[19]

Bridge Document 8.1 Use of Pesticides, A Report of the President's Science Advisory Committee, May 15, 1963

A major step in civilization was the domestication of food plants. With the birth of organized agriculture and the resultant concentration of crops and animals, the stage was set for outbreaks of pests. Until that time man had to search for food as did the pests. Afterword neither had to search; instead pest control became necessary. The welfare of an increasing human population requires intensified agriculture. This in turn enables the pests to increase, which necessitates the use of pesticides with their concomitant hazards. It thus seems inevitable that, as the population increases, so do certain hazards (p. 1).

…

The panel believes that the use of pesticides must be continued if we are to maintain the advantages now resulting from the work of informed food producers and those responsible for control of disease. On the other hand, it has now become clear that the proper usage is not simple and that, while they destroy harmful insects and plants, pesticides may also be toxic to beneficial plants and animals, including man.

13 *Id.* at 67.
14 *Id.* at 69.
15 *Id.* at 84.
16 *Id.* at 115.
17 *Id.* at 121.
18 *Id.* at 127.
19 *Id.*

The Act, which would become FIFRA, was significantly amended in 1972. FIFRA came into being in its current form after the nation's experiences with DDT and other toxic pesticides.[20] The effort to reform FFIRA responded in part to the delays EPA faced when it sought removal of certain pesticides from the market.[21] Although the 1972 FIFRA amendments brought environmental concerns into the purview of pesticide regulation, such concerns were more of an afterthought to an already established consumer protection licensing program. In fact, the legislative history of the 1972 FIFRA makes clear that the amendments were not seen primarily as environmental in nature, but instead were seen as a balancing between the importance of pesticides to securing the nation's food supply and the risks pesticides pose to man and the environment.[22] The Senate Report explicitly noted the concern that "some [pesticides] may have long lasting adverse effects on the environment. Some may be taken up in the food chain and accumulated in man and other animals. Improperly used they may endanger bees and other useful insects, birds and other animals and their food supply."[23] The Report provides a historical guideline to federal pesticide legislation, starting in 1910 with the Federal Insecticide Act. The Committee noted that the 1910 Act aimed to prevent the sale or manufacture of adulterated or mislabeled insecticides or fungicides. In 1947 the original FIFRA, which repealed and replaced the 1910 Act, focused on pesticide registration and warnings. In 1954, the so-called "Miller Amendment" was added, requiring manufacturers to test their pesticide before it was to be used on food crops, and to provide data on the toxicity of the chemical. The information allowed the Federal Food and Drug Administration to set a limit on the amount of chemical residue permissible on the food crop when sold. In 1959, Congress further amended FIFRA to include new forms of agricultural chemicals (e.g., defoliants and desiccants). The impact of pesticides upon wildlife was not addressed, however, until the 1972 amendments.[24] A 1963 Presidential Scientific Advisory Committee had recommended adding a pesticide's impacts on fish and wildlife as a factor for registration (prior to 1972, the impact upon "useful vertebrates and invertebrates was the only environmental consideration in the registration process").[25]

As is evident from FIFRA's legislative history, the Act started as a classic consumer protection act aimed at ensuring that pesticide products were not mislabeled or adulterated. The 1972 revisions that brought environmental considerations into the purview of the Act were never the primary focus of the Act. Moreover, it is clear from FIFRA itself, as well as its legislative history and

20 Senate Report 970, 1972 U.S.C.C.A.N. 4092, 4093 (discussing the need for reforming the federal schematics for pesticide regulation).

21 *Id.* at 4094.

22 Senate Report 938, 1972 U.S.C.C.A.N. 3993, 3996.

23 *Id.*

24 *Id.* at 3999–4000.

25 *Id.*

judicial interpretation, that economic and social considerations such as concerns for farmer profit, desire for cheap and safe food, and concerns over pest vector-borne public health diseases, are equally important to environmental considerations under the Act.[26]

FIFRA section 2(u) provides that the term "pesticide" means "any substance or mixture of substances intended for preventing, destroying, repelling, or mitigating any pest …."[27] Under FIFRA, all pesticides that are sold or distributed in the United States must be registered by EPA.[28] The primary standard for registration is that the pesticide be registered only if it will not cause an "unreasonable adverse effect on the environment."[29] Section 2(j) defines the term "environment" broadly to include "water, air, land, and all plants and man and other animals living therein and the interrelationships which exist among them."[30] Section 3(a) of FIFRA provides that EPA shall register a pesticide if it determines that, "when considered with any restrictions imposed, its composition is such as to warrant the proposed claims for it, its labeling and other material required to be submitted comply with the requirements of FIFRA; the pesticide will perform its intended function without unreasonable adverse effects on the environment; and when used in accordance with widespread and commonly recognized practice it will not generally cause unreasonable adverse effects on the environment."[31] FIFRA defines the term "unreasonable adverse effects on the environment" as "any unreasonable risk to man or the environment, taking into account the economic, social, and environmental costs and benefits of the use of any pesticide."[32] Accordingly, when making the determination of whether to register a pesticide, EPA must consider both the risks posed by the pesticide and the economic and social implications of using the pesticide. Although not expressly mandated by the statute, EPA has interpreted and consistently applied this standard as a cost/benefit balancing standard under which EPA weighs the costs or risks associated with the use of a pesticide against the economic and social benefits of the pesticide. It should be noted that cost/benefit terminology is used differently under FIFRA than in discussing most environmental regulation. Typically, in doing a cost/benefit analysis, the regulatory agency compares the costs of regulation (e.g., the cost of installing pollution controls) to the benefits of regulation (e.g., lives saved

26 *See* McGill v. EPA, 593 F.2d 631, 635 (1979) (providing that FIFRA is aimed not only at environmental goals, but also the economic interests of farmers and consumers).

27 7 U.S.C. §§ 136(u).

28 *Id.* § 136a(a). This subsection provides "Except as provided by this subchapter, no person in any State may distribute or sell to any person any pesticide that is not registered under this subchapter. To the extent necessary to prevent unreasonable adverse effects on the environment, the Administrator may by regulation limit the distribution, sale or use in any State of any pesticide that is not registered under this subchapter." *Id.*

29 *Id.* § 136a(c)(5).

30 *Id.* § 136(j).

31 *Id.* § 136a(c)(5).

32 *Id.* § 136(bb).

or cancers avoided). Under FIFRA, however, the "costs" are considered to be the costs of allowing the use of the pesticide (e.g., cancer deaths), whereas the benefits are considered to be the benefits of allowing the use of the pesticide (e.g., reduction in crop loss from pest insect damage).

A number of scholars have pointed out that although Congress did direct EPA to take into account economic factors, it did not explicitly mandate that EPA conduct a strict cost/benefit analysis.[33] In fact, as Professor William Rodgers has argued, the legislative history of FIFRA suggests that adverse effects were not intended to be tolerated unless there are "overriding benefits" from the use of the pesticide.[34] Despite the apparent intent of Congress in enacting FIFRA, for more than 30 years EPA has interpreted FIFRA to require a cost/benefit balancing, and this interpretation has been upheld by the courts.[35]

Although the registration standard requires EPA to determine that the pesticide "will perform its intended function" without unreasonable adverse effects on the environment,[36] FIFRA expressly states that EPA shall not make any lack of essentiality a criterion for denying registration of any pesticide, and that where two pesticides meet the requirements for registration, one should not be registered in preference to the other. Specifically, the statute states "the Administrator shall not make any lack of essentiality a criterion for denying registration of any pesticide. Where two pesticides meet the requirements of this paragraph, one should not be registered in preference to the other. In considering an application for the registration of a pesticide, the Administrator may waive data requirements pertaining to efficacy, in which event the Administrator may register the pesticide without determining that the pesticide's composition is such as to warrant proposed claims of efficacy."[37] Accordingly, there is no requirement to demonstrate that a pesticide is necessary to control a particular pest to obtain a registration and the availability of alternative pesticides for the same use does not preclude registration. Moreover, FIFRA expressly authorizes EPA to waive all

---

33 *See* Sidney A. Shapiro & Robert L. Glicksman, Risk Regulation at Risk: Restoring a Pragmatic Approach 29, 32 (2003) and Mary Jane Angelo, *Embracing Uncertainty, Complexity and Change to Protect Ecological Integrity: An Eco-Pragmatic Reinvention of a First Generation Environmental Law*, 33 Ecology L.Q. 177, 182 (2006).

34 *See* William H. Rodgers, Environmental Law 407, 453 (West, 2nd ed., 1994).

35 *See* Environmental Defense Fund, Inc. v. EPA (heptachlor-chlordane), 548 F.2d 998, 1004 (D.C. Cir. 1976), *cert. denied* 431 U.S. 925 (1977) (stating that "to evaluate whether use of a pesticide poses an 'unreasonable risk to man or the environment,' [EPA] engages in a cost-benefit analysis …"); In the Matter of Chapman Chemical Co., et al., FIFRA Dockets No. 246, et al. (EPA 1976) (stating that "before any pesticide can be cancelled under FIFRA [EPA] must be persuaded that the risks to man or the environment from continued use of the pesticide outweigh the benefits of its continued use."); In the Matter of Protexall Products, Inc., et al., FIFRA Docket Nos. 625, et al. (1989) (stating that "the risk-benefit assessment involves a balancing of the risks … against the benefits …).

36 7 U.S.C. § 136a(c)(5)(B).

37 *Id.* § 136a(c)(5).

data requirements pertaining to efficacy and in fact EPA has, by rule, done so.[38] Thus, as a practical matter in making registration decisions, EPA does not require any showing of the economic or social benefits to be derived from the pesticide, but instead assumes that such benefits will accrue.

One of the most significant aspects of FIFRA is that it requires an applicant for a pesticide registration to submit data to EPA. EPA evaluates the potential for risk for a particular pesticide using data, generally submitted by the registration applicant, specific to that pesticide. EPA's data requirements are designed to evaluate the two components of risk: hazard and exposure. Data requirements are found at 40 CFR Part 158, and provide for the submission of data on health and environmental effects. The applicant for registration must bear the cost of gathering and generating the necessary data. To avoid duplicative data generation, the statute encourages the joint development of data and provides that applicants seeking to reach agreement on the terms of a data development arrangement may seek binding arbitration.[39]

Data already submitted to the agency to support an existing registration may be relied upon to support a new registration application provided the applicant for the new registration offers to pay compensation to the registrant who originally submitted the data. Data submitted to support a registration the first time a particular active ingredient is registered, is protected by the "exclusive use" provisions of FIFRA and cannot be considered by EPA to support additional registrations for a period of 10 years. In addition, FIFRA section 10 generally governs the disclosure of information submitted to EPA pursuant to FIFRA requirements. Section 10(d) provides that health and safety data must be made available to the public, except that section 10(g) prohibits disclosure of health and safety data to multinational pesticide producers except during public proceedings under law or regulation. Sections 10(b) and 10(d) provide that other confidential business information ordinarily may not be released, and provide specific protection for the formula and information on inert ingredients. Exemptions from these confidentiality protections are provided to avoid imminent public health risks. Additionally, the Administrator may determine that disclosure is in the public interest during a proceeding to determine whether a pesticide causes unreasonable adverse effects. Any such release of information is subject to procedural protections involving prior notice and opportunity for district court review.

Under FIFRA, EPA may register products in certain situations even though all necessary data have not yet been generated. Such a premature registration is referred to as "conditional registration." Conditional registration can be used in a number of specified circumstances, including: (1) products with composition and proposed uses identical or substantially similar to currently registered pesticides; (2) products with proposed new uses; or (3) certain products with new

38 40 C.F.R. § 158.640(b)(1).

39 7 U.S.C. § 136a(c)(1)(F)(iii).

active ingredients.[40] For the first two types of conditional registration, EPA must determine that despite the lacking data, approval of the conditional registration would not significantly increase the risk of any unreasonable adverse effects on the environment.[41] To issue a conditional registration for new active ingredients, EPA must determine that the use of the pesticide during the period of conditional registration will not cause unreasonable adverse effects on the environment and that use of the pesticide is in the public interest.[42]

The vast majority of EPA's data requirements under FIFRA relate to human health effects. These data requirements include testing on residue chemistry to estimate human exposure to pesticides, acute human hazard, subchronic human hazard, chronic human hazard, mutagenicity, metabolism studies, reentry hazard, spray drift evaluation, as well as oncogenicity, teratogenicity, neurotoxicity, and reproductive effects in humans.[43] Unfortunately, EPA's data requirements for testing for wildlife and ecological effects are extremely limited.[44] EPA does require the submission of environmental fate data designed to "assess the presence of widely distributed and persistent pesticides in the environment which may result in loss of usable land, surface water, ground water, and wildlife resources," and "assess the potential environmental exposure of other non-target organisms, such as fish and wildlife, to pesticides."[45] EPA's data requirements related to wildlife impacts or other ecological effects are much less ambitious. EPA requires submission of some data designed to evaluate impacts to wildlife and aquatic organisms. The wildlife and aquatic organism data requirements include avian toxicity studies,[46] and freshwater fish and invertebrate acute toxicity studies[47] for most pesticides intended for outdoor use. Avian oral LD50 and dietary LC50s (the concentration at which 50 percent of the test animals die) are required when using the preferred test animal species, the mallard, and the bobwhite. Freshwater fish LC50 studies are required, with the preferred test species being the rainbow and bluegill fish, and acute LC50 studies are required on freshwater invertebrates, with the preferred test species being *Daphnia spp.*

Additional data are only required on a case-by-case basis depending on the result of lower tier studies such as acute and subacute testing, intended use pattern

40 7 U.S.C. § 136a(c)(7).

41 *Id.* § 136a(c)(7)(A).

42 *Id.*

43 *See* 40 C.F.R. paragraphs 158.202(a), (c), (e), (f), and (g) and *id.* §§ 158.240, 158.390, 158.440, and 158.340. *See also id.* § 158.34 (providing that certain human health effects data submitted to EPA must be flagged as indicating potential adverse effects).

44 *See* Leslie W. Touart & Anthony F. Macriowski, *Information Needs for Pesticide Registration in the United States*, 7 Ecological Applications 1083 (1997) (describing and evaluating EPA's ecological risk data requirements for pesticide registration).

45 EPA, *Data Requirement for Pesticide Registration, available at* http://www.epa.gov/opp00001/regulating/data_requirements.htm (last visited June 3, 2011).

46 40 C.F.R. § 158.490.

47 *Id.*

and environmental fate characteristics, or if certain specified criteria are met. Such conditionally required studies include mammal toxicity, avian reproduction, simulated and actual field testing of mammals and birds, acute toxicity to estuarine and marine organisms, fish early life stage, aquatic invertebrate life cycle, fish life cycle and aquatic organisms accumulation, and simulated or actual field testing of aquatic organisms for most outdoor uses.[48]

With regard to wildlife, EPA's main concern is with acute toxicity testing, and it does not typically require data submission on the potential adverse effects of pesticides on wildlife behavior, neurology, reproduction, birth defects, or other non-acute effects. EPA's data requirements do not contain any studies whatsoever aimed at evaluating effects on other species such as amphibians or reptiles, or other species not specifically identified in the rules. As to organisms other than birds, mammals and fish, EPA's requirements are even more limited. In fact, EPA rarely requires data submission related to adverse effects to non-target insects. Although EPA does conditionally require acute toxicity testing for honeybees and other pollinators if the proposed use will result in honeybee or other pollinator exposure, it does not have any data requirements related to pollinator subacute feeding studies,[49] non-target aquatic insects, or non-target predatory or parasitic insects.[50] In its data requirements rule, EPA identifies pollinator data requirements as "reserved pending development of test methodology" and data requirements for non-target predatory or parasitic insects as "reserved pending further evaluation to determine what and when data should be required, and to develop appropriate test methods."[51]

Once EPA evaluates submitted data, it must determine whether use restrictions are necessary to minimize risks sufficient to be outweighed by benefits, and thus, to meet the registration standard. However, EPA's ability to regulate pesticide use under FIFRA is very limited. Unlike many other environmental statutes, FIFRA does not establish a permitting system for pesticide use. Specifically, no EPA approval is required prior to using a pesticide, whether by permit or any other mechanism, even for very large-scale usage. Consequently, geographical and temporal factors are not evaluated under FIFRA prior to release of pesticides into the environment. In fact, FIFRA's regulation of pesticide "use" is achieved through labeling restrictions. It is the registration applicant's responsibility to propose all labeling with the registration application.[52] FIFRA defines the term "label" as the written, printed, or graphic matter on, or attached to the pesticide.[53] "Labeling," on the other hand is much broader and includes the label as well as all other written, printed, or graphic matter that accompanies the pesticide or to which

48 *Id.*
49 40 C.F.R. § 161.590 (2009).
50 *Id.*
51 *Id.*
52 7 U.S.C. § 136a(c)(1)(C).
53 *Id.* § 136(p)(1).

reference is made on the label.[54] All registered pesticide products must bear a label or labeling containing specified information including precautionary statements, warnings, directions for use of the product, and an ingredient statement. A product with a label or labeling which does not contain the information required by EPA or which sets forth false or misleading information is misbranded.[55] The primary means by which EPA regulates pesticide "use" under FIFRA is by requiring users of pesticides to follow all label directions. All pesticide product labels are required to state that it shall be unlawful for any person to use any pesticide in a manner inconsistent with its labeling.[56] This is the sole obligation placed by FIFRA on users of pesticides. Accordingly, directions for use are the only mechanism to regulate user behavior to accomplish risk-reduction goals. Unfortunately, pesticide users may not understand, or be willing to follow, the complex labeling instructions necessary to regulate use to prevent environmental harms. Moreover, it is virtually impossible for EPA to know who, where, when and how persons are using pesticides, much less to monitor each and every pesticide user in the country to assure the labeling instructions are followed.

FIFRA does authorize EPA to classify higher risk pesticides as "restricted-use pesticides." A restricted-use pesticide may be used only by or under the supervision of a certified applicator. These products may not be purchased by the general public.[57] However, such a designation is designed primarily to protect the users themselves and not to reduce risks to wildlife or ecosystems. EPA may classify a pesticide for restricted use if it would cause unreasonable adverse effects on the environment in the absence of such a restriction.[58] Certification of applicators is primarily conducted by the state. Each state must have a certification plan that conforms to certain standards enumerated in FIFRA. FIFRA provides that if any state, at any time, desires to certify applicators of pesticides, the governor of such state shall submit a state plan for such purpose. EPA shall approve the plan submitted by any state provided it meets certain specified conditions regarding the state's legal authority and funding mechanisms. Federal certification is required to be conducted by the Administrator, in consultation with the governor of any such state, in which a state plan for applicator certification has not been approved by the Administrator.[59] Over one half of all registered agricultural pesticides are restricted-use pesticides[60] and thus can only be applied only under the supervision of a certified applicator. Unfortunately, the certified applicator requirement does not mandate consideration of local ecological factors and consideration of lower-risk alternatives to address ecological risks posed by the use of a specific pesticide in

---

54 *Id.* § 136(p)(2).

55 *Id.* §§ 136(q) and 136j(a)(1)(E).

56 *Id.* § 136j(a)(2)(G).

57 *Id.* § 136a (d)(1).

58 *Id.*

59 *Id.* § 136i.

60 RODGERS, *supra* note 34, at 458.

a specific location at a specific time. The law does not require certified applicators to oversee directly the application of pesticides. Instead, certified applicators may simply be what are referred to as "arm chair" supervisors. They do not need to be present on site, but only need to be available via telephone during the time of the pesticide application.[61] Moreover, FIFRA does not require that certified applicators obtain any training in local ecological systems and their vulnerability to particular pesticides.[62] Finally, although FIFRA section 11 requires EPA and states to make available to certified applicators at their request instructional materials concerning Integrated Pest Management (IPM), the statute expressly states that certified applicators are not required to receive instruction on IPM and are not required to be competent with respect to such techniques.

Although there are no specific laws directing EPA to conduct comparative risk analyses or to promote one type of pesticide over another, it has undertaken certain policy initiatives to encourage the use of reduced risk pesticides. In 1992, EPA issued a *Federal Register* notice entitled "Incentives for Development and Registration of Reduced Risk Pesticides," which sought comment on potential approaches to encourage the development, registration, and use of pesticides or pest management practices that present lower risks to human health and the environment. Subsequently, EPA took steps to exempt from FIFRA registration requirements more than 30 specific naturally occurring pesticides that EPA found to pose little or no risk.[63] As described in more detail below, EPA has also taken steps to promote the use of lower risk biological pesticides and to streamline the registration process for pesticides that EPA classifies as posing reduced risk.

The 1972 revisions to FIFRA mandated that the agency go back and reexamine previously registered pesticides.[64] Congress mandated this reexamination or "reregistration" to ensure that previously registered pesticides meet current standards and to ensure that the data EPA had for these older pesticides was the same as that for newer pesticides. EPA's reregistration efforts moved extremely slowly and, as a result, in 1988 Congress imposed on EPA specific reregistration requirements intended to improve both the pace and the nature of reregistration.[65] These reregistration provisions establish a multiphased process with a number of deadlines that ensures that registrants submit required data for EPA review under current standards. Failure to meet the prescribed data submission deadlines may result in suspension or cancellation of registration.[66] In 1996, Congress passed the Food Quality Protection Act (FQPA), which among other things required EPA to go back and reassess the tolerances (safe level of pesticide residues in food)

61 *Id.* at 461.

62 For a description of certified applicator training programs, *see* Rodgers, *supra* note 34, at 461–3.

63 40 C.F.R. 152.25(f) (2010).

64 7 U.S.C. § 136a-l.

65 *Id.*

66 *Id.*

for registered pesticides to ensure they meet the current safety standard adopted in that year.[67] Although the FQPA is primarily focused on risks to human health from pesticides in or on food, the process of reevaluating tolerances led to the cancellation of some pesticides—sometimes merely because it was not worth it to the manufacturer to provide the data necessary for the evaluation or to make the reduction of amounts of pesticides that could be used. Thus, the FQPA indirectly affected ecological risks by eliminating some pesticides or restricting the use of others. In 2006, EPA initiated a new program, which it calls "registration review." The purpose of this process is to reevaluate all pesticides every 15 years, to ensure that as the ability to assess risks improves and as policies and practices change, registered pesticides continue to meet legal requirements with regard to human health and the environment.

Once a pesticide is registered, EPA maintains the authority to either cancel or suspend the existing registration based upon certain risk/benefit determinations. FIFRA section 6(b), which specifically addresses cancellation, provides that EPA may issue a notice of intent to cancel if a pesticide or its labeling does not comply with FIFRA or if, when used in accordance with widespread and commonly recognized practice, the pesticide generally causes unreasonable adverse effects on the environment. FIFRA requires a review of the proposed cancellation notice by the Secretary of Agriculture (USDA) and the FIFRA Scientific Advisory Panel (SAP). The statute dictates that the notice must be submitted to USDA and SAP 60 days prior to notification of the registrant or publication (whichever comes first). If USDA and the SAP do not submit comments within 30 days, EPA may publish the notice. If USDA and the SAP do submit comments, EPA may, after reviewing such comments, withdraw the notice, issue a final notice without modification, or modify the notice, as appropriate. Once the notice is published, persons adversely affected have 30 days to request a hearing. If no such hearing is requested, the notice of intent to cancel becomes final. If a hearing is requested, the hearing is considered a formal adjudicatory proceeding and is held before an Administrative Law Judge (ALJ). Such a proceeding is governed by the agency's rules at 40 CFR Part 164.[68]

Under section 6(b) there are two types of cancellation actions. Section 6(b)(1) authorizes EPA to issue a notice of intent to cancel or change classification. Section 6(b)(2) authorizes EPA to issue a notice of intent to hold a hearing to determine whether or not registration should be cancelled or classification changed. Section 6(b) provides that "if it appears to the Administrator that a pesticide or its labeling ... does not comply with the provisions of this subchapter or, when used in accordance with widespread and commonly recognized practice, generally causes unreasonable adverse effects on the environment, the Administrator may issue a notice of the Administrator's intent either: (1) to cancel its registration or to change it classification together with the reasons (including the actual basis) for

67 Food Quality Protection Act of 1996, P.L. 1070, 110 Stat. 1489.

68 7 U.S.C. § 136d(b).

the Administrator's action; or (2) to hold a hearing to determine whether or not its registration should be canceled or its classification changed."[69] This section goes on to provide that, in determining whether to issue any such notice, the Administrator shall include among those factors to be taken into account the impact of the action proposed in such notice on production and prices of agricultural commodities, retail food prices, and the agricultural economy. At least 60 days prior to sending such notice to the registrant or making public such notice, whichever occurs first, the Administrator shall provide the Secretary of Agriculture with a copy of such notice and an analysis of such impact on the agricultural economy. The proposed action shall become final and effective at the end of 30 days from receipt by the registrant, or publication of a notice, unless within that time either the registrant makes the necessary corrections, if possible, or a request for an administrative hearing before an ALJ is made by a person adversely affected by the notice. In taking any final action under this subsection, the Administrator shall consider restricting a pesticide's use or uses as an alternative to cancellation and shall fully explain the reasons for these restrictions, and shall include among those factors to be taken into account the impact of such final action on production and prices of agricultural commodities, retail food prices, and otherwise on the agricultural economy.

Regardless of the type of cancellation action initiated by EPA, the standard for cancellation is the risk/benefit balancing standard. Section 6(b)(2) is used when EPA's judgment concerning the risks and benefits of a pesticide is only tentative. There is no distinction between FIFRA section 6(b)(1) and 6(b)(2) hearing in the manner of conduct, burden of proof, or nature of initial decision by ALJ. One issue generally considered as part of the cancellation process is whether the agency should allow the continued sale and use of existing stocks of the pesticide. Before taking final action under section 6(b), EPA must consider whether any unreasonable risks posed by a pesticide's use can be sufficiently reduced by regulatory measures short of cancellation, such as additional labeling restrictions and/or the classification of the pesticide for restricted use. If EPA determines that sufficient risk reduction cannot be achieved by such measures, the registration of the pesticide for that use must be cancelled. Of the more than 60 pesticide cancellations and suspensions that occurred prior to 1994, only approximately one third have been judicially reviewed.[70]

In addition to authorizing cancellation, FIFRA authorizes EPA to suspend the registration of a pesticide based on certain findings. There are two types of suspension proceedings.[71] Ordinary suspension is used where necessary to prevent an imminent hazard during the time required for cancellation proceeding.

69 *Id.* § 136d.

70 Rodgers, *supra* note 34, at 480. And (D.C. Cir. 1972) EPA's refusal to initiate proceedings to cancel or suspend a registration is considered a final order reviewable in district court. *See* Envtl. Defense Fund v. EPA, 465 F.2d 528 (D.C. Cir. 1972).

71 7 U.S.C. § 136d(c) (2011).

"Imminent hazard" is defined as a substantial likelihood of serious harm during the duration of cancellation proceedings. A suspension action is merely to address the risks and benefits for the period involved, not an ultimate resolution of the cancellation issues.[72] In an ordinary suspension, EPA must provide notification to the registrant of the intent to suspend and an opportunity for a hearing. Only a registrant, not a third party such as an environmental organization, may request an adjudicatory hearing. The suspension order becomes effective either after a favorable decision following a hearing, or five days after notification if no hearing is requested.[73] If a registrant requests a hearing, an expedited administrative adjudicatory hearing is held before an ALJ. Interested third parties may intervene at this point. The sole issue to be decided at the hearing is whether an imminent hazard exists.[74] An emergency suspension order, on the other hand, is effective immediately. It is used where an emergency exists that does not permit even an expedited hearing before suspension takes place.[75] Registrants must request an expedited hearing within five days and the hearing must begin within five days of the agency's receipt of such a hearing request. The emergency order remains in effect until the issuance of a final suspension order following the hearing. Third parties do not have a right to request or intervene in an expedited hearing.[76] An emergency suspension order is subject to immediate review in district court.

In addition to regulating pesticides under FIFRA, EPA is responsible for regulating pesticide residues in human food or animal feed under the Federal Food, Drug, and Cosmetic Act (FFDCA).[77] FIFRA and FFDCA are closely linked in that a FIFRA registration for a food use pesticide may not be granted unless it complies with the standards in the FFDCA.[78] The Reorganization Plan No. 3 of 1970, which created EPA, granted EPA authority to establish tolerances for residues of pesticide chemicals in foods and animal feeds.[79] Regulatory authority over other non-pesticidal substances in foods and animal feeds was left within the jurisdiction of the Food and Drug Administration.

Pursuant to section 408(a) of the FFDCA, a pesticide chemical residue in or on food is not considered to be safe unless EPA has issued a tolerance for such residue and the residue is within the tolerance limits.[80] EPA may issue an exemption from the requirements of a tolerance if it determines that there is a reasonable certainty

---

72 *Id.* § 136d(c)(1).

73 *Id.* § 136d(c)(2).

74 *Id.*

75 *Id.* § 136d(c)(3).

76 *Id.*

77 21 U.S.C. § 346.

78 U.S.C. 136(bb) defining "unreasonable adverse effect on the environment" as including "a human dietary risk from residues that result from a use of a pesticide in or on any food inconsistent with the standard under section 346a of title 21."

79 Reorganization Plan No. 3 of 1970, 3 C.F.R. at 199 (1970 Comp.), *reprinted in* 5 U.S.C. app. 1 Reorg. Plan 3 1970 and *in* 84 Stat. 2086.

80 21 U.S.C. § 346a(a)(1).

that no harm will result from aggregate exposure to the pesticide chemical residue, including all anticipated dietary exposures and all other exposures for which there is reliable information.[81] Although the FFDCA only addresses dietary risks from pesticide residues in food, it can indirectly reduce ecological risks as a side benefit. For example, if pesticide use reductions are necessary to meet FFDCA tolerances based on dietary risk, the environment will be a third party beneficiary of the reduced amount of pesticide releases into the environment.

Although the majority of pesticides EPA regulates are synthetic chemicals, as described above, FIFRA defines the term pesticide very broadly to include naturally occurring substances as well as living organisms. EPA has expressed a strong interest in encouraging the development and use of many biochemical pesticides, whether derived from nature or created to mimic naturally occurring biochemicals, because in general they are considered to be more environmentally benign than conventional chemical pesticides. Many biochemical pesticides are important components of Integrated Pest Management (IPM) programs. It should be noted that as IPM is, by definition, an integration of a number of pest management tools (which can include cultural, biological and GMO pesticides, as well as chemical pesticides), no specific regulations apply. However, biochemical and biological chemicals, which are of critical importance to IPM, are regulated in a similar fashion as are chemical pesticides.

EPA has encouraged the use of biological pesticides by creating an expedited registration process and waiving certain data requirements for many of these products.[82] It has taken a number of steps to streamline the regulatory process with regard to pheromones, which are compounds produced by insects and other arthropods that modify the behavior of other individuals of the same species.[83] First, EPA has established an exemption for pheromones used in traps, based on the low potential for risk presented by these products. EPA has by regulation exempted from regulation or other regulatory requirements, "[p]heromones and identical or substantially similar compounds labeled for use only in pheromone traps (or labeled for use in a manner which the Administrator determines poses no greater risk of adverse effects on the environment than use in pheromones traps), and pheromone traps in which those compounds are the sole active ingredient(s)."[84] In addition, EPA has taken a number of regulatory actions to promote the development of pheromone pesticide products. For example, in January of 1994, EPA published in the Federal Register a final rule exempting the residues of arthropod pheromones resulting from the use of these substances in solid matrix dispensers with an annual application rate limitation of 150 grams/acre for pest control in or on all raw agricultural commodities from the requirement

81 *Id.* § 346a(c)(2)(A).

82 40 C.F.R. § 152.25(b).

83 40 C.F.R. § 152.25(b)(1) (2010).

84 40 C.F.R. § 152.25(b) (2010).

of a tolerance under the FFDCA.[85] On December 8, 1993, EPA published a final rule exempting certain inert ingredients used in solid matrix dispensers from the tolerance requirement.[86] In this notice, EPA announced that it was expanding the acreage cut-off for when an EUP is required under FIFRA from 10 acres to 250 acres for a certain pheromone products. The pheromone products covered by the notice are pheromones in solid matrix dispensers used at rates at or below 150 grams/acre/year. Tests conducted on these pheromones under the conditions specified in the notice would not require an EUP at acreages up to and including 250 acres. Tests conducted on acreages exceeding 250 acres, or with pheromone products other than those specified, would continue to require an EUP. The scope of pheromone products exempted under this notice is the same as the scope of products exempted from tolerance requirements.[87] In addition, in the January 1994 notice, EPA announced that it was considering whether to take additional actions to grant further regulatory relief for other types of pheromone products. Finally, EPA frequently grants waivers of the data requirements specific to biological pesticides and microbial pesticides, finding that data are not needed to demonstrate that the pesticides meet the criteria for FIFRA registration.[88]

85 59 Fed. Reg. 3681 (1994).
86 58 Fed. Reg. 64493 (1993).
87 *Id.*
88 *See* 40 C.F.R. 158.690, 40 C.F.R. 158.740, & 40 C.F.R. 158.45.

Chapter 9

Other Domestic Laws that Address the Ecological Impacts of Pesticides

In addition to FIFRA, the comprehensive domestic law that governs a range of issues associated with the sale, distribution and use of pesticides, there are several other US laws that, while not specifically geared toward pesticides, address certain ecological risks associated with them. These include the Plant Protection Act, the Clean Water Act, the Endangered Species Act, the Migratory Bird Treaty Act, and the Organic Food Production Act. Each of these statutes, as well as common law remedies that apply to certain harms caused by pesticides, is discussed in relation to pesticides in this chapter.

The Common Law

As potential harmful substances that are released into the environment, pesticides can cause harm to third parties actionable under common law torts. Most of the common cases involving pesticides address claims by non-pesticide users against pesticide users, typically neighboring landowners or aerial pesticide applicators, alleging property damage, crop damage, harm to commercial bees and/or personal injury from the negligent use of the pesticides. The use of state tort common law to address harm from pesticide use dates back to the early half of the twentieth century.

For example, in *Lenk* v. *Spezia*,[1] the court held that a defendant who is spraying a dangerous pesticide may be liable for damages to others if the spraying is done negligently, or if the pesticide drifts off the property. However, the court noted there is no duty to bees "trespassing" on the sprayed property other than to avoid malicious or wanton conduct. The *Lenk* court, while recognizing the importance of pesticides to agricultural productivity, nonetheless held that a pesticide user must act reasonably to avoid harming other landowner interests.[2] Reasonable action would include action such as ensuring proper weather and wind conditions and following directions by the county agent as well as what was considered reasonable in the community regarding pesticide spraying.[3]

---

1 213 P.2d 47, 51–52 (Cal. Ct. App. 1949).

2 *Id.*

3 *Id.*

Other cases from this pre-1972 period, however, looked to evidence of reasonableness tied somewhat more closely to the particular characteristics of the pesticide itself, as set forth in manuals and "circulars" that described the uses and risks of the pesticide. In *Lawler* v. *Skelton*,[4] the Mississippi Supreme Court held that the defendant aerial crop sprayer was liable for the plaintiff's illness when he sprayed pesticides in close proximity to the plaintiff who was working in the area. The court looked to the pesticide labels and the state aerial applicators' safety manual, which urged extreme caution in using pesticides because of the risk to human health, to support a finding that the defendant's actions were negligent.[5]

Similarly, in *Andreen* v. *Escondido Citrus Union*,[6] the court held that the defendants were negligent in fumigating a citrus orchard because they applied the fumigant when weather conditions were not appropriate and at excessive rates. The court further found that it was appropriate for the lower court to consider testimony showing that the defendants did not have the license required for engaging in fumigation services.[7] The court concluded that the lack of the license was legitimate evidence tending to show negligence because the foreman in charge had testified that he was not familiar with the fumigation regulations adopted by the county horticultural commissioner nor with any of the documents or circulars on the subject issued by the US Department of Agriculture, or the University of California.[8]

Subsequent to the passage of the 1972 amendments to FIFRA, state courts continued to apply the law of negligence in cases involving harm from pesticides. Because of the existence of the regulatory requirements of FIFRA, however, these courts began to move toward a negligence *per se* analysis based on compliance with FIFRA or state pesticide laws. In *J.L. Wilson Farms* v. *Wallace*,[9] one of the early, post-1972 FIFRA negligence cases, the court held it was proper to allow the jury to consider state plant board regulations restricting commercial aerial application of 2,4-D in determining whether the defendant was liable for using that pesticide without plant board authorization.[10] In relying on state regulation to set the standard of care, the court in essence employed a negligence *per se* analysis in the context of a common law negligence claim.

Bennett v. *Larsen*[11] was the first significant case to include a true negligence *per se* analysis and focus on the interrelationship between FIFRA, the EPA-approved pesticide label, and state common law claims. Bennett is significant for several reasons. First, the court made a clear distinction between common

4 130 So. 2d 565, 567–8 (Miss. 1961).

5 *Id.* at 568.

6 269 P. 556, 557 (Cal. Ct. App. 1928).

7 *Id.* at 558.

8 *Id.*

9 590 S.W.2d 42 (Ark. Ct. App. 1979).

10 *Id.* at 44.

11 348 N.W.2d 540 (Wis. 1984).

law negligence and negligence *per se*, and concluded that the only duty of care a landowner owed to bees or other animals was a duty to follow the EPA-approved pesticide label, which established the standard of care in a manner "tailored to the effects and use of the individual pesticide."[12] Second, although not explicit in the court's analysis, the court essentially held that the tort of negligence *per se* is the only negligence-based tort that is actionable where the FIFRA label provides a regulatory standard of care. Accordingly, the only other common law causes of action available for damages would be for intentional or wanton destruction of the bees or, presumably, other intentional torts such as trespass or nuisance.[13]

Three post-1972 cases from Minnesota have also used the FIFRA label as establishing a standard of care. In *Red River Spray Service* v. *Nelson*[14] the court noted that the label for the pesticide, Banvel, expressly prohibited aerial application if wind was in excess of five miles per hour whenever sensitive crops such as soybeans were in the vicinity.[15] Based on evidence that wind speeds at the time of spraying were in excess of five miles per hour, the court found the defendant was negligent because the manner in which the pesticide was applied "was specifically forbidden by the EPA."[16] The court in *Dosdall* v. *Smith*,[17] upheld a jury verdict of negligence where the defendant sprayer testified he did not read the pesticide label and applied the pesticide EVIK to corn during a stage of growth harmful to the corn and prohibited by the pesticide label. In contrast, in *Honek* v. *Kovar*[18] the court affirmed a verdict of no negligence on the part of the defendant and distinguishing Red River, because in *Honek* the evidence was that the defendant followed the pesticide label directions. More recently, in *Anderson* v. *Minnesota Department of Natural Resources*,[19] the Minnesota Supreme Court departed significantly from the approach of earlier courts that had replaced common law negligence with negligence *per se*. The court held that while the label provides the standard of care for a negligence *per se* claim, pesticide sprayers and landowners owe an independent common law duty of care toward trespassing bees or other animals when the sprayer or landowner is on notice of their presence and the potential danger.[20] Because the defendants had actual knowledge that there were bees in the area that might forage on their properties, "they may have come under a duty of reasonable care," and summary judgment on the common law negligence

---

12 *Id.*

13 *Id.*

14 404 N.W.2d 332 (Minn. Ct. App. 1987).

15 *Id.* at 333.

16 *Id.* at 334.

17 415 N.W.2d 332, 334 (Minn. Ct. App. 1987).

18 No. C7-96-480, 1996 WL 722085, at *2–3 (Minn. Ct. App., December 17, 1996).

19 693 N.W.2d 181 (Minn. 2005).

20 *Id.* at 187.

claims was inappropriate.[21] In reaching this decision, the court assumed, without explicitly deciding, that FIFRA did not preempt such claims.[22]

The case of *Hall* v. *Phillips*[23] is one of the few cases relying on private nuisance theory, rather than negligence. In this case, the court focused on whether the claim of private nuisance was actionable and whether the defendant's action was both intentional and unreasonable. On the issue of whether it was intentional, the court found that, based on the Restatement (Second) of Torts and state law, a nuisance is "intentional" if the actor: (1) acts for the purpose of causing the interference; or (2) knows that it will result or is substantially certain to result from his conduct.[24] As to whether the spraying was unreasonable, the court framed the question with reference to both the Restatement (Second) of Torts and Nebraska law and said the question was whether the gravity of the harm outweighs the utility of the defendant's conduct, or whether the defendant's conduct causes serious harm but payment of damages would render the defendant's continued conduct unfeasible.[25] The court then remanded the case to the trial court to determine whether the blowing of Atrazine on the plaintiff's property was substantially certain to follow from its application (i.e., intentional) and whether the invasion was unreasonable.[26]

As can be seen from the discussion of the cases above, common law negligence, and in some cases nuisance, has been used to address economic harm caused by pesticides for many decades. In many of these cases, the harms at issue include human illness and impacts to property such as commercial bees and neighboring crops. In determining liability, the courts have tended to rely heavily on whether the pesticide user complied with FIFRA or applicable state law and took appropriate precautionary measures to reduce the likelihood of harm. The courts in these cases have not been asked to address harm to non-commercial natural resources, and thus they have not had to engage in any real ecological assessment.

The National Environmental Policy Act

The National Environmental Policy Act (NEPA),[27] one of the earliest federal environmental laws, establishes a national policy of environmental protection and requires, among other things, that any major federal action significantly affecting the quality of the environment undergo an environmental impact assessment.[28] Although an EPA decision on whether to register a pesticide under FIFRA is a

21 *Id.*
22 *Id.* at 188.
23 436 N.W.2d 139 (Neb. 1989).
24 *Id.* at 142, 145.
25 *Id.* at 143–5.
26 *Id.* at 145–6.
27 42 U.S.C. §§ 4321–4370(h).
28 42 U.S.C. § 4332(C).

major federal action that could significantly affect the environment, the courts have found that a NEPA environmental impact assessment is not required for such action, because the environmental analysis required under FIFRA amounts to what is in essence the functional equivalent of a NEPA review.

In *Merrell* v. *Thomas*[29] the court found that the FIFRA unreasonable adverse effects determination is the functional equivalent of NEPA. In that case, the plaintiff filed suit against EPA for its failure to undertake environmental impact studies (EIS) when it registered pesticides under its FIFRA authority.[30] The district court granted summary judgment for the defendant, and the plaintiff appealed to the Ninth Circuit. The issue confronting the court was whether the 1970 NEPA and its subsequent requirements applied to FIFRA registrations that had been amended in 1972.[31] The court first noted that the 1972 FIFRA amendments did not address preparation of the environmental impact studies that are required of administrative agencies under NEPA. Of particular importance, the court noted the extensive nature of the FIFRA amendments, including a separate provision requiring EPA to consider environmental impact in its registration process, thereby making "NEPA superfluous." Provisions permitting limited public notice and participation in the registration procedure further influenced the court that these amendments indicated an intention on the part of Congress to not apply NEPA to FIFRA registrations. In assessing the relevance of the 1972 amendments, the court cited the limited notice requirements for registration, the limited timeframe requirements for registration and the restrictions on publicly released information regarding test data on pesticides as materially distinct from NEPA.[32] Furthermore, when EPA interpreted FIFRA's provisions in light of the 1972 amendments, the agency determined that it need not comply with NEPA. The court held Congress' failure to legislatively mandate a change in interpretation signaled an intention to permit that agency's interpretation and that application of NEPA's time-consuming requirements would "sabotage the delicate machinery" that Congress designed to register new pesticides.[33] In fact, in 1978, Congress amended FIFRA to encourage quicker registrations. FIFRA's registration standards for environmental impacts differed from NEPA's standard in terms of scope and balancing factors.[34] This balance between agricultural interests and environmental impact as sought by Congress further persuaded the court that NEPA and its corresponding EIS requirements did not apply to pesticide registrations under FIFRA. The Ninth Circuit's position that NEPA does not apply to pesticides registered under FIFRA has been taken by other

29 807 F.2d 776 (9th Cir. 1986).

30 *Id.* at 776–7.

31 *Id.* at 778.

32 *Id.* at 778–9.

33 *Id.* at 779.

34 *Id.* at 780.

courts as well. These courts have concluded that formal compliance with NEPA would be wasteful and redundant.[35]

The Plant Protection Act

Although biological pesticides, including living plants and animals, fall within the definition of pesticide under FIFRA, EPA has, by regulation, exempted all macroorganisms from FIFRA regulation.[36] EPA's rationale for this exemption is that these organisms are already adequately regulated by another federal agency, USDA APHIS, which regulates the introduction of non-indigenous species into the environment pursuant to the Plant Protection Act (PPA). The 2000 Plant Protection Act consolidated the authorities of two previously existing statutes under which APHIS asserted its regulatory jurisdiction over GMOs, the Plant Pest Act,[37] and the Plant Quarantine Act.[38] APHIS's mandate under the PPA is to prevent the release and spread in the environment of "plant pests," which are defined broadly as organisms which can directly or indirectly injure or cause disease or damage in or to any plants or plant parts.[39]

On January 26, 1995, under the authority of the precursor to the PPA, APHIS published a proposed rule in the Federal Register relating to the introduction of non-indigenous organisms into the environment.[40] The proposal would establish comprehensive regulations on the importation, interstate movement, and release into the environment of certain non-indigenous organisms.[41] APHIS stated that it believed this action was necessary because the plant pest regulations that regulated the movement of certain non-indigenous organisms at the time did not adequately address the introduction of the non-indigenous organisms that may potentially be plant pests.[42]

---

35 *Id.* at 781. *See, e.g.*, Wyoming v. Hathaway, 525 F.2d 66, 71–72 (10th Cir. 1975) (EPA need not prepare an EIS before cancelling or suspending registrations of three coyote poisons), *cert. denied*, 426 U.S. 906 (1976); Environmental Defense Fund, Inc. v. Environmental Protection Agency, 489 F.2d 1247, 1254–56 (D.C. Cir. 1973) (EPA need not prepare an EIS before cancelling registration of DDT with respect to nearly all uses); Environmental Defense Fund, Inc. v. Blum, 458 F.Supp. 650, 661–62 (D.D.C. 1978) (EPA need not prepare an EIS before granting an emergency exemption to a state to use an unregistered pesticide).

36 40 C.F.R. 152.20(a).

37 7 U.S.C. 150aa–150jj (1994).

38 7 U.S.C. 151–167 (1994).

39 7 U.S.C. 150aa–150jj (1994).

40 Introduction of Non-Indigenous Species, 60 Fed. Reg. 5288 (January 26, 1995) (to be codified at C.F.R. part. 305).

41 *Id.*

42 *Id.* David J. Bederman, *International Control of Marine 'Pollution' by Exotic Species*, 18 Ecology L.Q. 677, 678 (1991).

The proposed regulations would provide a means of screening current non-indigenous organisms prior to their introduction to determine the potential plant pest risk associated with the particular introduction.[43] The pre-1995 regulations for non-indigenous organisms were limited to the movement of known plant pests and did not address either the movement of non-indigenous organisms not previously known to present a plant pest risk or the release of such organisms into the environment.[44] A 1993 US Congress Office of Technology Assessment (OTA) report cited the loss of billions of dollars due to the negative effects of certain non-indigenous organisms and suggested that APHIS should revise its regulation to more adequately address such risks.[45] Accordingly, under the 1995 proposed regulations, persons wishing to import or move interstate a regulated non-indigenous organism would be required to obtain a permit from APHIS.[46] As part of its permit review process APHIS would be required to seek input from appropriate state agencies, as well as other federal agencies, such as the US Fish and Wildlife Service and Environmental Protection Agency.[47]

Under the proposal, a regulated organism of concern would fall into one of the following categories: (1) an organism of foreign origin that is not present in the United States; (2) an organism of foreign origin that is present in the United States but is capable of further expansion beyond its present established range; and (3) an organism of foreign origin that has reached its full range of potential establishment in the United States but is sufficiently biologically different from the organism that is present in the United States to warrant concern.[48]

The new regulation also proposed data requirements to assess the plant pest and environmental risks involved in a proposed introduction. Information required to be provided as part of the permitting process would include a description of the life cycle, biology, and ecology of the regulated organism.[49] In addition, information would be required on whether the regulated organism had been genetically modified, and if so, a description of the genetic modification would also be required.[50] If the regulated organism had been genetically modified through sexual recombination and selection for traits not typical of the organism in nature, through induced mutation and selection for special traits, or other classical techniques, APHIS would require a description of the modification in order to assess the biology of the modified regulated organism insofar as it differs from that of an unmodified organism of the same species.[51] If, on the other hand,

43 Introduction of Non-Indigenous Species, 60 Fed. Reg. 5288.

44 *Id.*

45 *Id.*

46 *Id.* at 5290.

47 *Id.* at 5291.

48 *Id.* at 5291.

49 *Id.* at 5292.

50 *Id.*

51 *Id.*

recombinant DNA techniques have been used to affect the modification, the permit application would be handled under the regulations for the genetically modified organisms.[52] Other information that would be required includes information on the geographic location where the regulated organism was originally collected and information on the established range of the regulated organism in the United States.[53] In addition, the permit applicant would be required to submit detailed information on the procedures, processes, and safeguards that would be used at the destination facility to prevent the escape and dissemination of the regulated organism and any material accompanying the regulated organism for a permit involving either the importation or interstate movement of a regulated organism.

Permits for the release of a regulated organism into the environment would require more information than permits involving importation or interstate movement with no intended release into the environment.[54] For release permits, information would be required regarding all testing and reviews that have been conducted to assess the effects of the regulated organism and the environment, the effect of the regulated organism on the environment in its established range, and the host specificity of the regulated organism under both artificial and natural conditions.[55] If APHIS issued a permit, the permit would specify the applicable conditions for the introduction of the regulated organism.[56] The proposal also provided a process for obtaining an exemption from regulation for organisms that are found not to present a significant plant pest risk.[57]

On June 16, 1995, less than six months after proposing the rule, APHIS withdrew the proposal.[58] APHIS stated that the reason for the withdrawal was that it had received numerous comments from a variety of stakeholder groups on the proposal, none of which supported the proposal. APHIS maintained that it would be impossible to reconcile the diverse views expressed in the comments regarding the proposed rule without making substantial changes to the proposal on which the public had an opportunity to comment. In the notice of withdrawal, APHIS stated that it intended to develop a new proposed rule to "address the inadequacies in our current plant pest regulation and to provide a means of screening organisms prior to their introduction …."[59] To date, APHIS has not proposed the new rule and continues to rely on existing regulations under the PPA[60] for the introduction of non-indigenous species including biocontrol agents.

52 *Id.*
53 *Id.*
54 *Id.* at 5294.
55 *Id.*
56 *Id.*
57 *Id.* at 5295.
58 60 Fed. Reg. 31647 (June 16, 1995).
59 *Id.*
60 7 C.F.R. 330.

The Clean Water Act

As described above, one of the most significant environmental concerns associated with pesticide application is pesticide residues getting into water bodies and harming aquatic life. The Clean Water Act (CWA) is the primary federal statute designed to address environmental impacts from discharges into waters of the US.[61] The core regulatory permitting program under the CWA is the National Pollutant Discharge Elimination System (NPDES) permitting program.[62] NPDES permits are required for discharges of pollutants from point sources into waters of the US.[63] Historically, the vast majority of discharges of pesticides into water bodies were not regulated under this program for two reasons. First, NPDES permits are only required for "point sources" discharges. The definition of point source in the CWA explicitly excludes agricultural stormwater runoff and irrigation return flows.[64] Thus, any pesticides that are washed away by rain from agricultural lands and runoff, via sheet flow into water bodies, are excluded from permitting requirements. Moreover, for decades, EPA did not require a NPDES permit even for direct discharges to water bodies from point sources such as point sources application from aircraft or watercraft. The controversy over whether permits were needed for these point-source discharges of pesticides centered on the issue of whether pesticides being applied near, on, or over water for a particular purpose (e.g., to control mosquito larvae or aquatic weeds) constituted "pollutants" under the Act. The definition of pollutant under the Act includes "biological materials" and "chemical wastes."[65] The use of the term, "chemical wastes," in contrast to the broader "materials" for biological matter, led many to argue that pesticides being intentionally applied to fulfill a particular purpose are not chemical "wastes" and thus are not pollutants requiring NPDES permits. This issue finally came to a head in 2002 with the decision in *Headwaters, Inc.,* v. *Talent Irrigation District,*[66] in which the court determined that an NPDES permit was required for the discharge of pesticides from point sources into waters of the US. Even after the *Headwaters* decision, EPA steadfastly refused to require NPDES permits for the application of pesticides directly to waters of the US for the purpose of controlling pests in or over such waters.[67]

---

61 33 U.S.C. § 1251 et seq. (1972).

62 *Id.* at § 1342.

63 *Id.* at § 1342(a).

64 *Id.* at § 1362(14).

65 *Id.* at § 1362(6).

66 Headwaters, Inc., v. Talent Irrigation District, 243 F.3d 526 (9th Cir. 2001) (holding that the application of an aquatic pesticide to irrigation canals in compliance with the registration and labeling requirements under FIFRA, did not eliminate the need for an NPDES permit).

67 *See* 68 Fed. Reg. 48385 (August 13, 2003).

In 2006, EPA issued a rule that it explained was to clarify and codify its past approach of not requiring NPDES permits for pesticides legally registered under FIFRA and legally applied to or near aquatic environments to control pests at those sites. Specifically, the rule would have made clear that NPDES permits would not be required in two circumstances: (1) the application of pesticides directly to water to control pests; and (2) the application of pesticides to control pests that are present over, on, or near water where a portion of the pesticides will unavoidably be deposited to the water to target the pests.[68] In 2007, environmental and industry groups petitioned for judicial review of the final rule. On January 7, 2009, the US Sixth Circuit Court of Appeals vacated the rule, finding that it was not a reasonable interpretation of the CWA and holding that NPDES permits are required for all biological and chemical pesticide applications that leave a residue in water when the pesticides are applied in, over, or near waters of the US.[69]

EPA estimated that the court's vacation of the rule would subject to NPDES permitting requirements approximately 365,000 pesticide applicators that perform 5.6 million pesticide applications annually. Thus, EPA sought a two-year stay of the issuance of the court's mandate to provide time for EPA to develop, propose, and issue a final NPDES general permit for pesticide applications. and to conduct outreach and education to the regulated community.[70] On June 8, 2009, the Court granted the two-year stay of the mandate, delaying the effective date of the court's ruling to April 9, 2011. On June 4, 2010, EPA issued a proposed Pesticide General Permit (PGP), which would authorize certain specified discharges of pesticides to waters of the US.[71] Pesticide applications not covered by the PGP would be required to obtain an individual NPDES permit, unless covered by an alternative NPDES general permit. The PGP would cover discharges to waters of the US from the application of: (1) biological pesticides; and (2) chemical pesticides that leave a residue. The use patterns covered by the PGP include mosquito and other flying insect control, aquatic weed control, and forest canopy pest control. The PGP would not cover discharge of pesticides or their degradates to waters already impaired by these specific pesticides or degradates, or discharges to outstanding national resource waters. EPA received approximately 750 sets of comments on the PGP.[72] At the time of this writing, EPA had not yet issued the final PGP.

The proposed PGP primarily consists of a narrative water-quality based effluent limitation, which provides "any discharge that causes or contributes to an excursion of any applicable numeric or narrative EPA-approved state, territory, or tribal or EPA-promulgated water quality standard is prohibited and is a violation of

68 40 C.F.R. § 122.

69 National Cotton Council, et al. v. EPA, 553 F.3d 927 (6th Cir. 2009).

70 EPA's Motion to Stay Mandate, April 9, 2009.

71 75 Fed. Reg. 31775 (proposed June 4, 2010).

72 EPA, Background Information on EPA's Pesticide General Permit, *available at* http://cfpub.epa.gov/npdes/home.cfm?program_id=414#decision (last accessed June 3, 2011).

this permit."[73] EPA's rationale for such a non-specific standard, which in essence merely reiterates that discharges must not violate existing water quality standards, is that it believes that compliance with both FIFRA and other permit conditions will generally ensure compliance with applicable water quality standards. The PGP would also require applicators to minimize pesticide discharges, including the prevention of leaks and spills, and calibration of equipment. Applicators would also be required to report any adverse incidents resulting from the pesticide's use. Applicators would also be required to not exceed specified annual pesticide use thresholds. If an applicator did exceed such a threshold, it would be required to submit of notice to EPA. Finally, applicators would be required to use IPM practices to minimize discharges of pesticides into waters of the US and to develop Pesticide Discharge Management Plans.

The Endangered Species Act

Although the Endangered Species Act (ESA) does not directly regulate pesticides, it applies to protect listed species, and thus applies to certain actions involving pesticides. The 1973 ESA was passed during the heyday of the environmental movement of the 1960s and early 1970s, against a backdrop of intense public concern over the health of the environment and fate of the dwindling populations of many wildlife species. During the 1960s, two more modest attempts to protect endangered species were enacted by Congress. The 1966 Act was limited to establishing a federal program for conservation and providing for species protection on federally owned land.[74] The 1969 Act focused on banning the importation of endangered species or endangered species' products.[75] These early Acts, although important in their own right, did not satisfy the clear public desire for strong species protection. Therefore, Congress passed a more comprehensive ESA in 1973 with a clear objective to "act early to save a vanishing species."[76] Accordingly, the 1973 ESA ambitiously sought to conserve, protect and encourage propagation of endangered species of both fish and wildlife through federal action and through the encouragement of state endangered species programs.[77] To carry out these ambitious goals, the ESA granted broad authority to the Secretary of the Interior to list species as threatened or endangered, to enforce the prohibition on taking, and to carry out the consultation requirements of section 7.

---

73 United States Environmental Protection Agency Office of Water, Office of Wastewater Management Water Permits Division, NPDES General Permit for Pesticides Discharges (April 2010), *available at* http://cfpub.epa.gov/npdes/home.cfm?program_id =410 (last visited May 7, 2011).

74 S. Rep. 93–307, at 2990, July 6, 1973 (Commerce Committee).

75 *Id.* at 2991.

76 *Id.*

77 *Id.* at 2989–90.

Currently, 1,973 species are federally listed as endangered or threatened by the US government.[78] These species are protected by regulation primarily under the Endangered Species Act,[79] the most far-reaching wildlife protection act in the United States. Although it is beyond the scope of this book, in recent years there has been considerable debate over whether the ESA is too restrictive or not restrictive enough and whether the ESA is unreasonably interfering with private property rights. [80] The purpose of the ESA is to conserve threatened and endangered plant and animal species and their habitat.[81] The vast breadth of the ESA is evident from Supreme Court descriptions of the Act as the "most comprehensive legislation for the preservation of endangered species ever enacted by any nation."[82] Under the ESA, species listed as endangered or threatened are protected in a variety of ways.

The term "endangered species" is defined as any species which is in danger of extinction throughout all or a significant portion of its range.[83] The term "threatened species" is defined as any species which is likely to become endangered within the foreseeable future throughout all or a significant portion of its range.[84] Although the statute creates distinct categories of endangered and threatened species, for the most part, species designated as either are subject to the same protections under the ESA. The agencies responsible for implementing the ESA are the United States Fish and Wildlife Service (FWS), which implements the ESA with regard to freshwater and terrestrial species, and the National Oceanic and Atmospheric Administration-Fisheries, also known as National Marine Fisheries Service (NMFS) (together, "the Services"), which implements the ESA with regard to marine and anadromous species. In addition to listing species as threatened or endangered, the Services are also required to designate critical habitat for each listed species. Critical habitat is defined as the specific areas within the geographic area occupied by the species which are essential to the conservation of the species and which may require special management considerations for protections.[85]

---

78 US Fish & Wildlife Service, Species Reports, *available at* http://ecos.fws.gov/tess_public/pub/boxScore.jsp (last visited June 3, 2011). Of the total listed threatened and endangered species, these numbers include both domestic and foreign listings, of which 1,178 are animals and the remainder are plants. *Id.* For a history of the ESA *see* Michael J. Bean & Melanie J. Rowland, The Evolution of National Wildlife Law (3rd ed., 1997).

79 16 U.S.C. § 1531–1599.

80 *See* Marcilynn A. Burke, *Klamath Farmers and Cappuccino Cowboys: The Rhetoric of the Endangered Species Act and Why it (Still) Matters*, 14 Duke Envtl. L. & Pol'y F. 441 (2004) (arguing that the push for ESA reform has been sold primarily on the basis of individual anecdotes, and as such these individual stories of "nightmarish" regulatory burdens on small landowners should be further examined).

81 16 U.S.C. § 1531(b).

82 Babbitt v. Sweet Home Chapter of Communities for a Greater Oregon, 515 U.S. 687, 698 (1995) (quoting TVA v. Hill, 437 U.S. 153, 180 (1978)).

83 16 U.S.C. § 1532(6).

84 16 U.S.C. § 1532(20).

85 16 U.S.C. § 1532(5).

Once a species is designated as either threatened or endangered under the ESA, several statutory protections apply. First, section 9 of the ESA prohibits the "taking" of listed species. The statute defines the term "take" broadly to include to "harass, harm, pursue, hunt, shoot, wound, kill, trap, capture, or collect or attempt to engage in any such conduct."[86] The US Supreme Court has upheld the Services' interpretation of the term "harm" to include acts that involve significant habitat modification or degradation which actually kills or injures wildlife by significantly impairing essential behavior patterns, including breeding, feeding, or sheltering.[87] Penalties for violations of the section 9 take prohibition vary depending on whether the violation involves a threatened or an endangered species and whether the violator knowingly violated the prohibition. Violations of the section 9 take prohibition can result in both civil and criminal enforcement, including penalties up to $50,000 and imprisonment up to one year for knowing violations resulting in takes of endangered species.[88] In addition, courts may award injunctive relief to prevent the takes from occurring or continuing.

Obviously, a government agency that directly kills or injures a listed species, such as through destroying an active nest during a federal construction project, would have section 9 liability. The more complicated issue is the extent to which regulatory agencies are liable for takes that occur as a result of action not taken by the agency, but instead taken by a party *with authorization* from the agency. Although in this situation liability will depend on the precise circumstances of the authorization, in general, federal regulatory agencies have been found liable under section 9 for authorizing activities that resulted in takes. For example, a Massachusetts state agency that issued licenses to use specific fishing gear was liable for taking endangered right whales because the gear entangled the whales[89] and, as discussed in more detail below, EPA was liable for a taking by allowing a pesticide to be marketed that was eventually ingested by endangered black-footed ferrets.[90]

The ESA contains a limited exception to the take prohibition for private parties. Section 10 of the ESA provides that one may obtain a permit to take a listed species, if the "taking is incidental to, and not the purpose of, the carrying out of an otherwise lawful activity," and "will not appreciably reduce the likelihood of the survival and recovery of the species in the wild."[91] To obtain an incidental take permit it is necessary for the permit applicant to develop a "habitat conservation plan" that minimizes and mitigates the impact of the taking to the maximum extent

86 16 U.S.C. § 1532(19).

87 Babbitt v. Sweet Home Chapter of Communities for a Greater Oregon, 515 U.S. 687 (1995).

88 16 U.S.C. § 1540(b).

89 Strahan v. Coxe, 127 F.3d 155 (1st Cir. 1997).

90 Defenders of Wildlife v. Administrator, Environmental Protection Agency, 882 F.2d 1294 (8th Cir. 1989).

91 16 U.S.C. § 1539.

practicable. Because many listed species occur primarily or exclusively on non-federally owned lands, FWS and NMFS have developed a policy to encourage and promote the voluntary management of non-federally owned lands for listed species through the use of "Safe Harbor Agreements." These agreements are entered along with the issuance of "enhancement of survival" permits to authorize any necessary future incidental takes and to provide the landowner with assurances that no additional restrictions will be imposed as a result of their activities.[92]

The second major regulatory program under the ESA is found in section 7,[93] which contains two mandates with regard to federal agencies. First, the federal agencies are required to use their existing authorities to conserve endangered and threatened species.[94] Second, section 7 mandates that federal agencies consult with the Services to "insure that any action authorized, funded, or carried out by such agency is not likely to jeopardize the continued existence of any endangered species or threatened species or result in the destruction or adverse modification of [critical habitat] of such species."[95] Federal agency action will "jeopardize the continued existence of a listed species" where the action can reasonably be expected, directly or indirectly, to appreciably reduce the likelihood of both the survival and recovery of a listed species in the wild by reducing the reproduction, numbers, or distribution of that species.[96]

The section 7 consultation process applies to any federal agency action that "may affect" listed species. The term "may affect" includes beneficial, as well as adverse impacts. If the agency makes a "may affect" determination, it may enter formal consultation or in the alternative, it may evaluate whether the action is "likely to adversely affect" listed species.[97] If the agency determines, with written concurrence of the Services, that the proposed action is "not likely to adversely affect" listed species, the consultation process is terminated."[98] If the agency determines that the action is "likely to adversely affect" listed species, the agency must engage in the formal consultation process.[99]

The product of the formal consultation process is the issuance of a Biological Opinion (BiOp), which will state whether federal agency action is likely to jeopardize listed species. If the Services conclude that the proposed agency action is likely to jeopardize the continued existence of listed species, the Services will include in the BiOp "reasonable and prudent alternatives" that if implemented

---

92 US Fish & Wildlife Service, For Landowners/Safe Harbour Agreements *available at* http://www.fws.gov/endangered/landowners/safe-harbor-agreements.html (last accessed June 3, 2011).

93 16 U.S.C. § 1536.

94 16 U.S.C. § 1536(a)(1).

95 16 U.S.C. § 1536(a)(2).

96 50 C.F.R. § 402.02.

97 *Id.* § 402.13.

98 *Id.*

99 50 C.F.R. § 402.13–14.

would avoid jeopardy.[100] The Services may also include in the BiOp an incidental take statement (ITS), which will identify actions that will not be considered a prohibited taking and will provide legal cover for harm that does occur to species if addressed in the ITS.[101] After the BiOp has been issued, the agency decides whether to proceed with action. However, if the agency's action results in a take, the agency will be liable under section 9, unless such a take is provided for by an incidental take statement.

Unfortunately, neither the statute nor the legislative history provides detailed guidance on the consultation process.[102] Although the Senate Report provided analysis of the section 7 consultation requirements, it did not elaborate or provide further insight as to what "steps [are required] 'to insure that actions authorized, funded or carried out' by it do not jeopardize the continued existence of any such species."[103] Similarly, one of the few ESA sections not discussed in the Conference Report[104] was section 7, and as a result there is little guidance from either the House or Senate as to the section's requirements.[105] Nevertheless, there was evidence in the legislative history that the ESA was intended to "substantially amplify the obligation" of federal agencies to use their authorities to carry out the purposes of the Act.[106] Moreover, the Supreme Court has interpreted the legislative history of section 7 of the ESA as giving greater importance to the protection of species than other agency missions.[107] Thus, although the ESA was clearly intended to provide very aggressive protection of threatened and endangered species, the exact intent of Congress in passing section 7 was not clearly articulated at the time of passage.

The Migratory Bird Treaty Act

In addition to the ESA, legislative authority for protecting wildlife is found in the Migratory Bird Treaty Act (MBTA).[108] As with the ESA, the MBTA can be triggered when the use of pesticides harms covered species. The MBTA implements four international treaties that are aimed at protecting migratory birds. The scope of the MBTA is quite broad and has been said to cover "almost all native North American

100 16 U.S.C. § 1536(b) (4).

101 16 U.S.C. § 1536(b) (4).

102 *Id.*

103 *Id.* at 2994–3000.

104 H.R. Conf. Rep. 93–740, December 19, 1973 (House Conference Report).

105 *Id.*

106 Representative John Dingell (D-Mich), 119 Cong. Rec. 42913 (Daily ed., December 20, 1973), cited in TVA v. Hill, 437 U.S. 153, 183–4 (1978).

107 TVA v. Hill, 437 U.S. at 185.

108 16 U.S.C. § 703. For the protection of marine mammals, the primary legislative authority, is the Marine Mammal Protection Act. 16 U.S.C. §§ 1361–1421h.

birds."[109] Some, but not all, migratory birds covered by the MBTA are also a listed species under the ESA and, thus, both Acts would apply to those species. Under the MBTA, it is unlawful at any time, by any means or in any manner to: hunt, take, capture, kill, possess, purchase, sell, barter, or transport any bird protected by the Treaty, any part, nest, or egg of a protected bird or any product composed of any part, nest, or egg of a protected bird, except as permitted by regulation of the Secretary of the Interior.[110] Violations of this prohibition can result in criminal penalties.[111] Violations of the Act are considered to be misdemeanors, which are punishable by a fine of not more than $15,000.00, imprisonment for not more than six months, or both.[112] To be guilty of a misdemeanor a violator need not have any intent or knowledge that it was violating the MBTA.

As with the ESA, the MBTA prohibits "takes." The MBTA does not define the term "take," however, regulations define it to mean "pursue, hunt, shoot, wound, kill, trap, capture, or collect, or attempt" any of the foregoing.[113] The state of the law on how broadly this definition extends is not clear. For example, although one court has found the language to be broad enough to include activities such as accidental poisoning by discharging pesticide waste into a storage pond,[114] in a more recent decision, another court has determined that habitat modifications such as logging activities are not considered to be a take under the MBTA.[115] Thus, although not well defined, the MBTA's definition of "take" appears to be narrower than the definition under the ESA, which, as described above, may include significant habitat modification or degradation where it actually kills or injuries wildlife. In a recent article, one author argues that the current protective aspects of the MBTA have been limited by narrow judicial interpretation of the Act, and as such there remains a need for legislative and administrative expansion of the Act to encompass the dangers posed by poisoning and habitat loss.[116] Nevertheless, the MBTA's narrow definition of the term "take" does not include habitat modification that cause death or injury to migratory birds as prohibited under the Act. However, at least with regard to accidental poisonings with pesticides, the MBTA appears to apply.[117] Because the MBTA protects species before they near extinction it can

109 Larry Martin Corcoran & Elinor Colbourn, *Shocked, Crushed and Poisoned: Criminal Enforcement in Non-Hunting Cases Under the Migratory Bird Treaties*, 77 Denv. U.L. Rev. 359, 381–85 (1999).

110 16 U.S.C. § 703.

111 16 U.S.C. § 707.

112 16 U.S.C. § 707(a).

113 50 C.F.R. § 10.12.

114 United Sates v. FMC Corp., 572 F.2d 902 (2d Cir. 1978).

115 Seattle Audubon Society v. Evans, 952 F.2d 297 (9th Cir. 1991).

116 Conrad A. Fjetland, *Possibilities for Expansion of the Migratory Bird Treaty Act for the Protection of Migratory Birds*, 40 Nat. Resources J. 47, 50–54 (2000) (tracing the history of MBTA cases where parties were found liable for indirect takings of migratory birds, including takings that occurred as a result of pesticide poisoning).

117 Corcoran & Colbourn, *supra* note 109, at 360.

be utilized more efficiently and quickly than the ESA (which requires a drawn-out listing process for species already nearing extinction).[118]

The Organic Food Production Act

The Organic Food Production Act of 1990 (OFPA)[119] established production certification and product labeling programs that require specific criteria be met before a farm or product can be called "organic." The link between the OFPA and pesticides is that to be a certified organic production facility or for a product to be labeled as organic, neither synthetic chemical pesticides nor GM pesticides can be used on the farm or product. Despite the technological and chemical transformation of agriculture during the Green Revolution, a small number of primarily small-scale farmers continued to grow their crops without synthetic fertilizer and pesticide inputs. This type of chemical-free farming, which gained a following during the heyday of the environmental movement of the 1970s became known as "organic farming." It was not until the 1990s, however, that organic farming saw significant growth and moved into the mainstream. As the organic foods market grew, farmers and consumers alike became concerned about whether foods that were marketed as "organic" truly were, and whether farming practices lived up to the "organic" label. Thus, in 1990, Congress passed the OFPA. After 10 years of controversy, the USDA organic foods regulations finally went into effect.[120] The focus of the regulations is on certification of farms that meet specified standards for organic farming and labeling of foods and other products as organic.

To obtain a USDA certification as an organic farm and to be able to label agricultural products as organic, food must be produced and handled without the use of: synthetic substances (except those specifically listed);[121] ionizing radiation;[122] GMOs; sewage sludge;[123] or most animal hormones and antibiotics (except vaccines). Accordingly, for a farm to be certified as organic or for agricultural products to be labeled as organic, synthetic chemical pesticides may not be used. In this way, the organic standards help to reduce overall chemical pesticide usage. Further, these standards encourage types of farming that rely on more ecologically sound practices such as cultural controls and biologic controls.

---

118 *See* Erin C. Perkins, Comment, *Migratory Birds and Multiple-Use Management: Using the Migratory Bird Treaty Act to Rejuvenate America's National Environmental Policy*, 92 Nw. U.L. Rev. 817, 841 (1998).

119 Organic Food Production Act of 1990, Pub. L. No. 101–624, 104 Stat. 3935 (1990); 7 U.S.C. § 6501 et seq. (1990).

120 7 C.F.R. Part 205.

121 *Id.* at § 205.105(a).

122 *Id.* at § 205.105(f).

123 *Id.* at § 205.105(g).

Under the USDA program, to become certified a grower must have, among other things, a number of management plans, which include an organic production plan[124] that includes monitoring practices; soil fertility and crop management practices to improve soil;[125] a crop rotation plan;[126] and a crop pest, weed, and disease plan.[127] Prohibited substances, including chemical pesticides, may not have been used on the land for at least three years prior to harvest.[128] Certification also requires the farm to have defined boundaries with buffers to prevent unintended contamination with prohibited substances.[129]

Despite the prohibition on synthetic chemical pesticide use, certified organic farms are not without recourse for pest control. Pests may be controlled through physical or mechanical methods such as augmentation or introduction of predators and parasites, development of habitat for natural enemies, lures, traps, and non-synthetic repellents.[130] Permitted weed control methods include mulching, mowing, livestock grazing, hand and mechanical weeding, burning, and plastic mulches as long as removed from the field at the end of the growing season.[131] Non-genetically modified biological pesticides such as microbial *B.t.* are allowed, as are certain "naturally occurring" pesticides derived from plants such as pyrethrum.[132] The certification standards contain an exemption for small growers selling less than $5,000 per year.[133]

In addition to certification requirements, the USDA regulations set forth standards for the labeling of organic agricultural products. To be able to label a product as "100 percent organic," the product must consist of 100 percent organically produced ingredients.[134] The term "Organic" and the USDA Organic seal may be used where the product consists of at least 95 percent organic ingredients.[135] For products consisting of at least 70 percent organically produced ingredient, the label may state that the product is "Made with [the] organic [ingredient]."[136] Products in this category may not display the "USDA Organic" seal or any organic logo, but can indicate which ingredients are organic in the ingredient statement.[137] The USDA regulations allow the application of otherwise prohibited pesticides and animal drugs in emergency situations. Such emergency

124 *Id.* at §§ 205.201, 205.400.
125 *Id.* at § 205.205.
126 *Id.*
127 *Id.* at § 205.206.
128 *Id.* at § 205.202(b).
129 *Id.* at § 205.202(c).
130 *Id.* at §§ 205.206(b), 205.271(b).
131 *Id.* at § 205.206(c).
132 *Id.* at § 205.206(e).
133 *Id.* at § 205.101(a)(1).
134 *Id.* at § 205.301(a).
135 *Id.* at § 205.301(b).
136 *Id.* at § 205.301(c).
137 *Id* at § 205.305(b)(1).

use does not result in a grower losing certification. However, products grown with emergency use of pesticides or animal drugs may not be labeled as organic.[138]

Although the USDA organic certification and labeling standards have served to clarify what is considered "organic" and reduce false organic claims, there is controversy over the appropriateness of the standards and debate about whether the standards actually encourage "true" organic farming.[139] On one hand, the standards clearly prohibit the use of synthetic pesticides and other synthetic substances, and thus have substantial environmental benefits over other forms of farming. On the other hand, the standards allow for large-scale industrialized monoculture farming practices, as long as prohibited substances are not used and the producers have certain specified plans in place. As the market for organic products has grown, large-scale industrial producers including some of the largest international food conglomerates and some of the largest international retailers such as Wal-Mart, have gotten into the "organic" business. Many purists argue that any large-scale industrialized agriculture operation is antithetical to the concept of "organic" farming, which in its purest form, is much more holistic and concerned with promoting healthy soil and making positive environmental and social contributions.[140]

138 *Id.* at § 205.672.

139 *See* Bryan Endres, *An Awkward Adolescence in the Organics Industry: Coming to Terms with Big Organics and Other Legal Challenges for the Industry's Next Ten Years*, 12 Drake J. Agric. L. 17 (2007).

140 *Id.*

Chapter 10
The Special Case of Genetically Modified Crops

Pesticidal GMOs Under FIFRA

Pesticidal GMOs are regulated under FIFRA in much the same way as traditional chemical pesticides. The decision to treat GMOs similarly to traditional pesticides is rooted in the early US biotechnology policies of the 1980s. The United States government's first systematic attempt to address the regulation of biotechnology in a comprehensive fashion was with the publication of the 1984 document entitled "Proposal for a Coordinated Framework for Regulation of Biotechnology."[1] The purpose of this document was "to provide a concise index to US laws related to biotechnology, to clarify the policies of the major regulatory agencies that will be involved in reviewing research and products of biotechnology, to describe scientific advisory mechanisms for assessment of biotechnology issues, and to explain how the activities of the federal agencies in biotechnology will be coordinated."[2] In 1986, the Office of Science and Technology Policy (OSTP) published in the Federal Register "Coordinated Framework for Regulation of Biotechnology: Announcement of Policy and Notice for Public Comment" (Coordinated Framework).[3] During the 1980s and early 1990s, the executive branch of the US government was focused on promoting biotechnology as the US's hope for strong economic future. The feeling at the time was that the US had allowed Japan to beat it in the electronics industry. The federal government was determined to not allow this to happen with the biotech industry. During this time, Vice President Quayle's Council on Competitiveness became intensively involved in planning for the commercialization of biotechnology. The message was clear that regulatory agencies were not to stand in the way of biotechnology and were not to develop any new regulatory programs.[4] For pesticidal GMOs this meant FIFRA. Moreover, a raging debate ensued over whether regulatory agencies should be regulating the "process" of genetic engineering or the "products" of genetic engineering. At that

---

1 Notice of Proposal for a Coordinated Framework for Regulation of Biotechnology, 49 Fed. Reg. 508566 (1984).

2 *Id.*

3 51 Fed. Reg. 23302 (1986).

4 *See* Vice President's Council on Competitiveness, *Report on National Biotechnology Policy* (February 19, 1991); *see generally* Emily Marden, *Risk and Regulation: U.S. Regulatory Policy on Genetically Modified Food and Agriculture*, 44 B.C. L. Rev. 733 (2003).

time, it was determined that from a risk standpoint, the process was irrelevant and that agencies should regulate only products of biotechnology. Under the coordinated framework, the regulatory approach taken by US regulatory agencies, including EPA, has been to rely on existing statutes and to focus on the "product" rather than the "process" used to create the product.[5]

Bridge Document 10.1 Coordinated Framework for Regulation of Biotechnology: Announcement of Policy and Notice for Public Comment, 51 Fed. Reg. 23302 (June 26, 1986)

Existing statutes provide a basic network of agency jurisdiction over both research and products; this network forms the basis of this coordinated framework and helps assure reasonable safeguards for the public. This framework is expected to evolve in accord with the experiences of the industry and the agencies, and, thus, modifications may need to be made through administrative or legislative actions.

The application of traditional genetic modification techniques is relied upon broadly for enhanced characteristics of food (e.g., hybrid corn, selective breeding), manufactured food (e.g., bread, cheese, yogurt), waste disposal (e.g., bacterial sewage treatment), medicine (e.g., vaccines, hormones), pesticides (e.g. *Bacillus thuringiensis*) and other uses. Federal agencies implement an array of laws which seek to ensure the safety of these products …. These laws are product-specific because they regulate certain product uses, such as foods or pesticides. This approach provides the opportunity for similar products to be treated similarly by particular regulatory agencies.

Biotechnology also includes recently developed and newly emerging genetic manipulation technologies, such as recombinant DNA (rDNA), recombinant RNA (rRNA) and cell fusion, that are sometimes referred to as genetic engineering. While the recently developed methods are an extension of traditional manipulations that can produce similar or identical products, they enable more precise genetic modifications, and therefore hold the promise for exciting innovation and new areas of commercial opportunity.

Concerns were raised as to whether products resulting from the recently developed techniques would pose greater risks than those achieved through traditional manipulation techniques. For example, what might be the possible environmental consequences of the many anticipated agricultural and environmental applications that will take place outside the physical constraints of a contained facility? In particular, the environmental application of genetically engineered microorganisms may elicit concern because they are of microscopic size, and some may be able to reproduce, proliferate, and become established.

5 *Id.*

> The underlying policy question was whether the regulatory framework that pertained to products developed by traditional genetic manipulation techniques was adequate for products obtained with the new techniques. A similar question arose regarding the sufficiency of the review process for research conducted for agricultural and environmental applications.
>
> The Administration, recognizing its responsibility to confront these concerns, formed an interagency working group under the former White House Cabinet Council on Natural Resources and the Environment in the spring of 1984. The working group sought to achieve a balance between regulation adequate to ensure health and environmental safety while maintaining sufficient regulatory flexibility to avoid impeding the growth of an infant industry.
>
> Upon examination of the existing laws available for the regulation of products developed by traditional genetic manipulation techniques, the working group concluded that, for the most part, these laws as currently implemented would address regulatory needs adequately.

EPA's primary authorities for regulating agricultural biotechnology products are found in two statutes: the Federal Insecticide, Fungicide, and Rodenticide Act (FIFRA)[6] and the Federal Food, Drug, and Cosmetic Act (FFDCA).[7] Under FIFRA, EPA has the authority to address all environmental and human health issues associated with pesticide use. Under FFDCA, EPA has the authority to set tolerances for pesticide residues in or on food. EPA also regulates biologicals and biotechnology products that are not pesticides, food, or drugs under the Toxic Substances Control Act (TSCA).[8] TSCA grants EPA the authority to screen new chemical substances and impose controls to prevent unreasonable risks and, through rulemaking, to acquire information and impose restrictions to prevent unreasonable risks on existing chemical substances. Although some agricultural biotechnology products may fall within the purview of TSCA, the majority of agricultural biotechnology products regulated by EPA are considered pesticides under EPA's broad definition of the term, and thus, are regulated under FIFRA and FFDCA. For pesticidal GMOs, this means using FIFRA to regulate the "pesticide" rather than targeting regulation at the process by which the pesticide is created. GMOs that are intended to kill, disrupt, repel, or mitigate pests are regulated in much the same way as traditional chemical pesticides under FIFRA. As described above, section 2(u) of FIFRA defines the term "pesticide" as: "(1) any substance or mixture of substances intended for preventing, destroying, repelling, or mitigating any pest, and (2) any substance or mixture of substances intended for use as a plant regulator, defoliator, or desiccant …." This definition is very broad and can include living organisms and substances produced by living organisms as well

6 7 U.S.C. § 136–136(y) (2004).
7 21 U.S.C. § 301 *et seq.* (2004).
8 15 U.S.C. §§ 2601–2697 (2011).

as traditional chemical pesticides. The definition of "pesticide" in FIFRA does not depend on the process by which a particular pesticide is produced. EPA has interpreted this definition to include biological pesticides and genetically modified pesticides. Thus, pesticidal GMOs must be registered under FIFRA prior to sale or distribution in the US. The standard for registration is the same for pesticidal GMOs as for traditional chemicals. EPA has developed some data requirements specifically geared to address potential risks from microbial pesticides, including microbial GMOs.[9]

The Coordinated Framework articulated two major policy choices that set the stage for at least the following 20 years of US regulation of biotechnology. First, the document stated that biotechnology could be adequately regulated under existing legal authorities and that new legal authorities were not necessary to address emerging technologies. The document stated: "Upon examination of the existing laws available for the regulation of products developed by traditional genetic manipulation techniques, the working group concluded that, for the most part these laws as currently implemented would address regulatory needs adequately."[10] Second, the document articulated a policy position that the "products" of biotechnology would be regulated rather than the "process" by which such products were created. The document provides that "[t]he manufacture by the newer technologies of food, the development of new drugs, medical devices, biologics for humans and animals, and pesticides, will be reviewed by FDA, USDA and EPA in essentially the same manner for safety and efficacy as products obtained by other techniques."[11] Specifically, the Coordinated Framework stated that "[t]echniques of biotechnology are not inherently risky in that biotechnology should not be regulated as a process, but rather that the products of biotechnology should be regulated in the same way as the products of other technologies." [12] This approach was based on a belief that rDNA technology in itself does not create risk.[13] Instead, certain types of products of biotechnology may pose risks that can be addressed in the same fashion that regulatory agencies address the risks posed by traditional chemical products. In other words, the Coordinated Framework set forth the position that the potential risks of genetic modification were not dependent on the process by which such modification was made, but instead only depended on the ultimate product that was produced regardless of the process or technology used.

9 The data requirement for microbial pesticides can be found at 40 C.F.R. §§ 158.2100–2174 (2011).

10 *Id.* at 23303, 23306.

11 *Id.* at 23302–4. The process versus product debate extends beyond the regulation of GMOs. For an interesting discussion of the issues, *see* Douglas A. Kysar, *Preferences for Processes: The Process/Product Distinction and the Regulations of Consumer Choice*, 118 Harv. L. Rev. 525 (2004).

12 *Id.*

13 *See* Celeste Marie Steen, *FIFRA's Preemption of Common Law Tort Actions Involving Genetically Engineered Pesticides*, 38 Ariz. L. Rev. 763 (1996).

In addition, the Coordinated Framework outlined the relationship and coordination between five federal agencies possessing legal authority in the regulation of biotechnology. These agencies include the Environmental Protection Agency (EPA), the United States Department of Agriculture (USDA), the Food and Drug Administration (FDA), the National Institutes of Health (NIH), and the Occupational Safety and Health Administration (OSHA).[14] The two federal agencies identified in the Coordinated Framework as having the primary authority to regulate environmental risks posed by genetically modified organisms are the EPA and the USDA.[15] The primary agency identified as having the authority to address risks from GMO food is the FDA.[16]

To date, EPA's regulation of GMOs has focused on three categories: (1) the regulation under FIFRA and FFDCA of genetically modified microbial organisms that have pesticidal characteristics; (2) the regulation under FIFRA and FFDCA of genetically modified plants that have pesticidal characteristics; and (3) the regulation under TSCA of genetically modified microorganisms that do not have pesticidal characteristics. EPA does not yet have any rules governing genetically modified animals.

Currently, EPA regulates pesticidal GMOs under FIFRA in much the same way as it does traditional chemical pesticides.[17] Thus, for pesticidal GMOs, EPA uses its authority under FIFRA to regulate the "pesticide," rather than targeting regulation at the process by which the pesticide is created.[18]

EPA has stated that it recognizes that many of the types of restrictive labeling that it relies on to regulate traditional chemical pesticides may not be appropriate for pesticidal GMOs.[19] For example, geographical limitations on the use of the GMO may not be meaningful if the organism that produces the pesticide can reproduce and spread in the environment beyond those geographical limits. Similarly, other use restrictions (e.g., "Do not use within 100 feet of a stream, river, or lake") may not be particularly useful if seeds from plants that produce the pesticide are saved and planted during subsequent growing seasons. Such seeds would not be labeled, and it is at least possible that farmers using these seeds would not even be aware that the seeds were from plants that had been engineered to produce a pesticide. If a GMO pesticide is found to pose an unreasonable adverse effect on the environment after it is registered and in commerce, FIFRA provides mechanisms for the cancellation of the pesticide registration, or in the case of

---

14 Coordinated Framework, *supra* note 1.

15 *Id.*

16 *Id.*

17 *See* Mary Jane Angelo, *Embracing Uncertainty, Complexity and Change to Protect Ecological Integrity: An Eco-Pragmatic Reinvention of a First Generation Environmental Law*, 33 Ecology Law Quarterly 105, 174 (2006).

18 *See id.* at n. 328.

19 59 Fed. Reg. 60496 (November 23, 1994).

imminent risk, for the immediate and temporary suspension of the registration.[20] As described above, EPA is authorized to cancel or suspend existing registrations based upon certain risk/benefit determinations.

During the 1990s, EPA attempted to develop a comprehensive regulatory program for GMOs. Unfortunately, these efforts were met with controversy, political pressure, scientific uncertainty, and bureaucratic delay, which together resulted in regulations for GMOs with very modest effect. The first EPA GMO final rule was the 1994 final rule on the regulation of GM microorganisms under FIFRA.[21] The other significant final regulation was the July 19, 2001 rule for the regulation of GM pesticidal plants, which EPA currently calls "plant-incorporated protectants," under FIFRA.[22] Each of these rules took approximately 10 years to develop. Countless public hearings, scientific advisory council meetings, Congressional hearings and interagency negotiations were held. Despite all of these efforts, however, the resultant rules are quite modest and do not really tackle the complex and novel risks of GMOs. The thrust of the rules is to define the scope of regulation—i.e., to outline what types of pesticidal GMOs EPA believes warrant regulation based on risk/benefit considerations. The rules do not, however, impose any new approaches to regulating pesticidal GMOs. Instead, at least for the foreseeable future, EPA has chosen to rely on the old standby of FIFRA regulation, with the cost/benefit analysis leading to the label restriction.

Microbial GMO Pesticides

The first category of pesticidal GMOs to be regulated by EPA under FIFRA was microbial GMOs. EPA had regulated naturally occurring microbial pesticides, such as *Bacillus thuringiensis* (*B.t.*), for many years.[23] In the early 1980s, when the pesticide industry began to develop microorganisms that had been genetically modified to impart or enhance a pesticidal characteristic, EPA began to regulate these organisms. Microbial pesticides are regulated in much the same way as traditional pesticides at the large-scale testing and registration stages. However, with regard to small-scale testing of microbials, EPA expressed concerns regarding the potential for adverse effects. Small-scale testing of most traditional pesticides is generally considered to pose very limited risks and thus is typically not regulated by EPA. Because microbial pesticides are living organisms that have the potential to reproduce and spread in the environment, even small-scale testing can present unreasonable adverse effects on the environment.[24] Thus, EPA promulgated a rule that requires notification prior to any small-scale testing of certain microbial pesticides, including microbial GMOs.[25]

---

20 7 U.S.C. §136d(b) (2006).

21 59 Fed. Reg. 45611 (September 1, 1994).

22 66 Fed. Reg. 37814 (July 19, 2001).

23 *See* R.E. Pfadt, Fundamentals of Applied Entomology 239 (3rd ed., 1978).

24 *See* 59 Fed. Reg. 45600 (1994) (codified at 40 C.F.R. § 172).

25 40 C.F.R. § 172.45 (2011).

Section 5 of FIFRA authorizes EPA to issue experimental use permits (EUPs) for the testing of new pesticides or new uses of existing pesticides if it determines that the applicant needs such a permit to accumulate information necessary to register a pesticide under section 3 of FIFRA.[26] EPA generally requires EUPs only for large-scale testing of pesticides.[27] A large-scale test under Part 172 includes any terrestrial application on a cumulative acreage of more than 10 acres of land or any aquatic application on more than 1 acre of surface water.[28] For traditional pesticides, EPA presumes that tests conducted on 10 acres or less of land or 1 acre or less of water ("small-scale tests") would not require EUPs. For certain GM microorganisms, however, EPA determined that even small-scale tests warrant an evaluation.

In October 1984, EPA published a policy statement entitled "Microbial Pesticides: Interim Policy on Small Scale Field Testing".[29] In June 1986, EPA reiterated the provisions of the Interim Policy Statement as part of the Office of Science and Technology Policy's Coordinated Framework for Regulation of Biotechnology.[30] These policy statements described EPA's concern about the potential for adverse effects associated with small-scale environmental testing of certain microbial pesticides. To address the situation, these statements specified that EPA be notified prior to initiation of small-scale testing of all non-indigenous and genetically modified microbial pesticides. The purpose of the notification was to allow EPA to conduct an assessment of these small-scale tests in order to make a determination as to whether or not the test should be carried out under an EUP that allows EPA oversight. In addition, the 1986 Policy stated EPA's plan for future rulemaking to codify the interpretation set out in the policy. Subsequent to the issuance of the 1986 Policy, a number of documents were issued by EPA or other parts of the federal government having relevance to this final rule, including a Federal Register notice issued by EPA in February 1989,[31] requesting comment on issues related to this final rule; Federal Register notices issued by the Office of Science and Technology Policy (OSTP) in July 1990[32] and February 1992[33] that addressed issues relating to the appropriate scope of federal oversight of introduction into the environment of modified organisms; and a "Report on National Biotechnology Policy" issued in February 1991 by the Council on Competitiveness. In addition, EPA made available to the public and to its FIFRA Scientific Advisory Panel (SAP) and Biotechnology Science Advisory Committee (BSAC) several draft proposals addressing the notification scheme for small-scale testing of certain genetically modified microbial pesticides. After almost 10 years of deliberation and a series of

26 7 U.S.C. § 136c.

27 40 C.F.R. § 172.3 (2011).

28 *Id.*

29 49 Fed. Reg. 40659 (October 17, 1984).

30 51 Fed. Reg. 23313 (June 26, 1986).

31 54 Fed. Reg. 7026.

32 55 Fed. Reg. 31118.

33 57 Fed. Reg. 6753.

EPA and federal government-wide policy statements that were made available to EPA's SAP and the BSAC, on September 1, 1994, EPA promulgated the final rule on experimental use permits and notifications for genetically modified pesticidal microorganisms.[34]

The rule codifies the early screening procedure first set forth in the Coordinated Framework by requiring notification before the initiation of small-scale field testing of certain microbial pesticides in order to determine whether an EUP is necessary. Under the rule, testing conducted in facilities designed and operated to adequately contain the microbial pesticide would not be subject to the notification requirements. Records describing containment, however, would be required to be developed and maintained.[35]

The most controversial issue that arose during the lengthy development of this rule was the appropriate scope of regulation. EPA decided to require notification for "microbial pesticides whose pesticidal properties have been imparted or enhanced by the introduction of genetic material that has been deliberately modified."[36] In other words, EPA rejected a "product"-based scope of regulation in favor of a "process"-based one.[37]

One other issue that was somewhat controversial was whether EPA should require notification for "non-indigenous" microbial pesticides. Under the 1984 Policy Statement and the 1986 Coordinated Framework, EPA had been requiring notifications to be submitted for all small-scale testing of non-indigenous organisms.[38] In the final rule, EPA opted to require small-scale notification only for those non-indigenous microbial pesticides that have not been acted upon by the US Department of Agriculture either by issuing or denying a permit or determining that

34 Microbial Pesticides, Experimental Use Permits and Notifications, 59 Fed. Reg. 45600 (September 1, 1994) (hereinafter Microbial Pesticides Rule). The proposed rule, Microbial Pesticides, Experimental Use Permits and Notifications, can be found at 58 Fed. Reg. 5878 (January 22, 1993) (hereinafter Microbial Pesticides Proposed Rule). FIFRA section 25(d) requires EPA to submit draft proposals and final rules to an advisory panel, the SAP, for comment as to the impact of the rule on health and the environment. The comments, evaluations, and recommendations of the SAP, and the response of the EPA Administrator, must be published in the Federal Register. Section 25(d) permits the chairperson of the panel, after consultation with the Administrator, to create temporary subpanels on specific projects to assist the full panel. Because of the unique issues associated with the regulation of biotechnology, specialized SAP subpanels have been convened from time to time to address biotechnology matters. In the 1986 Coordinated Framework, EPA announced that it was establishing a Science Advisory Committee for biotechnology to provide peer review of specific product submissions under FIFRA, TSCA, and other EPA statutes and scientific oversight of the agency's biotechnology programs.

35 Microbial Pesticides Rule, *supra* note 34.

36 58 Fed. Reg. 5882–3.

37 58 Fed. Reg. 5878 (codified at 40 C.F.R Part 172).

38 Notice of Coordinated Framework, 49 Fed. Reg. 50856 (1984); Coordinated Framework, 51 Fed. Reg. 23302 (June 26, 1986).

a permit is unnecessary. In the final rule, testing conducted in facilities designed and operated to adequately contain the microbial pesticide would not be subject to the notification requirements. Records describing containment, however, would be required to be developed and maintained.[39] EPA based this decision on its belief that to do otherwise and continue the imposition of the notification requirement on all non-indigenous microbial pesticides would constitute duplicative oversight because the USDA/APHIS already regulates small-scale testing of the vast majority of these organisms. The final rule also contains several provisions that were not very controversial and were not changed significantly from what was proposed.

The final rule also includes provisions that enable EPA to address situations where small-scale testing results in unanticipated and untoward effects. Section 172.57 requires persons using microbial pesticides in small-scale tests to submit any information they obtain concerning the potential for unreasonable adverse effects from the microbial pesticide,[40] and section 172.59 enables EPA to take immediate actions to prevent use of a microbial pesticide if such use would create an imminent threat of substantial harm to health or the environment. The final rule also amends 40 C.F.R. § 172.3 to clarify EPA's rationale for presuming that an EUP is not required prior to small-scale testing with most pesticides. The language of section 172.3 is modified to clarify that the determination of whether an EUP is required would be based on risk considerations, rather than on a definitional presumption about whether a substance is a pesticide. This clarification has general applicability to all pesticides and is not limited to microbial pesticides.[41] Although EPA has developed some data requirements geared to address potential risks from microbial pesticides in general, it has not yet developed any data requirements targeted specifically to microbial GMOs. The data requirement for microbial pesticides can be found at 40 C.F.R. § 158.740 (2005). These data requirements parallel the requirements for traditional chemical pesticides and do not specifically address potential risks caused by living organisms reproducing and spreading in the environment.[42]

GMO Plant-incorporated Protectants

Another category of pesticidal GMOs regulated by EPA under FIFRA are genetically modified pesticidal plants, or "plant-incorporated protectants" (PIPs). A PIP is defined as a pesticidal substance that is intended to be produced in a living plant, or in the produce thereof, and the genetic material necessary for its production.[43] In July 2001, EPA published its long-awaited rule for the regulation

39 40 C.F.R. § 172.45.

40 40 C.F.R. § 172.57.

41 40 C.F.R. § 172.59.

42 The data requirement for microbial pesticides can be found at 40 C.F.R. § 158.2110 (2011).

43 40 C.F.R. § 152.3.

of PIPs under FIFRA.[44] EPA's first attempt to describe its plans to regulate PIPs was in early 1994. On January 21, 1994, EPA held a joint meeting of a sub-panel of the agency's SAP and the BSAC to address certain scientific issues related to the regulation of pesticidal substances produced in plants. For the meeting, EPA made available to the public a draft proposal of a comprehensive policy and four draft proposed rules, together referred to as the "draft proposal," developed under FIFRA and FFDCA. On November 23, 1994, EPA published in the Federal Register somewhat modified versions of these draft documents, together referred to as "the proposal."[45] The proposal was intended to clarify the status of PIPs, referred to as "plant-pesticides" in the 1994 proposal and later renamed plant-incorporated protectants under FIFRA and FFDCA, and outline what types of PIPs EPA believed warranted regulation based on risk/benefit considerations. The final plant-incorporated protectant rule, promulgated in 2001, adopted some but not all of the exemptions proposed in 1994.

In the 1994 proposal, EPA identified several categories that it believed should be exempt from FIFRA regulation because they are low risk. The most significant proposed exemption was for PIPs that closely resemble plants that could be created naturally or through traditional plant breeding. EPA based this proposed exemption on the premise that new exposures would be unlikely if the genetic material leading to the production of the PIP is derived from a plant closely related to the recipient plant.[46] An example of this is using GM technology to insert a gene that is normally found in one variety of corn into another variety of corn. EPA posited that this type of GM plant would be exempt because it does not pose any new risks that would not have evolved naturally or through traditional breeding.[47] EPA presented three options for the exemption. All three options focused on the relationship between the source organisms and the recipient organisms. In other words, all three proposed options were based on the "product" rather than the "process" by which the product was created. Accordingly, in the proposed rule, no distinction was drawn between PIPs created through conventional plant breeding versus those created through genetic engineering. EPA proposed three categories of exemptions based on its proposed findings that plant pesticides in these categories did not warrant regulation either because they so closely resemble the types of

44 66 Fed. Reg. 37772 (2001) (codified at 40 C.F.R. Parts 152 and 174). *See* 40 C.F.R. Part 174. For an historical discussion of the Plant-Incorporated Protectant Rule, *see generally* Mary Jane Angelo, *Regulating Evolution for Sale: An Evolutionary Biology Model for Regulating the Risks Posed by Genetically Modified Organisms*, 42 Wake Forest Law Review 93 (2007).

45 59 Fed. Reg. 496 (1994); 59 Fed. Reg. 60519 (1994); 59 Fed. Reg. 60535 (1994); 59 Fed. Reg. 60542 (1994); and 59 Fed. Reg. 60545 (1994).

46 59 Fed. Reg. 60496 (November 23, 1994).

47 40 C.F.R. § 174.25.

plants that could be created through traditional plant breeding or that they had been found to be low risk.[48]

One category of pesticidal GMOs that EPA believed did not warrant regulation were plants that have been genetically modified to contain genes that are derived from closely related plants and thus will not cause new exposures to non-target organisms. Under this proposal, the *B.t.* delta-endotoxin would not be exempt when it is produced in corn, for example, because the delta-endotoxin is derived from a bacterium rather than from a plant that is closely related to corn.[49] A pesticidal substance that is naturally produced by a certain variety of corn and is introduced into another variety of corn, however, would be exempt. Another category that EPA proposed to exempt were those plant pesticides that would not be expected to adversely affect non-target organisms because they are less likely to be directly toxic due to their mechanism of action. This category consists of plant pesticides that act primarily by affecting the plant so that pests are inhibited from attaching to the plant, penetrating the plant's surface, or invading the plant's tissue. Under this proposed exemption, a substance that acts by causing a structural barrier to pest penetration in the plant would be exempt. The third category that EPA proposed to exempt was plants that have been genetically modified to contain the genes for coat proteins from a plant virus.[50] This type of genetic modification essentially acts as a vaccine protecting the plant from viruses. EPA proposed these GMOs for exemption based on the fact that plant viruses are ubiquitous in the human food supply and are not known to cause any adverse affects to humans or the environment.[51]

Between 1994 and 2001, when it published the final PIP rule, EPA held countless public hearings, scientific advisory council meetings, congressional hearings, and interagency negotiations.[52] Despite all of these efforts, however, the resultant rule is quite modest and does not really tackle the complex and novel risks of GMOs. The thrust of the new rule merely defines the scope of what types of pesticidal GMOs EPA believes warrant regulation. Under EPA's definition of PIPs, all substances produced by plants and intended for a pesticidal purpose are within EPA's jurisdiction, regardless of whether the plant is genetically modified. However, not all PIPs within EPA's jurisdiction warrant regulation under FIFRA. EPA believes that many PIPs do not warrant any regulation under FIFRA because they pose low probability of risk and will not cause unreasonable adverse effects on the environment. For example, in 1982 EPA promulgated a regulation under FIFRA § 25(b) that exempted all biological control agents from the requirements of FIFRA, except for certain microorganisms.[53] This exemption was promulgated

48 59 Fed. Reg. 60496 (November 23, 1994).

49 *Id.*

50 *Id.*

51 *Id.*

52 66 Fed. Reg. 37772 (2001).

53 40 C.F.R. §152.20(3).

because EPA found that microorganisms used as biological control agents were adequately regulated by other federal agencies, such as the USDA.

The final rule exempts from all FIFRA regulatory requirements, except for the requirement of reporting adverse effects information, PIPs derived through conventional plant breeding if the genetic material that encodes the pesticidal substance or leads to the production of the pesticidal substance is from a plant that is sexually compatible with the recipient plant and if the genetic material has never been derived from a source that is not sexually compatible with the recipient plant. The final rule requires any person who produces an exempt PIP for sale or distribution, or who obtains information regarding adverse effects on human health or the environment alleged to have been caused by the PIP, to submit such information to EPA. [54] EPA defines the term "sexually compatible" as when "a viable zygote is formed only through the union of two gametes through conventional breeding." [55] Because EPA has defined sexual compatibility as occurring only through conventional breeding, only conventionally bred crops are exempt from regulation. EPA's rationale for exempting the products of conventional plant breeding from FIFRA requirements is that conventionally bred plant varieties have been used by humans for thousands of years without ill effects. Because conventional breeding can only take place between plants that are sexually compatible, it is likely that such plants already share, or have shared in the past, genetic material; and therefore, exposure to the new plant variety, whether by humans or non-target organisms, will not likely be novel.[56] In other words, in the 2001 final rule, EPA rejected the "product"-based approach set forth in the 1994 proposed rule in favor of the "process"-based approach, which exempts PIPs based on the process by which they were created. If a PIP is developed through conventional plant breeding, it is exempt, whereas if the same PIP is developed through genetic engineering, it is subject to regulation. Thus, in the final rule, EPA departed from the "product-based" approach articulated in the Coordinated Framework in favor of the "process-based" approach that the US government had steadfastly avoided in the 1980s and 1990s.[57] The scaling-back of the exemption was in response to public comments received on the proposal, as well as to a joint FIFRA, SAP and BSAC meeting held in January, 1994, in which the joint panel considered and supported the use of exemption criteria based on the technology (i.e., process) used to produce the PIP.[58] The joint panel based its support on a combination of the uncertainties about how genes would function in the new genetic background and building public confidence in the products of genetic engineering.

54 40 C.F.R. § 174.71.

55 40 C.F.R. § 174.3.

56 66 Fed. Reg. 37794–5 (July 19, 2001).

57 40 C.F.R. § 152.20.

58 66 Fed. Reg. 37855, 37857–8 (July 19, 2001).

The 2001 final rule did not finalize the 1994 proposed exemptions for viral coat proteins, PIPs derived or modified through genetic engineering of sexually compatible plants, or PIPs that act primarily by affecting the plant. On the same day that the final PIP rule was published, EPA published a request for additional comment on the exemptions it proposed in 1994.[59] Specifically, EPA solicited comment on the two alternative approaches to PIPs derived from plants sexually compatible with the recipient plant: (1) whether all such PIPs should be exempt regardless of the technique used to introduce the PIP into the plant; and (2) whether EPA should establish a notification process that would implement a screening procedure to determine whether a PIP derived through genetic engineering from a plant sexually compatible with the recipient qualifies for exemption.[60] To date, EPA has not taken any action on either of the two alternative approaches for which it sought additional comment in 2001.[61]

EPA's final PIP rule merely draws a line between PIPs subject to regulation and those not subject to regulation under FIFRA.[62] The rule does not provide any provisions detailing how a PIP will be evaluated and regulated under FIFRA. Once it is determined that a substance is a pesticidal GMO subject to FIFRA regulation, the regulatory process is similar to the regulatory process for all pesticides with only very minor modifications—i.e., registration based on a cost/benefit analysis, labeling restrictions on use, and cancellation or suspension for registered GMOs found to pose unreasonable adverse affects. As described above, many of the risks posed by GMOs are of a different character than those posed by traditional chemical pesticides. Accordingly, existing FIFRA data requirements, labeling requirements, and regulatory approaches are not adequate to address these risks.[63] EPA has not yet developed any data requirements whatsoever for GM pesticidal plants, nor has it adopted any regulations addressing information requirements to support registration,[64] product labeling requirements,[65] or experimental use permitting for PIPs.[66] In the absence of any such new requirements, EPA relies on existing requirements that were crafted for traditional chemical pesticides to GM plants, regardless of the poor fit with GMOs. EPA does not yet have any rules governing GM animals. To date EPA has issued approximately 40 FIFRA registrations for PIPs, as well as a number of EUPs for field testing of PIPs. The

59 66 Fed. Reg. 37855 (July 19, 2001).

60 66 Fed. Reg. 37858–60 (July 19, 2001).

61 Laurel Celeste, US Environmental Protection Agency, Personal Correspondence, on May 23, 2006.

62 66 Fed. Reg. 37772 (2001).

63 *See generally*, 66 Fed. Reg. 37772 (2001).

64 40 C.F.R. § 174 Subpart H (reserving a subpart for future data requirements).

65 40 C.F.R. § 174 Subpart G (reserving a subpart for future labeling requirements).

66 40 C.F.R. § 174 Subpart U (reserving a subpart for future experiment use permit requirements).

majority of the registrations and EUPs issued are for crop plants, primarily corn, that produce *B.t.*[67]

Many products of genetic engineering have been modified to possess traits that increase their ability to reproduce and survive in the environment. Such traits include insect resistance, viral infection resistance, drought tolerance, and temperature tolerance in crop plants. By genetically manipulating microorganisms, plants, and animals to make them more "fit" from an evolutionary standpoint, science has altered the path of evolution to favor not those organisms that have evolved to be more fit for their natural environment, but instead those organisms that have become more fit at the hand of humans for commercialization and profit-making. US environmental law has not evolved to keep pace with these dramatic changes in the evolution of our biological systems. Thus, completely new approaches are needed to address these novel issues. US regulation of GMOs has occurred in a reactionary, haphazard fashion and has been fraught with political controversy and bureaucratic inertia. Moreover, regulatory agencies have been artificially constrained by early US policy on genetic engineering to rely on existing statutory authorities and to regulate based on the "products" of genetic engineering rather than the "process" by which they are created. Reliance on a mishmash of statutory authorities that predate the advent of genetic engineering, and the ensuing interagency turf battles and differing approaches to regulating similar products among different agencies with different missions, has resulted in profound regulatory gaps, overlaps, and inconsistencies.

GMO Foods Under the Federal Food, Drug and Cosmetic Act

In the 2001 final PIP rule, EPA adopted an exemption from certain FFDCA standards for PIPs. As with the FIFRA PIP exemptions, EPA's FFDCA exemption for PIPs focuses on sexual compatibility through conventional breeding. This exemption provides:

> Residues of a pesticidal substance that is part of a plant-incorporated protectant from a sexually compatible plant are exempt from the requirement of a tolerance if all the following conditions are met: (a) The genetic material that encodes for the pesticidal substance or leads to the production of the pesticidal substance is from a plant that is sexually compatible with the recipient food plant. (b) The genetic material has never been derived from a source that is not sexually compatible with the recipient food plant. (c) The residues of the pesticidal substance are not present in food from the plant at levels that are injurious or deleterious to human health.

67 EPA, *Current Plant Incorporated Protectant (PIP) Experimental Use Permits, available at* http://www.epa.gov/oppbppd1/biopesticides/pips/current_pip_eups.htm (last accessed June 3, 2011).

EPA has exempted from the tolerance requirement nucleic acids that are part of PIPs. In addition, EPA exempted inert ingredients from sexually compatible plants.[68] This exemption provides:

> An inert ingredient, and residues of the inert ingredient, are exempt if all of the following conditions are met: (a) The genetic material that encodes the inert ingredient or leads to the production of the inert ingredient is derived from a plant sexually compatible with the recipient food plant. (b) The genetic material has never been derived from a source that is not sexually compatible with the recipient food plant. (c) The residues of the inert ingredient are not present in food form the plant at levels that are injurious or deleterious to human health.

"Inert ingredient" is defined as "any substance, such as a selectable marker, other than the active ingredient, where the substance is used to confirm or ensure the presence of the active ingredient, and includes the genetic material necessary for the production of the substance, provided that genetic material is intentionally introduced into a living plant in addition to the active ingredient."[69] EPA has also exempted the residues of certain *B.t.* in specified crop foods from the tolerance requirement.[70]

FDA's primary authority governing the regulation of genetically modified foods is found in section 402(a)(2)(C) of the FFDCA.[71] This section provides that a food shall be deemed adulterated if it contains any food additives that are unsafe within the meaning of section 409.[72] Section 409 provides that a food additive is deemed unsafe unless the additive and its use or intended use complies with a properly promulgated food additive regulation.[73] The statute defines the term food additive to mean any substance that is intended for use in or which may be reasonably expected to become a component of or otherwise affect the characteristics of any food provided that the substance "is not generally recognized, among experts qualified by scientific training and experience to evaluate its safety, as having been adequately shown through scientific procedures (or, in the case of a substance used in food prior to January 1st, 1958, through either scientific procedures or experience based on common use in food) to be safe under the conditions of its intended use …."[74] Accordingly, commonly used natural substances that are

---

68 40 C.F.R. § 174.485.

69 40 C.F.R. § 174.3.

70 40 C.F.R. §§ 174.455 & 174.456 (exempting from the requirement for a tolerance *Bacillus thuringiensis* Cry1F protein and the genetic material necessary for its production in cotton and *Bacillus thuringiensis* modified Cry3A protein (mCry3A) and the genetic material necessary for its production in corn, respectively).

71 21 U.S.C. § 301, *et seq*. Pursuant to the Reorganization Plan No. 3 of 1970, FDA is responsible for the regulation of residues in food other than pesticide residues, which are regulated by EPA.

72 21 U.S.C. § 342(a)(2)(C).

73 21 U.S.C. § 348.

74 21 U.S.C. § 321(S).

added to foods, such as spices, in addition to certain chemical additives, are not considered food additives because they are considered to be generally recognized as safe (GRAS).[75]

In 1992 FDA published a policy statement on "foods derived from new plant varieties."[76] This policy provided guidance on how FDA would treat genetically modified foods in the regulatory process. The policy included within the definition of "genetic modification" alterations of the genotype that occur using any technique, whether conventional plant breeding or new biotechnology techniques.[77] Thus, under this definition virtually all cultivated food crops were considered to be genetically modified.[78] This approach of treating all cultivated food crops as genetically modified regardless of the process used to modify them is consistent with the Coordinated Framework's stated policy choice of regulating products rather than process.[79] However, the 1992 policy went further, establishing what is in essence a presumptive GRAS status for genetically modified foods as well as conventionally bred foods.[80] This presumption was based on FDA's conviction that, based on its experience, the likelihood of a significant risk from a genetically modified food is very low.[81] FDA believes that the traditional approach used by conventional crop breeders to ensure food safety has been successful in the past in identifying and eliminating food crops that exhibited unexpected, adverse traits prior to commercial use and that such processes would sufficiently screen out potentially risky genetically modified foods.[82] The 1992 policy statement explained that FDA believed that "in most cases, the substances expected to become components of food as a result of genetic modification will be the same as or substantially similar to substances commonly found in food, such as proteins, fats and oils, and carbohydrates …," and would thus qualify as GRAS.[83] FDA did acknowledge that some GM foods would not qualify as GRAS, including those that involve the transfer of genes coding for substances that can cause allergenic responses in humans, those that are known to be toxic, or those that are likely to become a macroconstituent in the human or animal diet, thereby affecting the nutritional value of GM foods.[84] FDA's position is, in essence, that if the GM food is "substantially equivalent" to a food product already in the human food supply with a history of safe use, the GM food will, in the vast majority of cases, be safe,

---

75 Statement of Policy: Foods Derived From New Plant Varieties, 57 Fed. Reg. 22984 (May 1992).

76 *Id.*

77 *Id.* at 22984.

78 *Id.* at 22984.

79 *See* Coordinated Framework, *supra* note 1 and accompanying text.

80 Statement of Policy: Foods Derived From New Plant Varieties, 57 Fed. Reg. 22990.

81 *Id.* at 22989.

82 *Id.* at 22991.

83 *Id.* at 22985.

84 *Id.* at 23000.

and therefore, no premarket evaluation of the safety of the GM food is necessary.[85] Nevertheless, FDA leaves it up to the producer of the new plant variety to determine the GRAS status of its product.[86] Thus, FDA's approach to regulating GM foods is to establish a presumption of safety and to leave it to the food producer, on a voluntary basis, to determine whether it is necessary to seek out an FDA review of the safety of its product prior to introducing the product into the market. This decision has been controversial and, in light of such controversy, in 2001 FDA published proposed regulations which would require manufacturers and importers of GM food to provide FDA with pre-market notification of their intent to market GM foods. In the January 2001 Federal Register, FDA proposed to require the submission of data and information regarding plant-derived bioengineered foods that would be consumed by humans or animals at least 120 days prior to the commercial distribution of such foods.[87] FDA stated that it was proposing this action to "ensure that it has the appropriate amount of information" and to "permit the agency to assess on an ongoing basis whether plant-derived bioengineered foods comply with the standards of the Federal Food, Drug, and Cosmetic Act."[88] In the Federal Register notice, FDA stated that the scientific community generally supports the regulatory approach articulated in FDA's 1992 policy, but that the proposal is a response to the many consumers and public interest groups, and some state officials, who have expressed concern regarding the lack of a requirement for pre-market review.[89] However, to date, FDA has not published a final rule addressing pre-market review.

Currently, FDA stands alone as the only federal agency following the policy of the Coordinated Framework to regulate based on product rather than process. FDA's 1992 policy does not distinguish based on the techniques used to produce the new plant variety. Instead, it relies on the standard of substantial equivalency, which applies equally to conventionally bred plant varieties and genetically engineered plant varieties. As described above, the lack of even pre-market notification had been widely criticized by those who believe that human health cannot be adequately protected without at least some level of evaluation of risk presented by new GM foods.

85 For a full discussion of the substantial equivalency doctrine, *see generally* Thomas O. McGarity, *Seeds of Distrust: Federal Regulation of Genetically Modified Foods*, 35 Univ. of Mich. J.L. Reform 403 (2002).

86 Statement of Policy: Foods Derived From New Plants Varieties, 57 Fed. Reg. 22985.

87 *Id.*

88 *Id.*

89 *Id.* at 4707.

Pesticidal GMOs Under the Plant Protection Act

The USDA's Animal and Plant Health Inspection Service (APHIS) has authority to regulate GMOs pursuant to the Plant Protection Act (PPA).[90] In 1993 APHIS published a final rule amendment to the regulations pertaining to the introduction of certain genetically engineered organisms and products to provide for a notification process prior to the introduction of certain GMOs.[91] The final rule was the outgrowth of a 1992 proposed rule[92] and followed the basic design of the proposed rule, with some modifications based on comments received.

APHIS also amended the regulation to provide for a petition process allowing for determination that certain GMOs are no longer considered "regulated articles,"[93] APHIS's term for GMOs that pose potential plant pest risk. The term "regulated article" is defined as "[a]ny organism which has been altered or produced through genetic engineering, if the donor organism, recipient organisms, or vector or vector agent belongs to any genera or taxa designated in § 340.2 and meets the definition of plant pest, or is an unclassified organism and/or organisms whose classification is unknown, or any product which contains such an organism, or any other organism or product altered or produced through genetic engineering," which APHIS determines is a plant pest or has reason to believe is a plant pest.[94] Excluded are recipient microorganisms which are not plant pests and which have resulted from the addition of genetic material from a donor organism where the material is well characterized and contains only non-coding regulatory regions.[95] In the final rule, APHIS stated that it believed these actions would relieve unnecessary restrictions on the introduction of regulated articles based on experience.[96] The notification procedure is allowed for the introduction of most GM plants that are considered regulated articles, provided that the introduction is conducted in accordance with specified eligibility requirements and performance standards.[97] This would alleviate the need to obtain a permit prior to the introduction of those regulated articles. The stated rationale for eliminating the permit requirement for most regulated articles and replacing it with a notification requirement is that APHIS believes that the notification process is sufficient for many regulated articles, based on the considerable experience APHIS gained in permitting GM plants since it established its permitting process

---

90 7 U.S.C. §§ 7701–7772.

91 58 Fed. Reg. 17044 (March 31, 1993).

92 57 Fed. Reg. 53036 (November 6, 1992).

93 *Id.*

94 7 C.F.R. § 340.1.

95 7 C.F.R. § 340.1.

96 *Id.*

97 Genetically Engineered Organisms and Products; Notification Procedures for the Introduction of Certain Regulated Articles; and Petition for Nonregulated Status 57 Fed. Reg. 53036–7 (November 6, 1992).

for regulated articles in 1987.[98] APHIS stated that it had issued over 300 permits for field tests and over 1,000 permits for the movement of regulated articles.[99] Based on this experience, APHIS stated that it has determined that introduction of many regulated articles can be conducted with little or no plant pest or environmental risk.[100]

To qualify for the notification process, six eligibility requirements must be met. These are as follows: (1) the regulated article must be one of a list of plants species which includes corn, cotton, potato, soy bean, tobacco, tomato or any additional plant species that APHIS has determined may be safely introduced in accordance with the performance standards; (2) the introduced genetic material is "stably integrated" in the plant genome; (3) the introduced genetic material is well characterized and does not contain genes whose expressions in the regulated article result in plant disease; (4) the introduced genetic material does not cause the production of an infectious entity or result in constituents that are new to the plant and are toxic to non target organisms; (5) the introduced genetic material does not pose a significant risk of the creation of any new plant virus; and (6) the plant has not been modified to contain functionally intact genes derived from human or animal pathogens.[101] The performance standards for introductions under the notification procedure include a number of requirements designed to prevent unintentional spread of the regulated article's genetic material in the environment.[102] These requirements are geared toward containing the spread of the organisms during field testing, but are not applicable to commercial release of the organism into the environment. The performance standard requirements include: (1) requirements that plants or plant materials be shipped in such a way that the viable plant material is unlikely to be disseminated while in transit; (2) that when released into the environment, the regulated plant must be planted in such a way that it is not inadvertently mixed with non-regulated plant materials which are not part of the environmental release; (3) that the plant and plant parts must be maintained in such way that the identity of all material is known while it is in use; and (4) that the plant's parts must be contained or devitalized when no longer in use, that there must be no viable vector agent associated with the regulated article, that when there is a significant probability that gene movement of the regulated article via pollen will result in viable progeny persisting in the environment,

98 *Id.* at 53036.

99 Genetically Engineered Organisms and Products; Notification Procedures for the Introduction of Certain Regulated Articles; and Petition for Nonregulated Status, 57 Fed. Reg. 53036. By 2001, USDA had issued 1,117 field test authorizations for more than 57,000 acres of GM crop field testing. 67 Fed. Reg. 50578.

100 Genetically Engineered Organisms and Products; Notification Procedures for the Introduction of Certain Regulated Articles; and Petition for Nonregulated Status, 57 Fed. Reg. at 53036.

101 7 C.F.R. § 340.3(b) (2006).

102 *Id.* at § 340.3(c).

such movement must be minimized, and upon termination of the field tests, no viable material shall remain which is likely to volunteer in subsequent seasons, or volunteers shall be managed to prevent persistence in the environment.[103]

For releases into the environment beyond controlled field testing, APHIS adopted a final rule which established a process for petitioning to determine non-regulated status.[104] For any organism for which such a petition is granted, that organism is no longer considered a "regulated article," and therefore is exempt from all APHIS regulation.[105] The petitioner must supply certain data regarding the organisms, including field test data.[106] APHIS then reviews the data for potential "plant pest" risk.[107] Plant pest risk is direct or indirect injury, damage to, or disease in any plant or plant product. The statute defines plant pest as "any living stage of any of the following that can directly or indirectly injure, cause damage to, or cause disease in any plant or plant product: (A) A protozoan (B) A nonhuman animal (C) A parasitic plant (D) A bacterium (E) A fungus (F) A virus or viroid (G) An infectious agent or other pathogen (H) Any article similar to or allied with any of the articles specified in the preceding subparagraphs."[108] The USDA regulations define plant pest to mean: "Any living stage (including active and dormant forms) of insects, mites, nematodes, slugs, snails, protozoa, or other invertebrate animals, bacteria, fungi, other parasitic plants or reproductive parts thereof; viruses; or any organisms similar to or allied with any of the foregoing; or any infectious agents or substances, which can directly or indirectly injure or cause disease or damage in or to any plants or parts thereof, or any processed, manufactured, or other products of plants."[109] Although the term "plant pest" is not defined to include organisms that cause harm to human health or environmental health in general, the 2000 PPA extended the authority of USDA to consider human health and broad environmental harm. The PPA gives the USDA the authority to prohibit or restrict the importation, exportation and interstate movement of plants, plant products, certain biological control agents, *noxious weeds* and plant pests. The term "noxious weed" means: "any plant or plant product that can directly or indirectly injure or cause damage to crops (including nursery stock or plant products), livestock, poultry, or other interests of agriculture, irrigation, navigation, the natural resources of the United States, the *public health, or the environment*." [110] If APHIS determines the organism poses no plant pest risk, it will grant the petition and the organism will

---

103 *Id.*

104 *Id.* at § 340.6.

105 *Id.*

106 *Id.*

107 A process for publication in the federal register and public comment is provided. 7 C.F.R. § 340.6(2)(d).

108 7 U.S.C. § 7702(14).

109 7 C.F.R. § 340.1.

110 7 U.S.C. § 7702(10) (emphasis provided).

be exempt from APHIS regulation. APHIS retains the authority to "reregulate" the organisms if it becomes a plant pest in the future. Organisms exempt from APHIS regulation may still be subject to EPA or FDA regulation.

As with EPA's GMO regulation to date, APHIS's regulations focus on which GMOs require submission of notification prior to field testing and which GMOs are completely exempt from APHIS regulatory oversight. APHIS's regulations do not address how to regulate GMOs that are released into the environment to minimize environmental risk. Moreover, APHIS's focus on plant pest risk does not adequately address the other types of unique risks that may be posed by GMOs.

Chapter 11

The Failure of US Law to Address the Ecological Considerations of Pesticide Use[1]

Although the basic regulatory framework of FIFRA dates back to 1972 some of its provisions, such as the misbranding prohibition, date back to the early consumer-protection statutes of the early twentieth century. Despite the fact that a statute containing the word "rodenticide" would hardly be expected to be cutting-edge, FIFRA surprisingly provides EPA with considerable legal authority to consider ecological concerns when registering pesticides. Certainly there is much that can be done to mold FIFRA into a more sophisticated ecologically based law. Nevertheless, ecologically based themes such as adaptive management and an ecological integrity baseline run throughout the existing statute.

FIFRA's Strengths

An Ecological Baseline

The concept of an ecological baseline can be found in a number of places in FIFRA. Specifically, the standard for registration under FIFRA is designed to address a large array of ecological, as well as human health, concerns. FIFRA's regulatory standard is aimed at preventing "unreasonable adverse effects on the environment."[2] The word "environment" is defined very broadly by FIFRA to include "water, air, land, and all plants and man and other animals living therein, and the interrelationships which exist among these."[3] Thus, FIFRA not only broadly includes land, water, air, and plant and animal resources within the definition of

---

1 Portions of this chapter are adapted from Mary Jane Angelo, *Corn, Carbon and Conservation: Rethinking U.S. Agricultural Policy in a Changing Global Environment* 17 George Mason Law Review 593 (April 2010); Mary Jane Angelo, *The Killing Fields: Reducing the Casualties in the Battle Between U.S. Endangered Species and Pesticide Law*, 32 Harvard Environmental Law Review 96 (2008); Mary Jane Angelo, *Regulating Evolution for Sale: An Evolutionary Biology Model for Regulating the Risks Posed by Genetically Modified Organisms*, 42 Wake Forest Law Review 93 (2007); and Mary Jane Angelo, *Embracing Uncertainty, Complexity and Change to Protect Ecological Integrity: An Eco-Pragmatic Reinvention of a First Generation Environmental Law*, 33 Ecology Law Quarterly 105 (2006).

2 *See, e.g.*, 7 U.S.C. § 136(l), § 136(x), § 136(ee)(2) (2011).

3 *Id.* § 136(j) (2011).

"environment," but also includes the "interrelationships," an important ecological consideration, among these resources within the definition.

Moreover, FIFRA contemplates an ecological baseline in its premarket evaluation. Under FIFRA, a registration is required prior to the sale, distribution, or commercialization of any pesticide product. Moreover, except for in limited circumstances qualifying for an exemption, some level of regulatory review is required prior to the release of any pesticide into the environment, even if only for experimentation purposes. Unlike some other environmental regulatory programs, such as the Toxic Substance Control Act (TSCA),[4] FIFRA imposes a somewhat precautionary approach in that pesticides, which are by definition intended to kill or disrupt living organisms, are presumed to pose unreasonable risks, and therefore cannot be sold or distributed without a premarket environmental review.[5] In contrast, under TSCA, new non-pesticide chemicals entering the marketplace do not require a premarket environmental review.[6] Instead, prior to manufacturing these new non-pesticide chemical substances under TSCA, all that is required is a 90-day notification to EPA.[7] During the premarket notification period, EPA conducts a cursory review of the proposed new chemical, but unless a determination is made that generation of new data are required for a particular chemical, EPA typically does not require environmental testing.[8] If a non-pesticide substance is later found to pose unforeseen risks, EPA can, by regulation, require additional testing to be conducted or impose regulations to reduce the risk from such a substance under sections 4 and 6 of TSCA, respectively.[9]

Moreover, FIFRA's precautionary approach is manifest in its allocation of the burden of proof. While this is not expressly stated in the language of FIFRA, pursuant to a series of administrative and judicial decisions, the burden of proof that a pesticide does not pose an unreasonable adverse effect of the environment remains at all times on the proponent of registration or continued registration.[10] Thus, a proponent of registration is required to demonstrate that it meets this burden prior to a pesticide being registered. Further, if EPA proposes cancellation of a pesticide or use of the pesticide, the burden of proof rests on the proponent of continued registration during any cancellation or suspension hearing that may ensue.

---

4 15 U.S.C. §§ 2607–2629 (2011).

5 7 U.S.C § 136a(a) (2011).

6 15 U.S.C. § 2604 (2011).

7 *Id.* § 2604(a) (2011).

8 *Id.* § 2604 (2011).

9 *Id.* § 2603 and § 2606 (2011).

10 Environmental Defense Fund, Inc. v. EPA (heptachlor-chlordane), 548 F.2d 998, 1004 (D.C. Cir. 1976), *cert. denied* 431 U.S. 925 (1977); Environmental Defense Fund, Inc. v. EPA (DDT II), 439 F.2d 584 (1971); Stearns Elec. Paste Co. v. EPA, 461 F.2d 293, 304 (7th Cir. 1972).

Adaptive Management: Flexibility and the Consideration of New Information

FIFRA also contains provisions that are compatible with the ecological concept of adaptive management. Many of FIFRA's provisions are specifically designed to seek or adapt to new information, or to tailor the level of regulation to the level of certainty of risks based on the sufficiency of available data. For example, FIFRA establishes two different levels of registration—full registration and conditional registration.[11] EPA may conditionally register a pesticide under certain circumstances despite the fact that sufficient data have not been generated to support full registration.[12] As described above, such circumstances may occur when a new use is proposed for a pesticide that is already registered for another use.[13] In such a situation, sufficient data exists to support the existing use but additional data may be required to support full registration of the newly proposed use. Under such circumstances, EPA may conditionally register the pesticide for the new use if EPA determines that the conditional registration would not significantly increase the risk of any unreasonable adverse effects on the environment.[14] Accordingly, through the conditional registration process a degree of flexibility is built into FIFRA, allowing products to be used in new ways prior to full data generation. Of course, should the new data demonstrate that the new use does not meet the standards for full registration, full registration will not be granted.

Other provisions of FIFRA that allow unique circumstances and changing information to be taken into consideration include the emergency exemption provisions,[15] the state registration provisions,[16] and the experimental use permit provisions.[17] The emergency exemption provision of section 18 of FIFRA authorizes EPA to grant an emergency exemption to any state or federal agency under emergency conditions—i.e., urgent non-routine conditions for which no economically or environmentally feasible alternative practices that provide adequate control are available.[18] Thus, section 18 provides flexibility to adapt to changing circumstances, which could include the outbreak and spread of a new pest or the spread of a public health disease. In such circumstances, EPA is authorized to act quickly to control the problem before the pest or disease vector is widely disseminated and to minimize the harm, without waiting for a full data set to be produced to support registration.[19]

11 7 U.S.C. § 136a(c)(7) (2011).
12 *Id.* at § 136a(c)(7)(B) (2011).
13 *Id.* at § 136a(c)(7)(A) (2011).
14 *Id.*
15 *Id.* at § 136p (2011).
16 *Id.* at § 136v (2011).
17 *Id.* at § 136c (2011).
18 *Id.* at § 136p (2011).
19 *Id.*

The state registration provision in section 24(c) authorizes states to issue registration to meet special local needs.[20] Accordingly, this provision allows states to take into consideration local circumstances that may warrant the use of pesticide products or uses that are not generally approved under FIFRA for nationwide use. Thus, in a state where a particular pest causes more severe harm than in other states, the cost/benefit analysis for the use of the pesticide in that state may have a different result from the nationwide cost/benefit analysis for that use, and accordingly, a special local needs registration may be granted for that state only. In this way, FIFRA's flexibility allows registrations to be tailored to the special agricultural, environmental, economic, or other needs of a state.

Finally, the experimental use permit provision contained in FIFRA section 5 is another example of the flexibility afforded by FIFRA in tailoring the amount of data necessary to the level of risk resulting from a particular use. FIFRA section 5 authorizes EPA to issue permits for the field testing of pesticides necessary to generate data to support full registration.[21] Thus, as the risk from exposure to a pesticide increases (i.e., as it moves from lab testing, to small-scale field testing, to full-scale use), progressively greater data requirements attach to ensure that sufficient data are available to make an unreasonable adverse effects determination for each level of use.[22] Similarly, the ability for EPA to classify a pesticide as either general or restricted use[23] allows EPA to adapt the amount of regulation required to the risks associated with the particular pesticide. By classifying a pesticide as restricted use, EPA ensures that users of the pesticide will have at least some level of training and supervision to reduce the risks associated with the use of that pesticide.

The ability to obtain and consider new information and to change course in response to new information is a cornerstone of adaptive management. FIFRA contains several provisions that authorize EPA to obtain, consider, and take action on new information. FIFRA contains some relatively forward-looking provisions that allow EPA to take advantage of new information as it develops, as well as to require new information as new testing methodologies become available or new risk scenarios are understood. Because hundreds of pesticides on the market were registered prior to the current registration data requirements being in place, FIFRA section 4 contains a detailed "reregistration" process designed to ensure that older pesticides are reexamined in light of more stringent regulatory standards and more sophisticated testing methodologies that have come into existence since the times of the early registrations of many pesticides.[24] The current reregistration provisions include a multi-phased process with a number of deadlines, which must be met to avoid suspension or cancellation.[25] This reregistration approach has been

20 *Id.* at § 136v(c)(1) (2011).
21 *Id.* at § 136c(a) (2011).
22 40 C.F.R. § 172.3(a) (2011).
23 7 U.S.C. § 136a(d)(1) (2011).
24 7 U.S.C. § 136a-1 (2011).
25 *Id.* at § 136a-1(b) (2011).

an extremely time-consuming, burdensome, and expensive process, but has also been important in evaluating pesticides that were registered under less stringent standards using outdated risk-assessment methodologies. Similarly, EPA's current "registration review" process provides a process for continual re-evaluation of pesticidal risk as new information and analytical methods become available.

While reregistration applies to all pesticides registered prior to 1984, there are a number of FIFRA data requirements that apply to previously registered pesticides in a more targeted manner. For example, FIFRA section 3(c)(2)(B) authorizes EPA to require holders of existing registrations to provide additional data to support the continued registration of a pesticide whenever EPA finds that "additional data are required to maintain in effect an existing registration of a pesticide."[26] Under FIFRA section 6(a)(2), if at any time after the registration of a pesticide the registrant obtains in any way factual information regarding unreasonable adverse effects on the environment of the pesticide, the registrant is required to submit such information to EPA. Unreasonable adverse effects information submitted to EPA may lead to EPA requesting additional data under section 3(c)(2)(B), EPA initiating a cancellation or suspension action, EPA reclassifying a general-use pesticide as a restricted-use pesticide, or some other form of regulation to ensure that the pesticide will not cause unreasonable adverse effects on the environment. Thus, FIFRA contemplates a process of ongoing evaluations of pesticide risks as new information becomes available. With regard to pesticidal GMOs, EPA has taken the section 6(a)(2) approach a step further to require ongoing reporting of adverse environmental effects data, not only for registered pesticides, but also for pesticides that qualify for a plant-incorporated protectant exemption.[27] In its plant-incorporated protectant rule, EPA requires that any person who produces, for sale or distribution, a plant-incorporated protectant exempt under the rule, and obtains any information regarding adverse effects on human health or the environment alleged to have been caused by the plant-incorporated protectant, must submit such information to EPA.[28]

Ecological Risk Assessment

Somewhat ironically, given the state of pesticide regulation today, the earliest applications of FIFRA were relatively adept at addressing ecological considerations. Even by today's standards, the 1972 DDT cancellation decision

26 *Id.* § 136a(c)(2)(B) (2011).

27 40 C.F.R. § 174.1 (2011).

28 40 C.F.R. § 174.71(b) (2011). This rule goes on to explain that for the purposes of plant-incorporated protectants, "[a]dverse effects on human health or the environment means at a minimum information about incidents affecting humans or other nontarget organisms where both (1) The producer is aware, or has been informed, that a person or nontarget organism allegedly suffered a toxic or adverse effect due to exposure to (e.g., ingestion of) a plant-incorporated protectant; and (2) The producer has or could reasonably obtain information concerning where the incident occurred."

represents a fairly sophisticated ecological analysis focused on the tendency of DDT and its metabolites to persist in the soil and aquasphere, the tendency of DDT to be readily transported by leaching, erosion, runoff and volatilization, DDT's fat-soluble characteristic, which enables it to collect and concentrate in animal fat tissue, the fact that DDT can bioaccumulate as it moves up the food chain, and the fact that once accumulated, DDT is toxic to animals and humans, and inhibits the ability of fish and other wildlife species to regenerate.[29] EPA did not shy away from tackling the difficult issue of how to take into account the risks that may occur in the future due to the persistent nature of the pesticide. EPA also was not paralyzed by a lack of certainty in data. For example, even though EPA recognized that the degree of transportability of DDT was unknown, it utilized data showing that DDT was found in remote areas of the world and in ocean species, such as whales, as providing enough evidence from which a logical reference could be drawn that DDT is readily transported in the environment.[30] Moreover, EPA concluded that persistence and biomagnification themselves were a "cause of concern, given the unknown and possibly forever undeterminable long-range effects of DDT in man and the environment."[31] Thus, even in 1972, EPA recognized the uncertainty surrounding ecological effects data, while at the same time employing a precautionary approach to prevent uncertain, possibly devastating effects that may occur in the long term.

Bridge Document 11.1 The 1972 EPA DDT Cancellation Hearing Final Order

[EPA has] established that DDT is toxic to non-target insects and animals, persistent in the environment, mobile and transferable and that it builds up in the food chain. No label directions for use can completely prevent these hazards. In short, [EPA has] established at the very least the risk of the unknown. That risk is compounded where, as in the case with DDT, man and animals tend to accumulate and store the chemical.

The risk-benefit equation is a dynamic one. Timing is a variable in that equation. What may, in the long run, be necessary to protect the environment could be a short-term threat to human health. This is exactly the case before me now. The benefits of using organophosphates are a long-range benefit and the risks of DDT result from continued long-term use. In the very short run, however, the equation balances out very differently.

Source: In the Matter of Stevens Industries, Inc, et al., I.F.& R. Docket Nos. 63, et al. (Consolidated DDT hearings, June 2, 1972).

29 In the Matter of Stevens Industries, Inc, et al., I.F.& R. Docket Nos. 63, et al. (Consolidated DDT hearings, June 2, 1972).

30 *Id.*

31 *Id.*

When the group petitioners who opposed cancellation argued that DDT was only one toxic substance in a polluted environment and that therefore whatever its laboratory effects, it could not be shown as the causative agent of injury to wildlife, EPA responded that this argument "does not redeem DDT, but only underscores the magnitude of effect that will be necessary for cleaning up the environment."[32] Persuaded by evidence showing that metabolites of DDT cause eggshell thinning in certain bird species despite some contradictory evidence, EPA found it sufficient that there was "laboratory data and observation data, and in addition, a scientific hypothesis, which might explain the phenomenon."[33] EPA further found that there were no label restrictions that could completely prevent effects on non-target organisms, persistence and transport in the environment, or biomagnification. Accordingly, cancellation was the only available risk-reduction measure.

EPA's analysis of the benefits of DDT in formulating its decision to cancel was also fairly sophisticated, and in some ways was a better job of considering benefits than many more recent EPA decisions regarding cancellation or suspension of pesticides. The benefits analysis focused on the availability of alternatives including non-chemical pest management programs and the fact that the crops protected by DDT were not in short supply. EPA found that DDT was not necessary to ensure an adequate supply of cotton at a reasonable cost, given that only 35 percent of the cotton-producing acreage at the time was treated by DDT.[34] In contrast to many recent EPA decisions, where non-chemical pest control alternatives and emphasis on availability of crops are conspicuously absent, or at least play a very minimal role in the decision-making process, EPA's 1972 DDT cancellation decision takes a broader look at the cost/benefit analysis rather than merely counting up dollars and cents on each side of the equation. In fact, EPA's 1972 DDT cancellation decision has other indicia of a sophisticated ecological approach. For example, in the final order, EPA's statement that "the risk/benefit equation is a dynamic one" can be viewed as a foreshadowing of the arguments for flexibility and adaptive management that are at the center of natural resources and environmental policy scholarship today.

FIFRA's Weaknesses

Notwithstanding all of these promising elements, FIFRA has not lived up to its early promise with regard to addressing ecological considerations. As described more fully below, a combination of statutory shortcomings, unfortunate interpretations, and problems with implementation and enforcement have resulted in FIFRA's lack of success as an environmental protection statute. Accordingly, there is considerable room for improvement.

32 *Id.*
33 *Id.*
34 *Id.*

The Cost/Benefit Standard: General Concerns with Cost/Benefit Balancing

One of FIFRA's most significant shortcomings that contributes to its failure to adequately protect ecological resources is its cost/benefit approach to registration and cancellation. The past 20 years of environmental law scholarship is characterized by a struggle between two opposite extremes—those who believe environmental values are paramount and those who believe economic interests are paramount.[35] Environmental regulation in the US is frequently criticized for being unrealistic, because it is based on protecting the public or environment from risks without considering the costs associated with such protection.[36] Of course it would be absurd to attempt to eliminate all environmental risks at all costs.[37] Opponents of pure risk-based approaches contend that environmental regulations seek to attain zero risk regardless of the costs of attaining such a goal. For the most part these criticisms are unfounded.[38] Despite the outcry of the opponents of risk-based regulation, the

35 Daniel J. Farber, Eco-Pragmatism: Making Sensible Environmental Decisions in an Uncertain World 35 (1999).

36 As Supreme Court Justice Breyer and Professor Cass Sunstein have described, some environmental risk reduction requires expenditures of money that go well beyond the bounds of common sense. *See* Stephen Breyer, Breaking the Vicious Circle: Toward Effective Risk Regulation 14 (1993) and Cass Sunstein, *Health-Health Tradeoffs*, 63 Chi. L. Rev. 1533 (1996).

37 Farber, *supra* note 35, at 3.

38 Although examples of environmental excesses certainly can be found, many if not most of such criticisms have been demonstrated to be without merit. Farber, *supra* note 35, at 22. In one of the most comprehensive analyses of the costs of environmental regulation to date, Lisa Heinzerling, demonstrated that many of the oft-cited examples of the excesses costs of environmental regulations are not accurate. Lisa Heinzerling, *Regulatory Costs of Mythic Proportions*, 107 Yale L.J. 1981 (1998). Heinzerling has shown that many of the examples used by cost/benefit proponents are either exaggerations or are examples of regulations that were considered but never adopted for a variety of reasons. *Id.* In this article, Heinzerling scrutinizes the "Morrall table," developed in 1986 by John Morrall, an economist at the Office of Management and Budget. *Id.* at 1983, 1987–9. The table includes information on the costs of various risk-reducing regulations per life saved. *Id.* at 1987–8. The numbers from this table have been relied on extensively by scholars, environmental policy makers, and politicians to support arguments that environmental regulations are not cost-effective and that taxpayer money could be better spent elsewhere. *Id.* at 1983. Heinzerling demonstrates that the estimates of the costs of environmental regulations developed by Morrall, and widely relied upon by proponents of cost/benefit analysis, are seriously flawed. First, Heinzerling shows how Morrall included a large number of regulations that were never adopted by the regulatory agency. *Id.* at 1999–2014. Heinzerling demonstrates that the alleged highest cost environmental regulations on Morrall's table have never taken effect. *Id.* at 1983, 1999–2014. In fact, some of these proposed rules were withdrawn by the agency specifically because of the high costs associated with them. *Id.* at 2000–10. Moreover, Heinzerling shows how Morrall's use of cost estimates skew the results by being up to 1,000 times higher than the agency estimates. *Id.* at 1983, 1991–1993.

vast majority of federal environmental laws are not purely risk-based and few, if any, are zero-risk statutes.[39] Instead, most environmental laws allow or require the consideration of other factors such as the costs of regulation or the economic feasibility of regulation. Even where zero-risk or absolutist risk-based environmental laws have been tried, the recognition of the potential illogical outcomes of these approaches have frequently led to either the interpretation of the laws to take other factors into account, or the amendment of the laws imposing these strict risk-based requirements.[40]

At the opposite end of the spectrum from the pure risk-based approaches to environmental regulation are those approaches that elevate economic efficiency above all other considerations. Cost/benefit analyses involve balancing the "social benefit" of a regulation against the cost to society to comply with the regulation.[41] Cost/benefit standards are based on the premise that maximum economic efficiency is the goal of any regulatory system.[42] Proponents of cost/benefit analysis often build

---

Further, Heinzerling shows how Morrall's estimates are misleading in that he used a high (10 percent) discount rate. *Id.* at 2018–24. Heinzerling's analysis shows that by using the agency's cost estimates and a lower discount rate, the cost per life saved for most of these regulations falls well within the accepted range. *Id.*

39 This is for good reason. It is impossible to eliminate all risk. First, as risks are reduced to lower levels, the costs of eliminating remaining risks may be excessively high. Second, in many cases it is impossible to attain zero risk because technology does not exist to measure very small amounts of pollutants or very small environmental disruptions that may cause risk. Third, it is not uncommon for risk-reduction measures that are employed to address one risk to cause some other unintended risk. Finally, risk is inherent in nature. FARBER, *supra* note 35, at 72–4.

40 Perhaps the most notable absolutist zero-risk-based law was the Delaney Clause of the Federal Food Drug and Cosmetic Act (FFDCA). 21 U.S.C. § 348 (1994). As one of the few environmental statutes to attempt to impose a zero-risk standard, the Delaney Clause was highly controversial. In fact, EPA attempted to interpret into the statute, a *de minimis* exception, despite the absolutist language contained in the law. EPA's *de minimis* interpretation of the Delaney Clause was struck down by the Ninth Circuit in Les v. Reilly, 968 F.2d 985 (9th Cir. 1992). After many years of controversy, Congress, in 1996, eliminated the zero-risk language in favor of the more flexible "safety" standard in the Food Quality Amendment Act. Pub. L. No. 104–170, 110 Stat. 1489 (codified at 7 U.S.C.A. §§ 136–136y and 21 U.S.C.A. §§ U301–381 (West Supp. 1996)).

41 For a general overview of cost/benefit analyses, *see* CASS R. SUNSTEIN, THE COST-BENEFIT STATE: THE FUTURE OF REGULATORY PROTECTION (2002).

42 The argument in favor of economic efficiency is that limited resources should be spent where they will do the most good. This makes sense in the context of choosing between two technologies to achieve the same level of risk reduction. For example, if, all else being equal, it would cost 10 million dollars to install an end-of-pipe technology that would reduce water pollutants, thereby saving five lives, but it would cost only 1 million dollars to use alternative raw materials in the manufacturing process that would achieve the same risk-reduction goal, no reasonable person would question the decision to utilize the more economically efficient risk-reduction method. However, the real world is rarely so clear-

their arguments in favor of economic efficiency by using examples of environmental regulations that achieved minimum risk reduction for enormous costs.[43]

The cost/benefit analysis measure of value is the consumer's willingness to pay for goods or services.[44] The concept of willingness to pay does not necessarily translate well to ecological values.[45] Moreover, relying on willingness to pay may

---

cut, and typically environmental decisions are not a choice between two equally effective risk-reduction methods. More often they involve complex trade-offs in risk reduction, costs and normative choices that are not easily monetized. Consequently, cost/benefit analyses reach far beyond the obvious and simple goal of achieving an environmental benefit for the lowest cost possible. *See* FARBER, *supra* note 35, at 7.

43 *Id.* at 22.

44 *Id.* at 36. The willingness to pay standard is a key component of welfare economics. For a good general discussion of the Pareto Criteria, as well as other theories of economic efficiency, including Kaldor-Hicks efficiency, *see generally* Jules Coleman, *Efficiency, Utility, and Wealth Maximization*, *in* FOUNDATION OF THE ECONOMIC APPROACH TO LAW (Avery Wiener Katz, ed., 1998). As Farber explains, welfare economics is based on the simple concept of the Pareto improvement. Under this standard, a decision that results in at least one person benefiting and no one being harmed is a sound decision. Of course, in real-world decision making, it is rare to find such a simple outcome. Thus, welfare economics has developed the concept of economic efficiency whereby, an outcome is considered to be more economically efficient than other outcomes if in an overall sense, the gains of the winners outweigh the losses of the losers. This is accomplished if a hypothetical transfer of some of the gains from the winners to offset the losses of the losers would result in a Pareto improvement (i.e., once the losers losses have been offset by some of the winners gains, the winners maintain at least some gain and the losers now have suffered no losses). FARBER, *supra* note 35, at 44.

45 It can be difficult to determine how much people would be willing to pay for a given ecological service or benefit. Although it may be a relatively easy task to determine how much people are willing to pay for certain types of values, such as how much more people are willing to pay for a safe product versus an unsafe product, in the environmental arena, many values that are sought to be protected by environmental laws are what is known as "non-use" values. These non-use values include "option value," which is how much people are willing to pay to leave open the option of receiving a benefit in the future. An example of option value is the value people are willing to pay to preserve tropical rainforests to preserve the option of obtaining plants with medicinal value in the future. FARBER, *supra* note 35, at 48.

Another type of non-use value is "existence value," which is how much people are willing to pay to protect something that they will never directly benefit from. An example of existence value is how much people are willing to pay to protect an endangered species even if they will never receive any direct benefit from that species, and in fact may never even *see* that species. *Id.* These non-use values are much more difficult to monetize in a cost/benefit analyses than other types of values that provide a more direct impact to the person who is willing to pay for them. *Id.* at 49. One way that economists attempt to monetize these values is through a survey technique called contingent valuation. Cass Sunstein has criticized this technique as unreliable because surveys asking people hypothetically how much they would pay to preserve a particular non-use value may be skewed by people

Bridge Document 11.2 Incorporating Emergy Synthesis into Environmental Law: An Integration of Ecology, Economics, and Law

Virtually all areas of environmental law are concerned in some way with both the ecological and the economic impacts of environmental decision-making. Unfortunately, existing environmental law statutes tend to incorporate ecological and economic considerations in a simplistic, piece-meal, and awkward fashion.

…

The interpretations of ecological realities on which existing statutes are based are outdated and in need of serious reexamination. Although many existing environmental laws pay lip-service to ecological science, they do not incorporate scientific understanding of the ecological world in any meaningful way.

…

One controversial issue in the cost/benefit debate is whether environmental values are significant only to the extent that consumers are willing to pay to preserve them. Many ecological goods and services are not assigned any value by neo-classical economic analysis despite the fact that they are integral for making economically valuable products and may even be essential for life on earth, leading to criticisms of economics as "pricing everything and valuing nothing." The value of many ecological goods and services is not readily quantified, and thus is rarely included in any meaningful way in traditional cost/benefit analysis. Consequently, human disruptions to ecological systems are rarely part of cost/benefit analyses. The values inherent in ecological integrity or biodiversity are particularly ill-suited to be reduced to a dollar value under neo-classical economics. Although, many ecological products and services have instrumental value as food, medication, fiber, etc., that can be valued in a market system, many goods and services provided by nature do not have direct instrumental value and are not traded in a market system. Moreover, most consumers do not have the information available to them, or the technical understanding of the value that many ecological goods and services have in sustaining life. For example, many species serve important roles as producers, consumers, decomposers, competitors, dispersers, or pollinators. Each of these roles provides value to other members of the ecosystem, including humans. However, due to a lack of information and technical understanding, a typical consumer's willingness to pay for these services probably has no relation to the true value that the good or service provides.

Source: Mary Jane Angelo & Mark T. Brown, *Incorporating Emergy Synthesis into Environmental Law: An Integration of Ecology, Economics and Law*, 37 Environmental Law 963 (2007).

---

answering in a way that makes them feel morally satisfied rather than stating their real valuation. *Id.* at 49 (citing Cass Sunstein, Free Markets and Social Justice 142–3 (1997). On the other side of the debate critics argue that contingent valuation is not an appropriate way to value natural resources. John Heyde, *Is Contingent Valuation Worth the Trouble?* 62 Chi. L. Rev. 331 (1995).

skew cost/benefit analyses away from protecting environmental values because embedded within the willingness to pay concept is an assumption that industry has the right to utilize resources, and that if environmentalists want to protect those resources, they have to be willing to pay for such protection.[46]

Proponents of cost/benefit analysis believe that sound decision making is based on market indicators, whereas, environmentalists believe that a sound decision should be based on political indicators (Farber refers to this as "willingness to pay" versus "willingness to vote").[47] This distinction has been described as follows: "social regulation expresses what we believe, what we are, what we

---

46 The question then becomes one of how much the environmentalists are willing to pay to protect the resources. If instead one assumes that environmentalists have the right to protect resources, and that industry must pay environmentalists, or the public, if it wishes to utilize those resources, the question becomes one of how much environmentalists or the public are willing to accept to allow the resources to be used. The amount of money environmentalists are willing to accept to allow a resource to be used is typically twice as high as the amount of money the same environmentalist is willing to pay to protect the resource. FARBER, *supra* note 35, at 100. Nevertheless, most cost/benefit analyses use willingness to pay. SIDNEY A. SHAPIRO & ROBERT L. GLICKSMAN, RISK REGULATION AT RISK: RESTORING A PRAGMATIC APPROACH 57 (2003). This is but one example of how cost/benefit analyses are not value-neutral. The value-based assumptions that underlie the analyses can dramatically alter outcomes. *See also id.* at 56, explaining that willingness to pay and willingness to accept are typically different due to the distribution of wealth in society. A person's wealth does not tend to limit willingness to accept in the same way it limits willingness to pay. As a result, willingness to pay tends to result in less regulation. *Id.* at 66.

An example of the difference between willingness to pay and willingness to accept provided in Farber's book is the example of whalers desiring to kill a pod of whales. If one assumes that the whalers have a right to kill the whales for their own economic benefit, the question becomes how much the environmentalists are willing to pay the whalers not to kill the whales. This willingness to pay then becomes the economic value of the whales. If instead, one assumes that the environmentalists have the legal right to protect the whales, the question becomes how much the whalers are willing to pay the environmentalists for the right to hunt the whales. Under this scenario, it is likely that the environmentalists will demand a much higher price from the whalers than they would be willing to pay in the first scenario. One explanation for this phenomenon is that by assuming the environmentalists have the right to protect the whales, and the consequent right to sell the hunting rights to the whales, the environmentalists have been made wealthier and more powerful. Thus, the demand curve is shifted in favor of a higher price to hunt the whales. Consequently, under the second scenario, the whaling that may have been economically efficient under the first scenario, may no longer be efficient. FARBER, *supra* note 35, at 99. Another explanation for the difference between willingness to pay and willingness to accept, may be that people tend to be attached to the status quo and that change can only be purchased at a high price. *Id.* at 100.

47 FARBER, *supra* note 35, at 42.

stand for as a nation, not simply what we wish to buy as individuals."[48] Moreover, relying solely on consumer choices leaves out the possibility of decision making based on community analysis, deliberation and vetting of ideas.[49] Critics of cost/benefit analysis focus their arguments on the distinction between private consumer choices and public choices based on public values.[50] In the pesticides arena, an economic argument could be made that the fact that most consumers purchase food that is treated with chemical pesticides means that consumers are making a conscious choice to do so, based on economic and other factors. While it is true that consumers may purchase such foods because they are generally significantly less expensive than foods that are grown without chemical inputs, to assume that this consumer choice means that consumers have consciously determined that the health and environmental risks of pesticides in food are outweighed by the increased cost of organically produced food is overly simplistic.

First, consumers may not be fully informed of the environmental and health risks associated with pesticide use. The time and thought that go into sorting out technical risk-assessment information may be more than the average consumer can, or desires to, commit. As described at length in this chapter, many of the risks of pesticides use are not well understood, even by the experts. Thus, it would be unreasonable for consumers to be able to make decisions based on preliminary or confusing data that experts are having difficulty grappling with. Moreover, the price of food treated with pesticides does not reflect the true cost of growing the food because of a failure to internalize what economists refer to as negative externalities. For example, the price of purchasing food treated with pesticides

48 *Id.*, quoting MARK SAGOFF, THE ECONOMY OF THE EARTH: PHILOSOPHY, LAW AND THE ENVIRONMENT 17 (1988). Professor Cass Sunstein made this point eloquently, when he explained human character as being more than simply the acts an individual chooses to do at any given point in time. *See id.* at 54–5.

49 *See* SHAPIRO & GLICKSMAN, *supra* note 46, at 60.

50 Nevertheless, economists do recognize that there is a difference between private consumer choices and environmental benefits. Economists consider environmental benefits to be a public good. Public goods cannot be provided solely through the marketplace because all members of the public benefit from the good and there is no way to charge all members of the public for the benefit they receive. The classic example of a public good in the environmental context is clean air. All members of the public breathe air, and thus, there is no way for a private enterprise to charge the public for the air it breathes. Accordingly, there is no way for the private enterprise to benefit financially by ensuring its operations result in clean air. In recent years, Professor Cass Sunstein has introduced a modified form of cost/benefit analysis, in which public values rather than private preferences govern environmental decision making. *See* FARBER, *supra* note 35, at 95 (describing the proposal in CASS SUNSTEIN, FREE MARKETS AND SOCIAL JUSTICE 139 (1996) involving a two-stage decision-making process. A traditional quantitative cost/benefit analysis would constitute the first stage. The second stage would take into consideration other non-quantifiable values). FARBER, *supra* note 35, at 95.

does not reflect the price that society may have to pay later to clean up pesticide contaminated sites, the long-term costs to society of the loss of insect pollinators, or the long-term costs to society of the development of pest resistance due to overuse of pesticides. Moreover, the price of the food is artificially low due to numerous government programs that encourage the use of chemical inputs in agriculture.[51] As described in detail in Chapter 4, government programs, such as commodity crop subsidy programs, provide economic incentives for growing high yield commodity crops that depend on chemical pesticides and no corresponding incentives for growing crops without chemical pesticides. This results in artificially low food prices, which in turn, results in consumers perhaps purchasing more food grown with pesticides than they would in a pure market.[52] Accordingly, in the agricultural pesticide arena as well as many others, a cost/benefit analysis that relies on market costs and benefits does not reflect consumers' true preferences. Moreover, the assumptions upon which cost/benefit analyses are based, including perfect information, no externalities, and a truly free market, deviate so far from reality that cost/benefit analyses, standing alone, are an unreliable guide.

One controversial issue in the cost/benefit debate is whether environmental values are significant only to the extent that they can be translated into economic terms.[53] Opponents of cost/benefit analysis maintain that economists "price everything and value nothing."[54] Cost/benefit analysts tend to limit their view of the benefits of environmental regulation to human lives saved or cancers averted.[55] The

---

51 For a detailed discussion of government policies that encourage pesticide use, *see* Kenneth A. Dahlbert, *Government Policies that Encourage Pesticide Use in the United States*, *in* THE PESTICIDE QUESTIONS: ENVIRONMENT, ECONOMICS AND ETHICS 281 (David Pimentel & Hugh Lehman, eds, 1993) (discussing a wide range of local, state, and federal programs including price supports, incomes supports, farm credit programs, crop insurance programs, export policies, and tax policies). *See also*, David Pearce & Robert Tinch, *The True Price of Pesticides*, *in* BUGS IN THE SYSTEM: REDESIGNING THE PESTICIDE INDUSTRY FOR SUSTAINABLE AGRICULTURE 50 (William Vorley & Dennis Kenney, eds, 1998).

52 For a good summary of agricultural price supports, *see* Robert L. Thompson, *Agricultural Price Supports*, *in* THE CONCISE ENCYCLOPEDIA OF ECONOMICS, *available at* http://www.econlib.org/library/Enc1/AgriculturalPriceSupports.html (last visited January 25, 2013).

53 FARBER, *supra* note 35, at 35.

54 *See generally*, FRANK ACKERMAN & LISA HEINZERLING, PRICELESS: ON KNOWING THE PRICE OF EVERYTHING AND THE VALUE OF NOTHING (2004). *See also*, David Pearce & Dominic Moran, *The Economics of Biological Diversity Conservation*, *in* CONSERVATION BIOLOGY FOR THE COMING DECADE (Peggy Lee Fiedler & Peter M. Kareiva, eds, 1998).

55 FARBER, *supra* note 35, at 88. Cost/benefit analyses typically rely on the concept of "statistical lives" saved, because under most risk assessments, it is virtually impossible to identify the actual person that would have died but for the environmental regulation. Typically people value a statistical life less than an actual life. This is illustrated by the lengths that people will go, and the money they are willing to spend, to save a child who

true benefits of environmental protection extend far beyond counting the numbers of lives saved. For example, non-lethal human health effects, including subchronic neurological, behavioral, or reproductive effects are not well understood, not easily quantified, and rarely included in any meaningful way in cost/benefit analyses.[56] Moreover, as described in detail below, ecological systems are not well understood by science and the value of such systems is not readily quantified. Consequently, human disruptions to ecological systems are rarely part of cost/benefit analyses. Further, ethical, religious, aesthetic, and other normative values of environmental protection are not typically included in cost/benefit analyses. Finally, even those components of the cost/benefit analysis that are more easily monetized, such as the cost of compliance with environmental regulations, are difficult to estimate. Frequently, regulatory agencies are asked to take industry at its word,[57] and often in retrospect the true costs of compliance prove to be not as high as the pessimistic estimates of industry would suggest.[58]

Another concern often raised by opponents of strict cost/benefit analyses is that even when costs and benefits are balanced in an overall sense, what is left out of the equation is who pays the costs and who gets the benefits. In other words, if a disproportionate share of the costs are borne by segments of the population who do not share proportionally in the benefits, is there truly a balance of the costs and the benefits? Environmental risks tend to be disproportionately borne by vulnerable populations—frequently people of color, low income, and/or weak political power—who do not share in the benefits gained by industrial pursuits.[59] The benefits of development and industry typically go to the advantaged segments of society who do not generally bear a great portion of the risks resulting from such activities. Environmental Justice[60] proponents point out that cost/benefit analyses and other market-based systems can lead to the creation of risk "hot spots," where the overall cost/benefit analysis may weigh in favor of an activity

---

falls into a well, as compared to how much society is willing to spend to save an anonymous statistical person. *See, e.g.*, Shapiro & Glicksman, *supra* note 46, at 64. In the ecological context, there are many examples of people going to great lengths to save a beached whale, while only being willing to spend a significantly lesser amount to save the statistical dolphin from the shrimper's net. It is even harder to predict how much people will be willing to spend to protect a statistical ecological service.

56 Farber, *supra* note 35, at 88.

57 *Id.* at 90.

58 *Id.* at 167–8.

59 *See generally*, Clifford Rechtschaffen & Eileen Gauna, Environmental Justice: Law, Policy and Regulation (2003).

60 Although a detailed consideration of Environmental Justice is beyond the scope of this chapter, it is a compelling issue in environmental policy, and it provides one example of how cost/benefit analysis does not adequately address ethical, moral or fairness issues that cannot be monetized. For an excellent discussion of how market-based approaches disproportionately impact vulnerable populations *see id.* at 33–7.

going forward, but within a given segment of society, typically economically and politically disadvantaged populations, the cost/benefit analysis weighs in the opposite direction—i.e., hot spots of risk with very little benefit. Accordingly, proponents of Environmental Justice reject strict cost/benefit analyses.[61]

A similar argument can be made with regard to ecological resources. Risk hot spots may emerge in situations where the overall cost/benefit analysis weighs in favor of allowing a pesticide to be used, but in a particular geographic area, or with a particular species or ecosystem, risk may be disproportionately high. This may occur with regard to sensitive species, including threatened or endangered species, or with ecosystems with low resistance and/or low resilience. One example of such a risk hot spot is the global amphibian crisis, in which pesticides are implicated.[62] Although the cost/benefit analysis for a particular pesticide may reflect that the benefits outweigh the costs overall, this does nothing to protect the highly sensitive and highly vulnerable amphibian population.[63]

Finally, the values inherent in ecological integrity or biodiversity are particularly ill suited to be reduced to a dollar value.[64] Of course many ecological products and services have instrumental value. Although, many species in nature can be eaten, made into medications, made into clothing or shelter, burned for fuel or

---

61 Proponents of cost/benefit analysis counter this argument by pointing out that disadvantaged segments of society will not necessarily be better off under an inefficient system. However, Environmental Justice advocates do not argue in favor of inefficiency for inefficiency's sake, but rather they are willing to tolerate some level of economic inefficiency to accomplish other important societal goals such as fairness.

62 *See* Carlos Davidson et al., *Spatial Tests of the Pesticide Drift, Habitat Destruction, UV-B, and Climate-Change Hypotheses for California Amphibian Declines*, 16 Cons. Biol. 1588 (2002).

63 Moreover, the amendment of FIFRA by the 1996 Food Quality Protection Act to include a "safety" standard for human food consumption may have resulted in a shift of risk away from food consumers to occupational workers and ecological resources. For example, organophosphate pesticides degrade quickly into benign breakdown products in the environment and on agricultural food products. Thus, the FQPA standard of "safety" for human food consumers may favor the use of organophosphate pesticides over other pesticides that do not break down as quickly, and thus leave residue on foods. However, as described above, organophosphates are acutely toxic to humans, other mammals, and birds and have been implicated in thousands of bird deaths. Accordingly, relying more heavily on organophosphates to meet the FQPA standard, while reducing risks to food consumers, may result in significantly increased risk to vulnerable farmworker populations and to ecological resources. The risk to farmworkers for pesticide exposure is extremely high. Ivette Perfecto & Baldemar Velasquez, *Farmworkers: Among the Least Protected*, in Rechtshaffen & Gauna, *supra* note 59, at 67–8. Because the vast majority of farmworkers are members of minority groups—primarily Latinos—theses risks raise significant Environmental Justice concerns. *Id.*

64 *See, e.g.*, J.B. Ruhl, *Working Both (Positivist) Ends Toward a New (Pragmatist) Middle in Environmental Law*, 68 Geo. Wash. L. Rev. 543 (2000) (book review).

otherwise used in a market-based economy,[65] what is considerably more difficult to value are the aesthetic, inspirational, religious, or spiritual reasons that many people value ecosystems.[66] And perhaps even more challenging, is attempting to reduce to dollars and cents the value that species have as members of ecosystems. For example, species in an ecosystem may serve important roles as producers, consumers, decomposers, competitors, dispersers, or pollinators.[67] Each of these roles provides value to other members of the ecosystem, including humans.[68] Another concern with the strict cost/benefit approach is that the lack of knowledge and uncertainty regarding biological systems, argues in favor of a cautious approach to the cost/benefits analysis. Our current, limited, understanding of ecological systems and inadequate methodologies for monetizing values, limit the ability of cost/benefit analyses to be sufficiently precise to control environmental decision making.[69]

In sum, the shortcomings of cost/benefit analysis are numerous. First, current data and methodologies are not adequate for accurate and precise analyses. Second, cost/benefit analyses do not adequately address ecological values, community values, and other normative considerations. It is impossible to reduce such values to monetary terms. Finally, far from being value-neutral, cost/benefit analyses are laden with biases in favor of those who seek to use resources and away from those who seek to protect resources. Nevertheless, economic analyses are considered to be useful factors to inform environmental decision makers.[70] In other words, cost/benefit analysis should assist rather than control environmental decision making.[71]

In a perfect world with perfect scientific data on ecological risks and perfect cost/benefit methodologies that adequately value all costs and benefits, a cost/benefit approach might be compelling. However, with the complexity pervasive in ecological systems, the dearth of good data, and the confused state of cost/benefit methodology, all coupled with the need to act now, strict cost/benefit balancing is simply inadequate and inappropriate when attempting to address ecological

---

65 Malcolm L. Hunter, Jr., Fundamentals of Conservation Biology 43 (1996).

66 *Id.* Noted zoologist E.O. Wilson has coined the term "biophilia" to describe the deep emotional and spiritual relationship that many people share with other living organisms and the related sense of awe that frequently accompanies experience with the natural world. *See generally*, E.O. Wilson, Biophilia: The Human Bond with Other Species (1984).

67 Hunter, *supra* note 65, at 52.

68 *Id.*

69 Farber, *supra* note 35, at 42. Farber estimated the costs and benefits of a number of environmental regulations from 1988 and determined that the costs ranged from 55 to 77 billion and the benefits ranged from 16 to 135 billion dollars. *Id.* at 167–8. Thus, he concludes, the return on every dollar spent ranged between 21 cents and $2.27. *Id.* As Farber puts it "we were either losing 80 percent of our investment or a more than doubling our money and we don't know which!" *Id.* at 168.

70 Farber, *supra* note 35, at 9; *see also*, Shapiro & Glicksman, *supra* note 46, at 63.

71 Ruhl, *supra* note 64, at 538.

concerns. All but the most staunch proponents of cost/benefit analysis seem to recognize that cost/benefit analysis involves so many assumptions, judgment calls, values and unquantifiable factors, that it is overly simplistic to view cost/benefit analysis as a clear-cut quantifiable test, where all one has to do is plug the numbers in and the "right" answer comes out. For all of the reasons discussed above, eco-pragmatism rejects pure risk-based approaches and cost/benefit analysis in favor of a hybrid approach based on the concept of feasibility.[72] Under the hybrid approach, environmental risks are reduced to the amount feasible, and cost/benefit analysis is used to assist, rather than control, decision making.[73] This type of approach has been described as "doing the best that we can."[74] In other words, while the goal is environmental risk reduction, there is recognition that at some point the costs of further risk reduction become too high to justify. Pragmatic risk regulation attempts to reconcile conflicting values by "striving to achieve the maximum level of protection consistent with reasonable costs."[75]

Although there has been extensive debate in the academic world over the appropriateness of utilizing cost/benefit analyses or pure risk-based approaches in environmental regulation, the truth is that for the most part, Congress has rejected both pure risk-based and cost/benefit standards and most of the substantial environmental regulatory programs involve "hybrid" standards—i.e., risk-based standards that take economic considerations into account but do not require strict cost/benefit balancing. Professors Shapiro and Glicksman have surveyed existing environmental regulatory statutes to determine which contain cost/benefit standards, which contain feasibility standards, and which are purely risk based. Their work demonstrates that the majority of existing statutes contain standards that require risk to be avoided to the extent feasible or to the extent that the best available technology can achieve. Accordingly, these statutes are referred to as technology-based statutes. The most common examples of environmental statutes that utilize technology-based standards include the Clean Water Act [76] and the Clean Air Act.[77] While these standards take costs into account, they are not cost/benefit balancing standards.

Shapiro and Glicksman describe constrained balancing as regulation whereby the legislature establishes a level of environmental protection to be achieved by

---

72 Farber has defined feasibility as where costs are "not grossly disproportionate to the benefits to be achieved." Farber, *supra* note 35, at 94.

73 *Id.* at 122–3.

74 Shapiro & Glicksman, *supra* note 46, at 22, 50.

75 *Id.* at 147.

76 *See, e.g.*, 33 U.S.C. § 1311(b) (setting forth technology-based standards for point sources of water pollution).

77 *See, e.g.*, 42 U.S.C. § 7502(c)(1) (specifying that existing stationary sources of air pollution in nonattainment areas implement all "reasonably available control technology").

identifying regulatory objectives based on some model technology.[78] In this way, costs are considered in choosing an appropriate technology as the model, but costs are not directly weighed against benefits.[79] This approach is what is commonly referred to as "technology-based"[80] standard setting and is also what Farber refers

78 *Id.* at 37.

79 Examples of this type of approach include: the Clean Air Act's requirement for non-attainment areas that existing stationary sources implement all "reasonably available control technology" as expeditiously as practicable, 42 U.S.C. § 7502(c)(1), the requirement of the Safe Drinking Water Act that EPA establish maximum contaminant levels as close as feasible to achieving the level at which no known or anticipated health effects will occur. 42 U.S.C. § 300g-1(b)(4).

80 Technology-based standards are credited with much of the pollution reduction that has occurred since the environmental movement of the 1960s and 1970s. *See* Wendy E. Wagner, *The Triumph of Technology-Based Standards*, 2000 U. Ill. L. Rev 83, 85 (2000). Technology-based standards are the dominant mode of controlling the amount of pollution released in to the environment in the US. *Id.* at 88. These standards typically are developed by EPA in response to a Congressional mandate for EPA to establish pollution limitations based on some prescribed standard such as the "best available technology." *Id.* at 89. Once Congress has mandated the use of a particular technology-based standard, EPA will conduct a review of currently available technologies for specific industrial sectors and chooses the technology that best fulfills the Congressionally-prescribed standard. *Id.* Typically, the specific technology chosen is not required to be employed. Instead, EPA determines the level of pollution control that can be achieved using the chosen standard. *Id.* Industry is then free to utilize any technology it chooses that complies with the numerical standard. *Id.* Accordingly, industry has an incentive to develop the most efficient technology to meet the numerical standard. One drawback to using feasibility as a standard is that feasibility will vary with the economic strength of the industry. Farber, *supra* note 35, at 83. For example, it may be feasible for a strong manufacturing sector to be able to install costly pollution control technology, while it may be completely infeasible for a weak sector of the economy to install the same technology. As a result, more economically successful industries are penalized for their success with the imposition of stricter rules that do not apply to less successful industries. *Id.* Many justifications have been expressed for using technology-based approaches to environmental protection. These approaches tend to better address non-economic values, while still taking economic concerns into account. *See* Shapiro & Glicksman, *supra* note 46, at 65. In addition, feasibility-based approaches recognize and account for bounded rationality. *See id.* at 65. A feasibility standard is pragmatic in the sense that it is consistent with widely shared beliefs that environmental protection is important to society. Shapiro & Glicksman, *supra* note 46, at 21. Feasibility approaches tend to prod industry toward environmental risk reduction, until the point is reached where additional protection would be technologically or economically infeasible. *See, e.g.*, Ruhl, *supra* note 64, at 537. Moreover, some commentators maintain that technology-based standards are the best standards available because we do not have enough data to conduct true cost/benefit analyses for most environmental decisions. *See* Adam Babich, *Too Much Science in Environmental Law*, 28 Colum. J. Envtl. L. 119 (2003). Finally, from a pragmatic perspective, technology-based approaches have worked well in our 30-year-history of

to as feasibility-based decision making. Under constrained balancing or feasibility approaches, risk reduction is sought to the point that additional reduction would result in extraordinary or disproportionate costs.[81]

Shapiro and Glicksman refer to the second category of balancing approaches as "open-ended balancing," which they describe as a type of balancing whereby the legislature prescribes the factors that the agency must consider in making a decision, but does not dictate the weight the agency must give to any particular factor.[82] Interestingly, Shapiro and Glicksman describe FIFRA as an open-ended balancing statute. And while it is true that an initial reading of the statutory standard of "unreasonable adverse effects" would lead one to believe that it is indeed an open-ended balancing statute, whereby the agency has been directed to consider a variety of risk and benefit factors but has not been directed as to how to weigh such factors, in practice, FIFRA has been interpreted and applied by EPA as a cost/benefit balancing statute such that for a pesticide to be registered, its benefits must outweigh its costs. This interpretation has been articulated in agency orders[83] and upheld in many judicial decisions.[84] Thus, despite the significance of pesticides in

---

environmental law implementation. *See generally*, Wagner, *supra* note 80. After conducting an extensive analysis of the role of technology-based standards in environmental law over the past 30 years, Professor Wendy Wagner concludes that technology-based approaches are one of the most important innovations in environmental law. *Id.* at 86. Wagner also lauds technology-based standards for being relatively easy to promulgate, readily enforceable, evenhanded, adaptable and efficient. *Id.* at 94–105 (finding that the rate of promulgation for technology-based standards outpaces other standards from three to ten times and describing how technology-based standards apply equally to all members of the regulated industry within a given category). Wagner explains that technology-based standards are easily enforceable because the numerical level of pollution allowed is clearly prescribed by rule. *Id.* Once emissions or discharges are sampled for pollutants, regulated entities either meet the numerical standard or they do not. *Id.* How they choose to meet the standard is irrelevant. For the same reason, technology-based standards are very predictable. *Id.* Regulated entities can readily determine what numerical standards they are expected to comply with by looking at the EPA rule for that industry category. *Id.* at 100. *See also*, Thomas O. McGarity, *Media, Quality, Technology, and Cost/Benefit Balancing Strategies for Health and Environmental Regulation*, Law & Contemp. Probs. 159–99, n. 194 (1983). Wagner demonstrates that technology-based standards generally are very efficient. *See generally*, Wagner, *supra* note 80. She also asserts that technology-based standards create incentives for regulated interests to develop more efficient technologies. *Id.* Wagner describes technology-based approaches as fulfilling a moral imperative for industry to "do the best it can" to reduce environmental harm. *Id.* at 92.

81 Shapiro & Glicksman, *supra* note 46, at 52.

82 *Id.* at 39.

83 In the Matter of Stevens Industries, Inc, et al., I.F.& R. Docket Nos. 63, et al. (Consolidated DDT hearings, June 2, 1972).

84 Environmental Defense Fund, Inc. v. Ruckelshaus, 439 F.2d 584, 594 (1971).

the development of environmental law, FIFRA virtually stands alone in its cost/benefit approach to environmental protection.[85]

The plain language of FIFRA, as it currently stands, does not mandate a strict cost/benefit balancing. Some scholars have noted that on its face, FIFRA could be read to contain what is known as an "open-ended balancing standard," which allows consideration of a variety of factors but does not require weighing costs against benefits to determine which one outweighs the other.[86] As described above, FIFRA's definition of "unreasonable adverse effects on the environment" means any unreasonable risk to humans or the environment, taking into account the economic, social, and environmental costs and benefits of the use of any pesticide. Congress' only direction to EPA was to "take into account" economic and social as well as environmental considerations. Nevertheless, EPA has interpreted and implemented this standard as more of a strict cost/benefit balancing.[87]

Failure to Adopt the "Overriding Benefits Standard"

Subsequent to the DDT cancellation, EPA brought forth a number of cancellation and suspension actions, through which the agency's interpretation of the statutory standard, "unreasonable effects on man and the environment," was further developed.[88] This series of FIFRA cancellation and suspension cases cemented the interpretation of FIFRA as containing a cost/benefit balancing standard, rather than the open-ended balancing standard that, at least arguably, it was intended to be.[89] As Professor William Rodgers has described, in the legislative history of FIFRA, the "unreasonable adverse effects" language was intended to be an environmentally stringent standard for registration.[90] The Senate Commerce Committee, which created the standard, described it as not tolerating any adverse

85 The only other major environmental statute that employs a strict cost/benefit balancing approach is TSCA Section 6, 15 U.S.C. § 2605, which has been rendered impotent by the Fifth Circuit Court of Appeal decisions in Corrosion Proof Fittings v. EPA, 947 F.2d 1201 (5th Cir 1991) (holding that EPA did not have a reasonable basis to conclude that asbestos presents an unreasonable risk and that EPA failed to choose the least burdensome alternative to protect against such risk).

86 Shapiro & Glicksman, *supra* note 46, at 14.

87 Environmental Defense Fund, Inc. v. EPA (heptachlor-chlordane), 548 F.2d 998, 1004 (D.C. Cir. 1976), *cert. denied* 431 U.S. 925 (1977); In the Matter of Chapman Chemical Co., et al., FIFRA Dockets No. 246, et al. (EPA 1976); In the Matter of Protexall Products, Inc., et al., FIFRA Docket Nos. 625, et al. (1989).

88 *Id.*

89 *See, e.g.*, In the Matter of Chapman Chemical Company (canceling certain uses of mercury in pesticides based on a finding that the risks of continued use outweighed the benefits); and In the Matter of Protexall Products, Inc., et al., FIFRA Docket Nos. 625, et al. (1989) (describing the registrant's burden in challenging a proposed cancellation as requiring a showing that the "benefits of continued use justify the risks").

90 William H. Rodgers, Environmental Law 407, 451 (2nd ed. 1994).

effects, "unless there are overriding benefits from the use of a pesticide."[91] Accordingly, it appears that the standard contemplated by the Senate drafters of the 1972 FIFRA amendments intended that, although economic and social factors should be considered and balanced against environmental risks, the balancing would not be a simple accounting of dollars and cents on two sides of the equation with the pesticide winning the right to registration as long as the scale was tipped, no matter how slightly, in favor of the benefits provided by the pesticide. Instead, the Senate drafters appeared to intend that, where environmental risks exist, the analysis would favor registration only where the risks were outweighed by "overriding benefits." An example of an overriding benefit might be where a particular pesticide is important to fighting a significant public health problem such as West Nile Disease, and where other less risky control alternatives are not available or are too costly. The importance that pesticides can play in preventing significant public health problems is specifically addressed by FIFRA. The definition of "unreasonable adverse effects on the environment" provides that:

> [t]he Administrator shall consider the risks and benefits of public health pesticides separate from the risks and benefits of other pesticides. In weighing any regulatory action concerning a public health pesticide under this subchapter, the Administrator shall weigh any risks of the pesticide against the health risks such as the diseases transmitted by the vector to be controlled by the pesticide.[92]

Other overriding benefits might include situations where a particular pesticide is necessary to the maintenance of a segment of agriculture, where nonchemical or less risky alternatives are not available, and where growing a particular crop without the pesticide would result in severe economic losses or dramatically increased food prices. However, an overriding benefit would not be found in a situation where, if the pesticide were taken off the market, the evidence showed only that chemical companies would lose money or farmers would have to switch to other existing alternative pest control practices, which might involve some additional cost. If FIFRA were interpreted to mean that only overriding benefits could outweigh significant environmental risks, then potential registrants would face a more stringent standard and pesticides that posed significant risk would not routinely be registered. Unfortunately, EPA's longstanding interpretation of FIFRA's "unreasonable adverse effects" standard as a strict cost/benefit balancing standard has resulted in the registration of pesticides where all that is demonstrated is that the economic benefits outweigh risks despite the fact that lower risk alternatives may be widely available and may even be less costly.

---

91 *Id.* (quoting Senate Comm. on Commerce, Federal Environmental Pesticide Control Act of 1972, S. Rep. No. 970, 92d Cong., 2d Sess. 11 (1972)).

92 7 U.S.C. § 136(bb) (2011).

Failure to Require Demonstration of Benefits

Another significant shortcoming of EPA's approach to cost/benefit balancing, at least with regard to the registration of pesticides, is that EPA does not really engage in a true cost/benefit analysis because it does not require applicants to demonstrate the benefits of the pesticide. Moreover, as described above, EPA generally does not require efficacy data prior to registering a pesticide. EPA has, by rule, waived all requirements to submit efficacy data unless the pesticide product bears a claim to control pest microorganisms that pose a threat to human health or a claim to control vertebrates (such as rodents, birds, bats, canids, and skunks) that may directly or indirectly transmit diseases to humans.[93] Accordingly, at the time of a registration decision, EPA does not know how efficacious a particular pesticide is. Further, at the time of registration, EPA does not conduct an analysis to determine whether more efficacious alternatives, including non-chemical alternatives, exist. Thus, at the time of registration, EPA does not know the extent of the benefits of the pesticide. Instead, benefits are assumed. In determining whether to register a pesticide, EPA assumes that a manufacturer would not invest the resources necessary to support registration and commercialization of the pesticide unless the pesticide was efficacious and thus has benefits. Of course, one need only look at the glut of weight loss products, anti-balding products, and anti-wrinkle cream on the market, and the billions of dollars consumers spend each year on such products to conclude that marketability and efficacy are not necessarily one and the same. When coupled with the limited cost data that EPA considers, it can be seen that EPA's cost/benefit analysis to support registration is not a true cost/benefit analysis and is flawed in many respects.

It is not until EPA begins to consider whether to cancel the registration of a pesticide that it evaluates the benefits of the pesticide and whether there are viable alternatives available. In determining whether it should proceed with cancellation, EPA necessarily makes a threshold determination that the risks posed by a pesticide are significant. Once that determination is made, EPA conducts a full cost/benefit analysis. It is only at this point that EPA takes a look at the economic benefits of the pesticide.

Limited Consideration Relative Risks of Alternative Pesticides

When conducting a benefits analysis, EPA's analysis of alternatives is typically limited to looking at other registered pesticides for the same use (which are assumed to be efficacious if they are registered). EPA's consideration of non-chemical alternative pest control techniques such as cultural control, biological

93 40 C.F.R. § 158.400 (2011) (containing product performance data requirements for antimicrobial agents, products for treating water systems, and pesticides intended to kill or repel rodent, avian, and bat vectors). However, EPA has reserved the right to require, on a case-by-case basis, submission of efficacy data for other pesticides.

control, or organic farming practices has proven to be extremely limited and cursory. Moreover, when evaluating whether existing chemical alternatives are available for the target pest, EPA does not conduct a comparative risk analysis, which leads to the situation where a less risky pesticide may be cancelled because other alternatives exist, but as more and more pesticides are cancelled, the benefits of the remaining pesticides grow. Thus, the benefits of the "last pesticide standing" will be very high because no alternatives exist. Accordingly, the last pesticide standing will have benefits that outweigh the risks, even if the risks are relatively high. This problem could be solved by requiring a true benefits analysis for each registered pesticide, including a consideration of non-chemical alternatives, and requiring relative risk analysis, which compares the risks of pesticides targeted toward a particular pest, for each registered pesticide.

Typically, EPA does not routinely consider the relative risks of alternative pesticides when making registration or cancellation decisions. However, EPA has implemented certain policies to encourage the development and registration of lower risk pesticides. In 1997, EPA issued Pesticide Registration (PR) Notice 97-3, which sets forth policy for the expedited review of conventional pesticides under the reduced risk initiative and of biological pesticides.[94] This policy was developed partially in response to the 1996 Food Quality Protection Act (FQPA) mandates to develop procedures and guidelines for expedited pesticide review. The policy supersedes EPA's prior reduced-risk criteria.[95] The goal of the policy is to encourage the development, registration, and use of lower-risk pesticide products "which would result in reduced risks to human health and the environment, when compared to existing alternatives." To accomplish this goal, EPA offers the incentive of an expedited registration review for qualifying products. Qualifying pesticides include those that may reasonably be expected to accomplish one or more of the following: "(i) Reduce the risks of pesticides to human health; (ii) Reduce the risks of pesticides to non-target organisms; (iii) Reduce the potential for contamination of groundwater, surface water or other valued environmental resources; and (iv) Broaden the adoption of integrated pest management strategies."[96] EPA has further interpreted these criteria to develop a list of factors that will most significantly contribute to EPA's decision to grant reduced risk status. These factors include, in descending order of importance, very low mammalian toxicity; toxicity generally

94 US Environmental Protection Agency, Pesticides: Regulating Pesticides, Pesticide Registration (PR) Notice 97-3: Guidelines for the Expedited Review of Conventional Pesticides under the Reduced-Risk Initiative and for Biological Pesticides [hereinafter PR Notice 97-3], *available at* http://www.epa.gov/pesticides/PR_Notices/pr97-3.html (last visited May 26, 2011).

95 Incentives for Development and Registration of Reduced Risk Pesticides 57 Fed. Reg. 32140 (July 20, 1992); Incentives for Development and Registration of Reduced Risk Pesticides Program Update 58 Fed. Reg. 5854 (January 1993); and PR Notice 97-3, *supra* note 94.

96 7 U.S.C. § 136a(c)(10) (2011); PR Notice 97-3, *supra* note 94, at 3–4.

lower than alternatives (10–100X); displaces chemicals that pose potential human health concerns [e.g., organophosphates, probable carcinogens (B2s)]; reduces exposure to mixers, loaders, applicators and reentry workers; very low toxicity to birds; very low toxicity to honeybees, significantly less toxicity/risk to birds than alternatives; not harmful to beneficial insects, highly selective pest impacts; very low toxicity to fish; less toxicity/risk to fish than alternatives; potential toxicity/risk to fish mitigatable; similar toxicity to fish as alternatives, but significantly less exposure; low potential for groundwater contamination; lower use rates than alternatives, fewer applications; low pest resistance potential (i.e., new mode of action); highly compatible with IPM; efficacy."[97]

In 1997, EPA also issued Pesticide Registration Notice 97-2 (PR 97-2), which among other things established a priority for the registration of pesticides as follows: (1) methyl bromide alternatives; (2) reduced-risk candidates; (3) USDA-EPA identified potentially vulnerable crops; (4) minor use priorities; (5) non-minor use priorities; and (6) addressing trade irritants.[98] In 1998, EPA issued Pesticide Registration Notice 98-7, which changed the registration priority system involving organophosphate and reduced-risk pesticides, and amended PR 97-2 to have the following priority scheme: (1) methyl bromide alternatives; (2) OP alternatives that pass the reduced-risk screen; (3) other reduced-risk candidates; (4) OP alternatives that are submitted to the reduced-risk committee, judged to be significant OP alternatives, denied reduced-risk status, but recommended by the reduced-risk committee for expedited review; (5) USDA-EPA identified potentially vulnerable crops; (6) minor use priorities; (7) non-minor use priorities; and (8) addressing trade irritants.[99] These PR notices form the foundation of EPA's "Conventional Reduced Risk Pesticide Program," which is intended to expedite the review of and regulatory decision making on conventional pesticides that pose less human health and environmental risk. EPA describes the advantages of Conventional Reduced Risk Pesticides over existing conventional pesticides as having lower impact on human health, lower toxicity to non-target organisms such as birds, fish, and plants, lower potential for groundwater contamination, lower use rates, lower pest resistance potential, and more compatibility with IPM practices.[100]

---

97 PR Notice 97-3, *supra* note 94, at 3–4.

98 US Environmental Protection Agency, Pesticides: Regulating Pesticides, Pesticide Registration (PR) Notice 97-2: New Chemical, New Use, EUP, Non-Fast Track Amendments and Inert Ingredient Registration Priorities for Conventional Pesticides (1997), *available at* http://www.epa.gov/PR_Notices/pr97-2.pdf.

99 US Environmental Protection Agency, Pesticides: Regulating Pesticides, Pesticide Registration (PR) Notice 98-7: Changes to Registration Priority System Involving Organophosphate (OP) Alternatives and Reduced Risk Candidates, *available at* http://www.epa.gov/pesticides/PR_Notices/pr98-7.html (last visited May 26, 2011).

100 US Environmental Protection Agency, Pesticides: Regulating Pesticides, What is the Conventional Reduced Risk Pesticide Program? *available at* http://www.epa.gov/opprd001/workplan/reducedrisk.html (last visited May 26, 2011).

Despite the various reduced-risk policies adopted by EPA in the past 20 years, EPA has never gone so far as to adopt the commodity specific pesticide use reduction goals that then-Administrator Carol M. Browner proposed in her 1993 testimony before Congress.[101] Nor has FIFRA been amended as Administrator Browner proposed to state a clear policy goal favoring reduced use of pesticides and to direct federal agencies to lead an effort to reduce pesticide use and promote IPM.[102]

Failure to Consider all Ecological Costs

With regard to the costs side of the equation, EPA's analysis, although more complete than for benefits, does not fully address the suite of environmental risks posed by pesticides. It is not clear why EPA has focused its attention on human health risks and given short shrift to ecological concerns. While EPA certainly has the legal authority to address ecological concerns, its focus continues to be on regulating human health concerns. Nowhere is this more apparent than in the area of pesticide regulation. Although a full evaluation of this issue is beyond the scope of this chapter, the history of the EPA pesticide program may be a factor in creating such a bias. Prior to the creation of EPA, pesticides, to the extent they were regulated, were under the jurisdiction of the US Department of Agriculture. When EPA was formed, many USDA employees working on pesticide issues moved over to the EPA pesticide office. Perhaps this created a culture focused on promoting agriculture even at the expense of ecological harm. Another explanation for this institutional bias may simply be that, from a political standpoint, it is easier to justify imposing costs on industry or farms to protect human health than it is to protect ecological resources.

Most of the indirect environmental and economic costs of pesticide use are not considered.[103] Environmental and economic costs not typically addressed in any meaningful way in pesticide cost/benefit analyses include: domestic animal poisonings and contaminated products, destruction of beneficial natural predators and parasites, honeybee and wild bee poisonings and reduced pollination, crop and product loss, groundwater and surface water contamination, fishery losses, adverse non-lethal effects on wild birds and mammals, adverse effects on microorganisms and invertebrates, and adverse effects on ecosystem services. In 1993, Professor David Pimentel of Cornell estimated that if the full environmental and social costs of pesticide use, including indirect effects, were taken into account, the environmental and social costs of pesticide use would be significantly greater than

101 *See* Review of the Administration's Pesticide Policy; Hearing Before the Subcomm. on Department Operations and Nutrition, 103rd Cong. 57 (1993).

102 *See id.*

103 *See generally*, David Pimentel et al., *Assessment of Environmental and Economic Impacts of Pesticide Use*, *in* THE PESTICIDE QUESTION: ENVIRONMENT, ECONOMICS AND ETHICS 47, 68–9 (David Pimentel & Hugh Lehman, eds, 1993).

$8 billion per year.[104] Pimentel estimated that of this $8 billion per year, users of pesticides in agriculture pay only approximately $3 billion, leaving the remaining $5 billion to be borne by society.[105] Moreover, Pimentel points out that many of the true costs of pesticide use are either not well understood or difficult to quantify. Thus, the true cost of pesticide use many be even higher. Unfortunately, very few of these costs are considered by EPA when conducting the cost/benefit analysis it relies on to register a pesticide. Moreover, in the manner that EPA has interpreted FIFRA, even a very small risk may warrant cancellation of a pesticide, provided that the benefits are relatively low.[106] Presumably, even a very high risk may not trigger cancellation if the economic benefits to be achieved are very high. By this logic, if a pesticide poses a great economic benefit, high risks to vulnerable species or ecosystems will be tolerated.

FIFRA's cost/benefit approach also raises concerns with regard to pesticidal GMOs. EPA's cost/benefit approach raises particular concerns with regard to pesticidal GMOs that are exported, or that spread naturally, to other parts of the world. For example, one risk concern with plant-incorporated protectants is whether a favorable ecological niche exits for the modified plant in its new environment that will enable the plant to thrive, and perhaps become a pest or disrupt the balance of ecosystems. EPA may evaluate a particular pesticidal GMO under FIFRA, and may find that it will not pose an unreasonable environmental risk in the US because a suitable niche does not exist for the modified plant in the US that would allow the creation of a pest. To determine whether the pesticidal GMO poses an unreasonable risk in another country, however, the exporting company (or EPA) would be required to identify the types of ecosystems that exist in that country and the likelihood that the modified plant would thrive or spread in that environment. To do this, the exporting company (or EPA) would have to address a wide array of issues to determine whether the pesticide produced by the plant gives it a selective advantage in the new environment. Such a determination would depend on a number of considerations such as whether the modified plant has wild relatives in the new environment, how the modified plant is affected by factors such as climate, what selective pressures (e.g., viruses or other pathogens that normally keep the plant population in check) exist in the new environment, and how the modified plant interacts with the types of species present in the various ecosystems of the importing country. It would seem to be too burdensome to require such a site-specific risk assessment for every country that imports US pesticides. Nevertheless, international concerns could be averted if the United States were to participate in international efforts to provide advanced informed consent prior to exporting GMOs to other countries. Such a process is contemplated by both the

104 *Id.* at 72.

105 *Id.*

106 In the Matter of Protexall Products, Inc., et al., FIFRA Docket Nos. 625, et al. (1989) (upholding cancellation of sodium arsenate ant baits which posed a small risk of minor illness to children because the benefits of the pesticide were minimal).

Rio Convention on Biodiversity and the Cartagena Biosafety Protocol, to neither of which the United States is a party.

Limitations in Data Requirements

As described in more detail above, EPA requires applicants for registration to submit a considerable amount of data to support registration. However, these data requirements are extremely limited with regard to understanding the impacts to individual members of wildlife species, populations or ecosystems. Data requirements geared toward understanding impacts to wildlife are limited primarily to acute toxicity concerns, with limited data requirements to address reproductive or other sublethal effects. For example, EPA's data requirements are not designed to address issues related to animal growth or subacute conditions, which may adversely affect the ability of the individual to find food or to find and compete for mates. Moreover, EPA's data requirements rely on a few surrogate species on which tests are conducted. Data results from tests conducted on these surrogate species do not necessarily translate well to other species. Thus, although EPA's data requirements may be sufficient to determine acute toxicity and some other limited impacts to individual organisms, they do not include tests designed to assess population survival. Even more significant from an ecological perspective however, is that EPA's risk assessment does not typically include population models, let alone ecosystem-level models. Not only does EPA not consider population- or ecosystem-level concerns in its typical risk assessment for purposes of determining whether to register a pesticide for a particular use, but even where EPA has determined that a particular pesticide use may affect a threatened or endangered species and conducts an "Effects Analysis," it does not typically utilize population models. By not considering population effects, and instead focusing only on a limited number of individual effects, EPA is only looking at one aspect of ecological effects at one level of organization and is not conducting a true ecological risk assessment, which would consider population-level effects at minimum. Perhaps most important in this regard, EPA's current data requirements are not adequate to evaluate ecological risks posed to wildlife species, and in particular ESA listed species, or ecosystem services such as pollination, decomposition, nitrogen fixation, and other ecological services provided by organisms that may be affected by pesticides. Currently, EPA's data requirements for pesticide registration address only a limited number of these concerns.[107]

The minimum data requirements for registration, experimental use permits, and reregistration are set forth in 40 C.F.R. § 158. More detailed standards for conducting tests, guidance on evaluation and reporting of data and additional guidance is provided in a series of advisory documents that EPA makes available to applicants and the public.[108] In its data requirement rules, EPA identifies some

107 *See generally* 40 C.F.R. § 158 (2011).

108 *See id.* § 158.20(c) (2011).

data as required and other data as "conditionally required." Conditionally required data are required only if the product's proposed pattern of use, results of other tests, or other factors meet the criteria specified in the rules.[109] EPA's rules also allow certain data requirements to be waived if they are not applicable to the particular pesticide or use.[110] In addition, EPA's rules set forth varying data requirements for minor use of a pesticide (i.e., used on a minor crop) and biochemical and microbial pesticides.[111] EPA's data requirements for testing for ecological effects are limited, and many such data requirements have not yet been developed. In addition to substantial data requirements related to product chemistry,[112] much of EPA's requirements regarding data necessary to support the registration of a pesticide are focused on human health effects. For example, EPA has extensive data requirements related to residue chemistry to estimate human exposure to pesticides, acute human hazard, subchronic human hazard, chronic human hazard, mutagenicity, metabolism studies, reentry hazard, spray drift evaluation, as well as oncogenicity, teratogenicity, neurotoxicity, and reproductive effects in humans.[113]

EPA's environmental fate data are designed to "assess the presence of widely distributed and persistent pesticides in the environment which may result in loss of usable land, surface water, ground water, and wildlife resources, and assess the potential environmental exposure of other nontarget organisms, such as fish and wildlife, to pesticides."[114] Environmental fate studies include studies to determine the rate of pesticide degradation,[115] metabolism studies to determine the nature and availability of pesticides to rotational crops and to aid in the evaluation of the persistence of a pesticide,[116] mobility studies pertaining to leaching, adsorption/desorption and volatility of pesticides,[117] dissipation studies,[118] and accumulation studies.[119] Perhaps because environmental fate data are used to evaluate human exposure to pesticides, as well as wildlife exposure, these data requirements appear to be fairly comprehensive. However, EPA's data requirements regarding effects on wildlife and non-target organisms are considerably less comprehensive.

---

109 *See id.* §§ 158.25(a) and 158.101 (2011).

110 *See id.* § 158.25(b) (2011) (setting forth policy on flexibility and waiver); 40 C.F.R 158.35 (describing the flexibility in data requirements) and § 158.45 (regarding waiver of data requirements).

111 *See id.* §§ 158.60 and 158.65 (2011), respectively.

112 *See id.* §§ 158.150, 158.153, 158.155, 158.160, 158.162, 158.165, 158.167, 158.170, 158.175, 158.180, and 158.190 (2011).

113 *See id.* paragraphs 158.202(a), (c), (e), (f), and (g) and *id.* §§ 158.240, 158.390, 158.440 and 158.340 (2011). *See also id.* § 158.34 (2011) (providing that certain human health effects data submitted to EPA must be flagged as indicating potential adverse effects).

114 *Id.* § 158.202(d)(1) (2011).

115 *Id.* § 158.202(d)(2) (2011).

116 *Id.* § 158.202(d)(3) (2011).

117 *Id.* § 158.202(d)(4) (2011).

118 *Id.* § 158.202(d)(5) (2011).

119 *Id.* § 158.202(d)(6) (2011). *See also id.* § 158.290 (2011).

With regard to ecological effects, EPA has, by rule, developed limited data requirements designed to evaluate wildlife and aquatic organisms, environmental fate, and non-target insects. EPA's data requirements for effects on wildlife and aquatic organisms include avian toxicity studies,[120] and freshwater fish and invertebrate acute toxicity studies[121] for most pesticides intended for outdoor use. For most outdoor uses, wild mammal toxicity, avian reproduction, simulated and actual field testing of mammals and birds, acute toxicity to estuarine and marine organisms, aquatic invertebrate life cycle, fish life cycle and aquatic organisms accumulation, and simulated or actual field testing of aquatic organisms are only conditionally required. Conditionally required studies are conducted only on a case-by-case basis depending on the results of lower tier studies, such as acute and subacute testing, intended use patterns and environmental fate characteristics, or whether certain specified criteria are met. [122] As can be seen by EPAs primary focus on acute toxicity testing, EPA does not generally require data related to potential adverse effects of pesticides on wildlife behavior, neurology, reproduction, birth defects, or other non-acute effects. Moreover, EPA's data requirements do not contain any studies aimed at evaluating effects on other species, such as amphibians or reptiles, or other species not specifically identified in the rules. Accordingly, EPA's wildlife data requirements should be revisited to determine what additional data requirements should be included to more fully address the wide range of potential effects on fish and wildlife that may result from exposure to pesticides.

Likewise, EPA's data requirements for non-target insects are limited. EPA does conditionally require acute toxicity testing for honeybees and other pollinators if the proposed use will result in honeybee or other pollinator exposure, but EPA does not have any data requirements related to honeybee subacute feeding studies,[123] non-target aquatic insects, or non-target predatory or parasitic insects.[124] Moreover, EPA does not have any data requirements whatsoever related to soil microorganisms, which provide critical ecological services such as decomposition and nitrogen fixation, or any data requirements designed to evaluate the effects of pesticides on any other ecological services.

---

120 Avian oral LD50 and dietary LC50s are required using the preferred test animal species, the mallard and the bobwhite. *Id.* § 158.490 (2011).

121 Freshwater fish LC50 studies are required, with the preferred test species being the rainbow and bluegill fish and acute LC50 studies are required on freshwater invertebrates, with the preferred test species being Daphnia. *Id.* § 158.490 (2011).

122 *Id.* § 158.490 (2011).

123 In its data requirements rule, EPA identifies this type of requirement as "reserved pending development of test methodology." 40 C.F.R. § 158.590 (2011).

124 In its data requirements rule, EPA identifies these types of requirements as "reserved pending further evaluation to determine what and when data should be required, and to develop appropriate test methods." *Id.* § 158.590 (2011).

The Limitations of Regulating Use Through Labels

Although EPA's data requirements do include some studies designed to evaluate risks to fish, wildlife, aquatic organisms and nontarget insects, EPA's primary purpose in requiring such studies is not to determine *whether* to register a pesticide product, but instead is to "provide data which determines the need for (and appropriate wording for) precautionary label statements to minimize the potential adverse effects to nontarget organisms."[125] However, label requirements do not always provide sufficient protection against the environmental harms resulting from pesticides use.

Despite all of the testing and labeling that EPA imposes, large numbers of birds, insects, amphibians, and aquatic species, including threatened and endangered species, continue to be harmed by EPA-registered pesticides.[126] While many label requirements that EPA imposes are aimed at protecting human users of pesticides,[127] as well as other humans, such as children,[128] from accidental poisonings, EPA does require certain environmental hazard information to appear on pesticide labels. For example, if a pesticide intended for outdoor use contains an active ingredient with a specified level of acute mammalian or avian toxicity, the label must bear a precautionary statement, such as "This pesticide is toxic to wildlife." [129] If either accident history or field studies demonstrate that the use of the pesticide may result in fatality to birds, fish or mammals, the pesticide label must bear the precautionary statement, such as "This pesticide is extremely toxic to wildlife (fish)."[130] Similarly, if a product intended for certain uses contains an active ingredient that is toxic to pollinating insects, the label must bear an appropriate caution.[131] Finally, if a product is intended for outdoor use other than aquatic applications, the label must bear the precautionary statement "Keep out of

---

125 *Id.* § 158.202(h)(1) (2011).

126 *See* Brian Litmans & Jeff Miller, Silent Spring Revisited: Pesticide Use and Endangered Species, Center for Biological Diversity Report (2004), *available at* http://www.biologicaldiversity.org/swcbd/PROGRAMS/science/pesticides/REPORt.pdf.

127 *See* 40 C.F.R. §§ 156.10 (general labeling requirements); 156.60 (human hazard and precautionary statements); 156.62 (human hazard toxicity categories); 156.64 (signal words for human hazard toxicity categories); 156.68 (first aid statement); 156.70 (precautionary statements for human hazards); 156.78 (precautionary statements of physical or chemical hazards); and §§ 156.200–156.212 (worker protection statements).

128 *See id.* § 156.66 (child hazard warning).

129 The specified level of acute toxicity for mammals warranting such a statement is an oral LD50 of 100mg/kg or less. The specified level of acute toxicity for fish warranting such a statement is an LC50 of 1ppm or less. The specified level of acute toxicity for birds warranting such a statement is an oral LD50 of 100 mg/kg or less or a subacute dietary LC50 of 500 ppm or less. *Id.* § 156.85(b) paragraphs (1), (2) and (3) (2011), respectively.

130 *Id.* § 156.85(b)(4) (2011).

131 *Id.* § 156.85(b)(5) (2011).

lakes, ponds or streams. Do not contaminate water by cleaning of equipment or disposal of wastes."[132]

Although EPA does require these precautionary statements on labels, it is unclear what effect they have in practice. For example, if a farmer intends to apply a particular pesticide to combat a particular pest of her crops, and the pesticide label indicates it is toxic to wildlife, how will this information influence the farmer's behavior? It is unlikely that the farmer will choose not to apply the pesticide, because virtually all of the major chemical pesticides used in agriculture today are acutely toxic to at least some non-target organisms. It is difficult to imagine that a statement on a label indicating that a product is toxic to wildlife will have any significant influence on user behavior. Without more specific directions about when, where or how it is appropriate to apply the pesticide to minimize risks to wildlife, the farmer is left with an essentially useless warning. The lack of useful directions to minimize risk is likely due to the fact that because most chemical pesticides are acutely toxic to at least some wildlife, it is impossible to release them into the environment in large amounts without creating the possibility for harm to wildlife.

EPA's Limited Ability to Consider Local Conditions

Currently, there is no federal system in place, and only very limited state or local systems in place, that regulate the "uses" of pesticides registered under FIFRA other than the label instructions on each registered pesticide. These label instructions are generally the same nationwide. Thus, there is not currently a widespread system for considering local factors to determine what pest control method should be used in what location under what circumstances.

Although FIFRA provides a regulatory system that applies to any pesticide sold or distributed in the United States, FIFRA does not generally preempt state or local government regulation of pesticide use. In 1991, the right of a local government to regulate pesticide use was clearly established by the US Supreme Court in the case of *Wisconsin Public Intervenor* v. *Mortier*.[133] In that case, a Wisconsin local government had adopted an ordinance which required a permit from the local government prior to certain types of pesticide use. Prior to the *Mortier* decision, only a small number of states had laws in place that preempted local governments from regulating pesticide use. After *Mortier*, all but 11 states have laws preempting local regulation of pesticides.[134]

Justice White, writing for the majority in the *Mortier* decision, recognized the benefit of local decision making for the actual use of pesticides when he wrote: "FIFRA nowhere seeks to establish an affirmative permit scheme for the actual use

---

132 *Id.* § 156.85(b)(6) (2011).

133 501 U.S. 597 (1991).

134 Laura A. Haight, *Local Control of Pesticides in New York: Perspectives and Policy Recommendations*, 9 Alb. L. Envtl. Outlook J. 37, 39 (2004).

of pesticides. It certainly does not equate registration and labeling requirements with a general approval to apply pesticides throughout the Nation without regard to regional and local factors like climate, population, geography, and water supply." [135] And yet, this is in practice what FIFRA does. Once a pesticide receives a FIFRA registration, unless a particular state actively seeks to further regulate such a pesticide, it can be used anywhere in the United States with the only limitation being that it must be used in accordance with the FIFRA label instructions. As is described more fully below, most states do not have detailed environmental permitting requirements for pesticide use. Although EPA attempts to impose risk-reducing measures on users through detailed labeling requirements, a set of instructions on a container that have been drafted to apply to the entire United States is a poor substitute for a site-specific, circumstance-specific decision on what pesticide to use where, when, and how.

More recently, the US Supreme Court in *Bates* v. *Dow Agrosciences*,[136] soundly rejected a broad view of FIFRA preemption. In this case, the Court clearly stated that nothing in FIFRA itself would prevent a state "from making the violation of a federal labeling or packaging requirement a state offense, thereby imposing its own sanction on pesticide manufacturers who violate federal law."[137] The Court acknowledged that pursuant to a prior Court decision, the term "requirements" in section 24(b) of FIFRA "reaches beyond positive enactments" such as statutes or regulations to embrace judge-made rules or jury verdicts, but that it is crucial to determine the scope of that preemption.[138] According to the Court, in order for a "requirement" to be preempted it must be a requirement for labeling or packaging and must be "in addition to or different from" those required under FIFRA.[139] In reviewing the plaintiffs' state law claims, the Court held that the common law claims for defective design, defective manufacture, negligent testing, and breach of express warranty were not requirements for "labeling or packaging" and thus were not preempted.[140] Even if a verdict in favor of the plaintiffs on such claims might induce the manufacturer to alter its label, the Court rejected such an "effects-based" test, choosing instead to focus on whether the elements of the common law claim impose labeling or packaging requirements more burdensome than federal law.[141] The Court concluded by confirming that FIFRA did preempt competing state labeling standards as well as "any statutory or common-law rule that would impose a labeling requirement that diverges from those set out in FIFRA and its implementing regulations."[142]

---

135 501 U.S. 597 (1991).

136 125 S. Ct. 1788 (2005).

137 Bates, 544 U.S. at 1797.

138 *Id.* at 1798.

139 *Id.* (citing 7 U.S.C. § 136v(b) (2000)).

140 *Id.*

141 *Id.* at 1799.

142 *Id.* at 1803.

The issue of localized decision making regarding pesticide use has arisen in another context in recent years. Beginning with the case *Headwaters, Inc., v. Irrigation District*[143] in 2002, the courts, as well as EPA have been grappling with the issue of whether and under what circumstances a National Pollutant Discharge Elimination Systems (NPDES) permit is required under the Clean Water Act for the application of pesticides into waters of the United States.[144] Historically, EPA had not required NPDES permits for such pesticide applications. Thus, although EPA, at least arguably, had the legal authority to require permits for applications of pesticides to waters, which would take into consideration localized environmental factors, it had chosen not to do so. After the *Headwaters* case, it appears that EPA is required to impose NPDES requirements on pesticides applications to waters of the US. Unfortunately, EPA has chosen to fulfill this obligation by simply issuing a nationwide "Pesticide General Permit," which eliminates EPA's ability to take into account local circumstances when permitting such pesticide application.

Nationwide decision making regarding pesticides can lead to disproportionate risks being placed on vulnerable populations of people, as well as on vulnerable species or ecosystems. While an overall cost/benefit analysis for a particular pesticide may weigh in favor of use of the pesticide, geographic or ecological "hot spots" may occur where the risks outweigh the benefits on those localized geographic areas or for those particular species. Without some way to bring localized concerns into the decision-making process, risks will continue to fall disproportionately on the most vulnerable human and ecological resources.

Although over one half of all registered agricultural pesticides are restricted-use pesticides,[145] which must be applied only under the supervision of a certified applicator, the certified applicator requirement does not provide the level of oversight, consideration of local ecological factors, and consideration of lower risk alternatives that would be needed to adequately address ecological risks. First, certified applicators do not necessarily directly oversee the application of pesticides, but instead typically serve as "arm chair" supervisors. Second, certified applicators are not required to receive any particular training in local ecological systems and their vulnerability to particular pesticides.[146]

---

143 243 F.3d 526 (9th Cir. 2001) (holding that the application of an aquatic pesticide to irrigation canals in compliance with the registration and labeling requirements under FIFRA, did not eliminate the need for an NPDES permit).

144 For a detailed discussion of the judicial decisions and EPA's position on the issue of requiring NPDES permits for aquatic pesticide application, *see* Kelly C. Connelly (case note), *Pesticides and Permits: Clean Water Act v. Federal Insecticide, Fungicide and Rodenticide Act*, 8 Great Plains Nat. Resources J. 35 (2003) and Paul Herran (case note), *Headwaters, Inc. V. Talent Irrigation District: Application of Aquatic Pesticides to Irrigation Canals, a Discharge, Which Requires a Clean Water Act Permit?* 25 Haw. L. Rev. 629 (2003).

145 Rodgers, *supra* note 90, at 458.

146 For a description of certified applicator training programs, *see* Rodgers, *supra* note 90, at 462–3.

Finally, although FIFRA section 11 requires EPA and states to make available to certified applicators at their request instructional materials concerning Integrated Pest Management (IPM), the statute expressly states that certified applicators are not required to receive instruction on IPM and are not required to be shown to be competent with respect to such techniques.[147] Thus, certified applicators are not required to know about less risky pest control techniques, let alone to consider them in making decisions regarding which options to choose to control a particular pest. In fact, a certified applicator's job is not to decide what approach to take to control a pest, but is merely to ensure that once a particular pesticide is chosen, it is applied properly in accordance with label instructions.

Unfortunately, despite compelling reasons for considering local conditions in determining what pesticides to use in what locations and under what conditions, and the fact that FIFRA allows state and local governments to regulate pesticide use, state regulation of pesticides tends to be minimal. Most states do not have regulatory systems in place that are any more environmentally protective than FIFRA's basic nationwide protections. Most states, in fact, do not have any significant regulations addressing the use of pesticides under localized conditions. For example, most states do not require site-specific permits to be obtained before a pesticide can be applied, even for large-scale agricultural pesticide application into the environment. Likewise, most states do not require anyone with specialized knowledge of the presence of threatened or endangered species or rare or sensitive ecosystems to make any evaluation prior to the release of pesticides into the environment. Some states, however, do require certified applicators to have knowledge of how pesticides impact wildlife, water resources, and other ecological resources. For example, the state of Nevada requires certified applicators to have an understanding of the ecological impacts of pesticides. For agricultural pesticide application, Nevada law provides "[p]ractical knowledge is required concerning soil and water problems, preharvest intervals, phytotoxicity and potential for environmental contamination, non target injury and community problems resulting from the use of restricted use pesticides in agricultural areas."[148] For aquatic and mosquito pest control, Nevada requires certified applicators to demonstrate a "practical knowledge of the potential effects of pesticides on plants, birds, fish and other organisms in aquatic environments, and methods of application that avoid unacceptable contamination of the habitat and exposure to people and animal life." Interestingly, Nevada law also requires a consideration of the comparative risks associated with pesticides when deciding which pesticide to use. Specifically, the law requires a certified applicator, when choosing between two or more equally effective pesticides, to select the "least toxic material if possible."[149] However, it is not clear whether certified applicators have the knowledge to determine

147 7 U.S.C. § 136i(c) (2011).

148 Nevada Admin. Code Chapter 555(B)(1)(a). *See also* http://www.unce.unr.edu/publications/files/ag/other/sp8707.pdf (last visited June 5, 2011).

149 *Id.*

which pesticides are "equally effective" or to determine which are "least toxic," particularly when one considers that there are many different forms of toxicity and risk that are not easy to compare.[150] This type of directive, while well-intentioned, is not easy to enforce.

In addition, some states do have limited permitting requirements for pesticide use, however these requirements generally apply only to aerial application of pesticides and a permit is not typically issued for each application. For example, in Hawaii, a permit is required prior to aerial application of pesticides.[151] However, the permit can be issued for repeated uses or for a specified length of time.[152] Consequently, changing local environmental conditions are not likely to be adequately addressed for each application. In Massachusetts, a permit is required for the aerial application of pesticides, however, the permit is for a one-year duration and is not specific to the date or time of application.[153] Nevertheless, a site inspection is required prior to permit issuance, which presumably means that local conditions are assessed prior to issuing the permit.[154] In addition, in Massachusetts, a special permit is required for application of restricted-use pesticides to an area greater than 25 acres.[155] Similarly, in Vermont, one-year duration permits are required for aerial application of pesticides.[156]

What state pesticide regulatory programs generally do is treat farmers as "private applicators," who are required to obtain a state certification that demonstrates that they have attained a level of practical knowledge of pesticide use and hazards, before such farmer is allowed to apply pesticides. Some states have more stringent requirements for chemigation, fertigation and aerial application,[157] but generally,

---

150 *Id.*

151 *See* Haw. Admin. Rules § 4-66-64 (2004).

152 *Id.* at § 4-66-64 (a)(4).

153 *See* Mass. Regs. Code tit. 333, § 13.05(3)(b) (2004).

154 *Id.*

155 *See id.* § 13.03(18).

156 *See* Vt. Code R. 20 031 012 § IV (5) (2003).

157 *See, e.g.*, Fla. Admin. Code Ann. r. 5E-9.019(2)(a) (2005) (aerial applicators require a separate certification); Haw. Admin. Rules § 4-66-64(6) (2004) (mandating a notice requirement prior to any aerial application); Idaho Admin. Code 02.03.03.310, 02.03.03.320 (2004) (placing specific restrictions on aerial application geared primarily toward minimizing spray drift); 302 Ky. Admin. Regs. 27:050 § 2(8) (2003) (requiring separate certification for aerial applicators); La. Admin. Code tit. 7, Part XXII § 145 (A) (iii-vii) (2005) (detailing specific requirements regarding aerial application, including the requirement to adjust flight patterns to avoid sensitive areas and prohibitions on aerial application if wind velocity exceeds 10 mph or if it is raining or rain is imminent); Mass. Regs. Code tit. 333, § 13.05(3)(b,c) (2004) (requiring a permit for aerial application, as well as prohibiting aerial application within 400 feet of a water body and requirements that aerial spraying cease if spray drift occurs); Mich. Admin. Code r. 285.637.10 (2005) (detailing specific requirements regarding aerial spray drift, including a requirement for applicators to have a written spray draft plan in place and restrictions to protect sensitive areas including

once a farmer has obtained the required training and certification, he is free to apply pesticides at will provided it is done in accordance with FIFRA or state label directions and restrictions. What's more, generally, the certified private applicator can "supervise" the application of pesticides by non-certified persons. In most states, "direct supervision" means that the certified private applicator is within a telephone call away from his supervisees.[158] Some states do have specific pesticide regulations aimed at protecting groundwater or surface water bodies. However, even these states do not generally require any form of site-specific permitting evaluation prior to pesticide use.

---

all water bodies); NEB. ADMIN. CODE tit. 25, chp. 2 005.03 (2002) (mandating separate certification requirements for application of restricted-use pesticides via chemigation); N.C. ADMIN. CODE tit. 2, r. 9L.1002, 9L.1003 (2004) (detailing spray drift control regulations and a prohibition on aerial application to restricted areas, including water bodies); VT. CODE R. 20 031 012 § IV (1)(b) (2003) (detailing rules regulating pesticide drift); WASH. ADMIN. CODE §§ 16-202-2001–16-202-2021 (2004) (detailing rules regulating fertigation); WIS. ADMIN. CODE § 29.50(2) (2003) (listing restrictions to prevent significant spray drift).

158 See, *e.g.*, FLA. STAT. ANN. § 487.021(64) (2004) ("under direct supervision" means that the licensed applicator is available "if and when needed"); IDAHO CODE § 22-3401(34) (2004) (stating that a supervisory certified applicator must be available as needed, even though the "certified applicator is not physically present at the time and place the pesticide is applied"); NEB. REV. STAT. § 2-2642(5)(c)(iv) (2002) ("direct supervision" defined as being available by voice or electronic means and the ability to be physically present if needed); VA. CODE ANN. § 3.1-249.27 (2004) ("under the direct supervision of" means the certified applicator is responsible for supervisees' actions); and WIS. STAT. § 94.67(10m) (2003–2004) (allowing private applicators to "direct the use of" restricted-use pesticides, provided they are available if and when needed during the application); MD REGS. CODE tit. 15.05.01.01(b)(23)(b) (2005) (stating that the supervisory certified applicator must be available is needed); N.J. ADMIN. CODE tit. 7, § 30-1.2(2005) (defining "available if and when needed" to mean that the direct supervisor is in constant voice contact and no more than 3 hours away via ground transportation); OR. ADMIN. R. 603-057-0001(10) (2004) ("immediate supervision" defined as no more than five minutes away). Nevertheless, a few states do have more stringent supervisory requirements for certified applicators. *See, e.g.*, IND. ADMIN. CODE tit. 355, r. 4-2-3 (2004) (stating that, if an applicator has never worked with a pesticide before or is not a "registered technician," the certified applicator must be both physically present and in direct voice contact during the application; for registered technicians, the certified applicator can be available through telephone or walkie-talkie); MASS. REGS. CODE tit. 333, § 10.07 (1996) (requiring that the supervision of non-certified applicators increase based upon the hazard of the given pesticide being used); N.Y. COMP. CODES R. & REGS. tit. 6, § 325.7 (2004) (defining "under the direct supervision" to mean that the certified applicator must be physically present and within voice contact); and VT. CODE R. 20 031 012 § I(7)(17) (2003) (certified applicators must be physically present and actively supervise non-certified applicators).

Chapter 12
The Conflict between Pesticide Law and Wildlife Protection Law

The problem with EPA's inability to effectively regulate the use of pesticides under specific geographic and temporal conditions is highlighted by the conflict that exists with regard to FIFRA and wildlife protection statutes, such as the Endangered Species Act (ESA) and the Migratory Bird Treaty Act (MBTA). In the decades since the ban of DDT and its relatives, pesticides have caused the deaths of literally millions of birds, fish, and other wildlife, and have placed hundreds of threatened and endangered species at risk of extinction. The Center for Biological Diversity recently released a report that found that EPA has approved registrations for pesticides that put more than 375 Endangered Species Act (ESA) listed species at risk.[1] In the past several years, EPA has come under considerable attack for its failure to adequately address risks to certain species listed under the ESA[2] from pesticide uses. Unfortunately, the laws governing pesticides conflict in a number of significant ways with the laws designed to protect wildlife, including threatened and endangered species. Specifically, the species protection laws and pesticide laws differ dramatically in their goals, standards, focus, and methods, making it virtually impossible for EPA, as the agency charged with implementing the pesticide laws, to comply with species protection laws. When coupled with EPA's institutional bias in favor of registering pesticides, its generic bureaucratic inertia, and the George W. Bush Administration's hostility towards species protection, these conflicting laws have thrown EPA into a chaotic mixture of defensive entrenchment and regulatory paralysis, resulting in an utter failure to comply with the species protection laws such as the ESA. The unwitting casualties of EPA's failure are the countless species, including threatened and endangered species, that have been placed in harm's way.

Although the legal wrangling over the pesticide/species protection conflict has simmered for decades, in the past several years, the battles have intensified. Recently, a number of environmental organizations have sued EPA over its failure to comply with the ESA. In response to such suits, EPA has evoked every legal defensive strategy and become even more entrenched in its position of non-compliance with the ESA. All three branches of the federal government have

---

1 Brian Litmans & Jeff Miller, Silent Spring Revisited: Pesticide Use and Endangered Species, Center for Biological Diversity Report 19–44, (2004), *available at* http://www.biologicaldiversity.org/swcbd/PROGRAMS/science/pesticides/REPORt.pdf.

2 15 U.S.C. §§ 1531–1599 (2011).

entered the fray, with the judiciary attempting to resolve the conflicts inherent between the ESA and FIFRA, the regulatory agencies attempting to regulate their way out of ESA compliance, and Congress attempting to make the problem go away by exempting pesticide regulatory decisions from the ESA. To complicate matters more, in July 2007 the Supreme Court ruled in *National Ass'n of Home Builders* v. *Defenders of Wildlife*,[3] that federal agencies are not required to undergo the consultation process provided for in section 7 of the ESA unless their action is a discretionary one. Thus, the relationship between federal pesticide law and federal species protection law has been thrown into even greater disarray.

Starting in the late 1970s, the US brought a number of lawsuits involving the issue of pesticides harming protected species. The earliest cases involved liability under the MBTA. In *United States* v. *FMC Corporation*,[4] the Second Circuit considered whether the defendant corporation's manufacture and subsequent release of toxic pesticides that killed 92 migratory birds violated the MBTA.[5] The government pursued 36 individual violations pursuant to specific dates of bird kills, of which the jury found the corporation guilty on 18 counts.[6] The MBTA makes it unlawful to kill migratory birds and makes such unlawful killings a misdemeanor.[7] The key issue was whether a violation of the MBTA could be predicated on an intentional action without specific intent to kill protected birds, or if a violation required that the defendant possess *mens rea* to kill migratory birds.[8]

FMC owned a pesticide-manufacturing facility in upstate New York, producing and storing dithiocarbamate pesticides on site.[9] A 10-acre pond on site served as a wastewater conduit for pesticide production.[10] Although pesticides were to be treated before entry into the pond, pesticides were being dumped directly into the pond; as a result the water body had a concentration of carbofuran 200 times greater than the amount required to cause a significant probability of bird kills.[11] Unfortunately, the size of the water body attracted migrating birds, and as a result of the high concentration levels, numerous bird deaths were found over a two-month span in 1975.[12]

The Second Circuit acknowledged that hunting issues had predominated MBTA cases, and that these were fairly inapposite to the instant case.[13] The court noted further, however, that even though FMC did not know that its release of

---

3 Nat'l Ass'n of Home Builders v. Defenders of Wildlife, Slip opinion No. 06-340, 127 S.Ct. 2518 (Decided June 25, 2007).

4 572 F.2d 902 (2d Cir. 1978).

5 *Id.* at 903.

6 *Id.*

7 *Id.*

8 *Id.* at 904.

9 *Id.*

10 *Id.*

11 *Id.* at 905.

12 *Id.*

13 *Id.*

carbofuran was the cause of the bird deaths, FMC's product directly caused the deaths.[14] FMC argued that it not only lacked the intent to kill the birds, it also took no affirmative act causing the deaths and, as a result, did not violate the MBTA. The court disagreed, noting that the "term 'act' itself is ambiguous," and citing cases in which even a failure to act in light of a duty itself creates liability as a criminal action.[15] The court stressed that FMC had acted when it manufactured a toxic pesticide and that FMC had failed to prevent the pesticide's release in spite of its affirmative duty.[16] The court analogized the release to a strict liability offense in tort law, citing the Restatement (Second) of Torts, which made individuals carrying on abnormally dangerous activities liable for injury regardless of due care shown.[17] Thus, the court reasoned, FMC's manufacture of pesticides constituted an abnormally dangerous activity, and FMC's failure to prevent a toxic release leading to bird deaths made it strictly liable regardless of the care it might have shown.[18]

In another MBTA case, *United States* v. *Corbin*,[19] the Eastern District of California held that the Migratory Bird Treaty Act's prohibition on bird killings applied to poisonings even in the absence of actual intent to kill protected bird species.[20] In a lengthy opinion, the court considered whether the terminology of FIFRA with regard to certain provisions was unconstitutionally vague, and whether the appropriate basis for an individual violation of MBTA was each bird death or each act resulting in a bird death, regardless of actual birds killed.[21] The court held that FIFRA clauses pertaining to user liability were not unconstitutionally vague.[22] In light of the statutory ambiguities of the MBTA, the court agreed with the defendants that the appropriate number of charged counts should be predicated on the number of actual applications of the pesticide, not the number of birds killed as a result of that one act.[23]

The court first addressed whether the MBTA prohibited only hunting deaths by noting that while the Act was primarily intended by Congress to address hunting of migratory birds, section 703 of the MBTA, in broad language, made it illegal to kill a migratory bird in any manner.[24] Furthermore, the court noted, song birds were among the protected species even though such birds were not commonly

---

14 *Id.* at 906.
15 *Id.*
16 *Id.* at 907.
17 *Id.*
18 *Id.* at 908.
19 444 F.Supp. 510 (E.D. Cal. 1978).
20 *Id.* at 536.
21 *Id.* at 515–31.
22 *Id.* at 517.
23 *Id.* at 531.
24 *Id.* at 532.

hunted, indicating an intention by Congress to broadly safeguard the protected species.[25]

The defendants had also contended that violations of the MBTA required an intent on their part to kill migratory birds.[26] The court examined prior case law outside of the hunting context but found little of precedential value. The court briefly reviewed *United States* v. *Union Texas Petroleum*,[27] and *United States* v. *FMC Corp*[28] but found neither truly applicable, as *Union Texas* was settled with guilty pleas, whereas *FMC Corporation* involved a bench determination.[29] The court then focused on bait-hunting, noting that prior courts did not require knowledge of the poisoned bait itself as an element of the crime.[30] The defendants argued that bait-hunting materially differed from the instant conduct as placing baits inherently involved a party with intent to kill protected birds whereas the defendants claimed they lacked such intent.[31] The defendants predicated their argument on two fronts: first, by reference to the Bald and Golden Eagle Protection Act's (BGEPA) scienter requirement and, secondly, by claiming the government's MBTA rationale would require some individuals to be liable upon the death of *any* protected bird as a result of human contact.[32]

The court rejected both defenses. The court found that while the BGEPA has a scienter requirement, the MBTA is silent on this element. A defendant must have acted "knowingly or with wanton disregard for the consequences of (their acts)."[33] The second argument was easily rejected, as the court noted that the defendant, as a pesticide applicator, had far more control in terms of preventing the death of a bird than a random automobile driver who had struck a bird with his car.[34] The court analogized the scienter requirement to that of FIFRA, and stated "the guilty act alone [was] sufficient to make out the crime," and that requiring reasonable care was appropriate under the Constitution.[35] Because the reasonableness of the defendants' actions had not yet been decided at trial, the court did not address whether a defendant who had acted with reasonable care could be held liable without intent to kill.[36] The court then held that the MBTA could be applied constitutionally against defendants lacking intent to kill migratory birds.[37]

---

25 *Id.*
26 *Id.*
27 73-CR-127 (D. Colo. 1973).
28 572 F.2d 902 (2d Cir. 1978).
29 444 F.Supp. at 533.
30 *Id.*
31 *Id.* at 534.
32 *Id.* at 534–5.
33 *Id.* at 534.
34 *Id.* at 535.
35 *Id.* at 536.
36 *Id.*
37 *Id.*

While both *FMC* and *Corbin* imposed liability for pesticides that caused harm to protected species under the MBTA, neither addressed the issue of whether EPA, the regulatory agency responsible for registering pesticides, could be liable as a non-pesticide user based on its conduct in approving the registration of a pesticide later used by another party that caused harm to protected species. As described above, in *FMC* the liable party was the manufacturer of the pesticides that handled the waste products in such a way as to cause harm, and in *Corbin* the liable party was the user of the pesticide product. It was not until 1989 that a court addressed the issue of whether EPA could be liable for approving a pesticide registration that ultimately resulted in harm to a protected species.

In 1989, EPA was dealt a stunning blow when the Eighth Circuit found it liable for a take under ESA section 9 for allowing the continued registration of a pesticide that caused harm to a listed species. In *Defenders of Wildlife* v. *Administrator, Environmental Protection Agency*,[38] the court considered a citizen suit brought under the ESA against EPA for harm caused to listed species by the application of strychnine, a FIFRA-registered pesticide.[39] In the 1970s EPA reviewed the above-ground use of strychnine and consulted with the FWS about the impact of the pesticide on listed species.[40] The consultation culminated in a 1979 FWS Biological Opinion with findings that the continued use of above-ground strychnine would jeopardize listed species.[41] EPA initiated a cancellation process for several registered uses for strychnine after several environmental groups intervened in the FIFRA process. The groups, Defenders of Wildlife and the Sierra Club, intervened, as did the Farm Bureau, FWS and the United States Department of Agriculture.[42] The discussions among the parties continued from 1984 until 1986.[43] Most of the intervenors settled with EPA, but Defenders of Wildlife and the Sierra Club refused, and along with Friends of Animals filed suit under the ESA's citizens' suit provision, alleging that the continued registration of strychnine was a take under the ESA.[44]

EPA argued that because the plaintiffs sought cancellation of a pesticide, the plaintiff's suit had to be brought under FIFRA.[45] While the Eighth Circuit acknowledged that an action for pesticide cancellation alone should be sought under the FIFRA legislations,[46] the court held that FIFRA did not permit EPA to ignore the ESA when regulating pesticides.[47] The Eighth Circuit then considered

38 882 F.2d 1294 (8th Cir. 1989).
39 *Id.* at 1296.
40 *Id.* at 1296–7.
41 *Id.* at 1297.
42 *Id.*
43 *Id.*
44 882 F.2d at 1298 (8th Cir. 1989).
45 *Id.*
46 *Id.* at 1299.
47 *Id.*

whether EPA's continued registration of strychnine constituted an illegal taking under the ESA.[48] EPA did not dispute that the distribution of strychnine had caused the death of endangered species.[49] Noting that the definition of a take is quite broad and that distribution of strychnine could only occur upon registration of the pesticide, the court held that EPA action had caused the deaths of endangered species and, as a result, an illegal taking had occurred.[50]

It is important to note that EPA is not the only federal agency that has been reluctant to comply fully with section 7 of the ESA. For example, there has been an ongoing battle between environmental organizations and the Federal Emergency Management Agency (FEMA) regarding FEMA's ESA obligations in administering the National Flood Insurance Program (NFIP). In *Florida Key Deer* v. *Brown*,[51] the National Wildlife Federation, Florida Wildlife Federation and Defenders of Wildlife sued Michael Brown, in his official capacity as the head of FEMA, and Gale Norton in her official capacity as Secretary of the Interior, seeking a judicial order requiring FEMA to consult with the FWS concerning its NFIP for the Florida Keys. In 1994, the court ordered FEMA to consult with the FWS, and in 1997 the FWS issued a BiOp that the FEMA program "was jeopardizing" several endangered species. Among the nine jeopardized species were the Key deer, Key Largo cotton mouse, Key Largo woodrat, Key tree-cactus, and the Lower Keys marsh rabbit. After a series of subsequent consultations, each of which was challenged by environmental organizations, the court ultimately granted the plaintiffs' request to enjoin FEMA from issuing flood insurance for any new developments in the listed species' suitable habitats in Monroe County.[52]

As described in detail below, EPA's loss in the 1989 *Defenders of Wildlife* case did little to prod EPA into ESA compliance. Despite decades of noncompliance with the ESA, it was not until the early 2000s that a number of environmental organizations started to bring or threaten suit against EPA. Although most of these cases settled, the cases in which the courts rendered decisions demonstrate the courts' frustration with EPA's noncompliance with the ESA. Indeed, the courts have attempted to impose the types of geographic use restrictions on pesticide usage that FIFRA is so poorly designed to accommodate and which EPA has resisted implementing. In fact, as further described below, in one of the most extensively litigated cases, the court recognized that species protection must occur on a geographic basis and took it upon itself to impose very site-specific buffer requirements for spraying certain pesticides near water bodies containing certain listed salmon species.

In addition to the lawsuits against EPA for its failure to comply with the ESA when making FIFRA regulatory decisions, the early 2000s also saw a rush of

48 *Id.* at 1300.

49 *Id.* at 1301.

50 882 F.2d at 1301 (8th Cir. 1989).

51 364 F.Supp.2d 1345 (S.D. Fla. 2005).

52 Florida Key Deer v. Brown, 386, F.Supp.2d 1281, 1294 (S.D. Fla. 2005).

litigation against federal agencies that used pesticides that put listed species at risk as part of their federal land management practices. In *Oregon Natural Resources Council* v. *Keys II*,[53] plaintiffs filed suit against the Bureau of Reclamation (BOR), claiming that its application of acrolein and copper-containing pesticides in areas near the Lost River jeopardized the shortnose sucker and thus violated section 7(a)(2) of the ESA. After a series of consultations and lawsuits, BOR reinitiated consultation with the FWS mooting the ongoing litigation.[54] In *San Juan Audubon Society* v. *Veneman*,[55] plaintiffs filed suit against the Secretary of Agriculture and the Services (defendants), alleging that the defendants' use of sodium cyanide ejectors violated restrictions set by EPA as the ejectors were frequently misused and led to the deaths of the California condor and other endangered species. The sodium cyanide ejectors, labeled M-44s, are designed to kill predatory animals for the protection of livestock, but because of the nature of the device it also poisoned scavenging condors.[56] As a result, EPA regulated the use of M-44s and required the defendants to follow maps prepared by the FWS to prevent poisonings of listed species. The plaintiffs alleged that the defendants failed to follow these maps when applying M-44s in the condor's area.[57] The defendants filed a 12(b)(1) motion, arguing that the court lacked subject-matter jurisdiction over the dispute because there had not been a final agency action for the plaintiffs to appeal.[58] The court denied the motion, stating that the "particular agency actions at issue in this case are the defendants' approvals and individual decisions to place the M-44s" in areas causing harm to listed species without consulting the FWS maps as required by EPA. Furthermore, the placement of M-44s in the areas in question constituted the climax of the "defendants' decision-making process," and the failure to review the FWS maps when using the M-44s provided the plaintiffs with the right to sue.[59] Because the plaintiffs only challenged these discrete agency actions and not the M-44s program as a whole, the court held that the plaintiffs could sue under the APA and that the motion to dismiss was improper.[60]

The current wave of litigation over the wildlife impacts to protected species from pesticide use started in 2002, when 40 environmental groups, including the American Bird Conservancy and Defenders of Wildlife, sent the EPA a Notice of Intent to Sue for violations of the ESA, MBTA, and Administrative Procedures Act (APA) concerning the registration of the pesticide fenthion due to the high risks fenthion posed to a number of bird species. Later in 2002, the US Fish and Wildlife Service recommended that EPA cancel existing registrations for fenthion

53 2004 WL 1048168, page 1 (D. Or. 2004).
54 *Id.* at 2.
55 153 F.Supp.2d 1–3 (D.C. District 2001).
56 *Id.* at 3.
57 *Id.* at 4.
58 *Id.* at 5.
59 *Id.* at 6.
60 *Id.* at 7.

immediately, due to the unreasonable adverse effects fenthion posed to avian species protection under ESA[61] and MBTA.[62] EPA failed to take action to reduce the risks as requested by the plaintiffs and as recommended by FWS. Consequently, in October of 2002 Defenders of Wildlife, the American Bird Conservancy and the Florida Wildlife Federation filed suit against EPA in federal district court alleging EPA had violated the ESA and MBTA. EPA was let off the hook when in 2003 the manufacturer of fenthion voluntarily canceled its registration of fenthion, rendering the lawsuit moot.[63]

In September of 2004, environmentalists won a significant victory when the Ninth Circuit Court of Appeals issued a decision affirming a January 2004 US District Court for the Western District of Washington order, which found that EPA had violated the ESA because it had failed to take steps to ensure that the registration of 54 pesticides would not jeopardize the survival of listed salmon species. The court's ruling upheld the district court's injunction, which imposed detailed buffer zones restricting the use of more than 30 pesticides along listed salmon-supporting waters in California, Oregon, and Washington states.[64] The buffer zones adopted by the court came directly from the reasonable and prudent alternatives recommended by the Services in the 1989 BiOps. The attorney representing the environmental organizations in the case recommended to the judge that the minimum buffers from the 1989 BiOp would be appropriate because the Services had already evaluated the pesticides at issue and determined that use of these buffers would avoid jeopardy.

In its brief to the Ninth Circuit, EPA argued that the district court had erred in its decision that created mandatory buffer zones for application of the 55 active ingredients in question, and had required the EPA to provide written notifications to accompany pesticides sold in urban areas.[65] The notification was headed "Salmon Hazard," with the following text: "This product contains pesticides that may harm salmon or steelhead. Use of this product in urban areas can pollute salmon streams." EPA claimed several grounds for appellate relief, focusing mainly on the relationship of its FIFRA duties with its responsibilities under the

61 16 U.S.C. §§ 1532–1544 (2011).

62 16 U.S.C. §§ 703–711 (2011).

63 *See* Fenthion: Product Registrations Cancellation Order, 68 Fed. Reg. 55609–55611 (September 26, 2003). EPA approved the manufacturer's request to cancel Fenthion in May 2003. *See* 68 Fed. Reg. 32495–32497 (May 30, 2003).

64 *See* Wash. Toxics Coalition v. EPA, Case No. C01-013132C, Order issued January 22, 2004. This Order was the third in a series of orders granting injunctive relief to the environmental plaintiffs in this matter. *See* Wash. Toxics Coalition v. EPA, Case No. C01-013132C, Orders issued July 16, 2003 and August 8, 2003. All of these Orders are available on EPA's website at http://www.epa.gov/espp/litstatus/wtc/index.htm (last visited May 26, 2011).

65 EPA Brief to the Ninth Circuit in Washington Toxics vs. Environmental Protection Agency, 2004 WL 1763203, pages 12–13 (2004).

ESA.[66] EPA first argued that since it had already granted a FIFRA license, any action that would result in a cancellation or modification of that license must be according to the statutory requirements of FIFRA.[67] Furthermore, EPA claimed that FIFRA, when read in conjunction with the ESA, already took into account any concerns that registration might affect listed species.[68] EPA argued that although the ESA had a citizen's suit provision, it should not be read so as to provide citizen plaintiffs' greater ability to enjoin pesticide registration than EPA itself possessed. EPA also sought a determination that the plaintiffs had not fully exhausted their administrative remedies,[69] and that the district court, under the doctrine of primary jurisdiction, should have deferred to EPA as the agency with the necessary experience to fashion appropriate orders regarding the complexities of pesticide regulations. Under the doctrine of primary jurisdiction a court should abstain from ruling on certain issues that fall within the primary responsibilities of the acting agency.[70] EPA proposed that the citizen plaintiffs should have first petitioned EPA to suspend registration of the offending active ingredients, and only upon an EPA decision of that petition should a lawsuit have been allowed.

The Ninth Circuit rejected EPA's position, and agreed with the Eighth Circuit's logic in *Defenders of Wildlife* v. *EPA*,[71] stating that FIFRA does not allow EPA to exempt itself from the requirements of the ESA, and that EPA must comply with the ESA if its registration of pesticides will affect listed species.[72] The court held that while the statutes have different purposes and different calculations, EPA could not avoid its duties under the ESA simply "because it is bound to comply with another statute that has consistent, complementary objectives."[73] The ESA, on the other hand, provides a virtual blanket prohibition against the takings of endangered species. The court then summarily dismissed another of EPA's arguments; namely that EPA lacked discretion to cancel registration except under the statutory requirements of FIFRA.[74] The court simply noted that the cases EPA cited involved completed agency actions in which there was "no ongoing regulatory authority."[75] These cases were not dispositive, however, as the court noted that EPA under FIFRA maintains continuing discretion to register pesticides, or in the alternative, modify or even cancel registration.[76] The court similarly dismissed EPA's argument that the plaintiffs had not adequately exhausted their

66 *Id.* at 14–15.
67 *Id.* at 14.
68 *Id.* at 15.
69 *Id.*
70 *Id.* at 16, 27–8.
71 882 F.2d 1294 (8th Cir. 1989).
72 *Wash. Toxics Coalition*, 413 F.3d at 1032.
73 *Id.*
74 *Id.* at 1032–3.
75 *Id.* at 1033.
76 *Id.*

administrative remedies, noting that FIFRA does not require a plaintiff to exhaust a FIFRA remedy before seeking relief under another statute.[77]

EPA primarily argued that it was bound only to apply the provisions of FIFRA, which had its own statutory language relating to endangered species. Under FIFRA section 6(c)(1)-(2), EPA may suspend registration of a pesticide for an immediate hazard, which per FIFRA section 2(l) can include its effect on endangered species.[78] As such, EPA argued that it did not have an independent duty under section 7(a)(2) to consult with the FWS or the NMFS.[79] The Ninth Circuit upheld the district court's injunctive relief, noting that because it was the "maintenance of the 'status quo' that [was] alleged to be harming the endangered species,"[80] the injunction was appropriate pending EPA compliance with the ESA. Furthermore, the court placed the burden of proof on EPA to show that its action was non-jeopardizing to the listed species, finding that such burden-shifting was appropriate under the ESA for agency actions that have violated section 7(a)(2).[81]

In September 2005, CropLife America petitioned for writ of certiorari to the United States Supreme Court seeking to have the Ninth Circuit's decision in *Washington Toxics Coalition* v. *EPA* reversed.[82] The petitioner argued that the Ninth Circuit had misapplied the scope of judicial review pertaining to agency actions under the APA, FIFRA and section 7 of the ESA. Principally, the petitioner argued that the district court and Ninth Circuit misapplied the judicial standard of review by applying a *de novo* standard to the challenged pesticide regulations.[83] In conjunction with its main argument, the petitioner argued that since the EPA violation was procedural in nature, EPA was not substantively violating the ESA and therefore APA guidelines should have governed the judicial review.[84] Furthermore, the petitioner asserted that until EPA had made a determination that the registration of pesticides "may affect" a listed species, there was no duty to consult with the NMFS.[85] The petitioner's final argument criticized the lower court rulings for failure to apply FIFRA requirements to a pesticide suspension.[86] The majority of the petition argued, much like the petitioner's brief to the Ninth Circuit, that the lower court failed to apply the pre-set judicial standards and deference to the acting agency throughout the judicial review process.[87] The petitioner first noted that other circuits addressing the standard of review for the ESA's citizen suit provisions had

77 *Id.*
78 413 F.3d at 1031.
79 *Id.*
80 *Id.* at 1035.
81 *Id.*
82 CropLife America Petition for Writ of Certiorari No. 05-395, 1 (September 2005).
83 *Id.* at 11.
84 *Id.* at 19.
85 *Id.* at 23.
86 *Id.* at 25–6.
87 *Id.* at 11–20.

held that section 706 of the APA still applied (rather than a *de novo* standard).[88] Because the district court applied a *de novo* standard, the court did not review the administrative record with regard to each pesticide registration, which the petitioner argued was relevant with regards to the "unreasonable adverse effects on the environment" as required by FIFRA.[89] In addition, the refusal to apply the APA also meant that the APA remedies were not available—instead of simply compelling the EPA to make the appropriate effects determination, the petitioner argued, the court improperly enjoined the registration process and supplanted FIFRA.[90] Not applying the APA was improper, the petitioner claimed, because the ESA "does not specify a standard of review … and therefore the agency action continues to be governed by the arbitrary and capricious standard of the APA."[91] The petitioner further argued that the court did not appropriately delineate between a substantive and procedural violation by EPA of the ESA as it held EPA to be a regulated party rather than an administrative agency.[92] The petitioner argued that because EPA was only in procedural violation of the ESA (through its unreasonable delay in making effects determinations), the APA should have governed the cause of action.[93] The petitioner's non-APA argument first stated that the decision to consult with the NMFS must be predicated upon a decision by EPA that its actions may affect listed species.[94] Since there had been no such determination, consultation should not have been required. In its second non-APA argument, the petitioner claimed that FIFRA's substantive registration procedures, which provide for suspension due to an unreasonable hazard to a listed species, should not be bypassed by ESA injunction.[95] The petitioner argued that FIFRA provides the necessary due process projections and procedures that reflect the balancing concerns Congress meant for when it enacted the statute. As a result of these concerns, the petitioner argued, the Supreme Court should review the below case.

The Supreme Court declined to hear the Washington Toxics Coalition case. However, in June 2007 the Supreme Court decided another pivotal ESA case, which has potential implications for FIFRA. In *National Ass'n of Home Builders et al.* v. *Defenders of Wildlife*,[96] the issue presented to the Court was whether EPA must undergo consultation with the Services under section 7 of the ESA when determining whether to approve transfer of permitting-authority to a state

---

88 *Id.* at 11–12 (discussing the D.C., Fifth, Sixth, Eighth, Tenth and Eleventh Circuits' holdings that APA standards apply to ESA citizen suits).

89 *Id.* at 15.

90 *Id.* at 15–16.

91 *Id.* at 16.

92 *Id.* at 19.

93 *Id.*

94 *Id.* at 23.

95 *Id.* at 25–6.

96 National Ass'n of Home Builders v. Defenders of Wildlife, Slip opinion No. 06-340 (Decided June 25, 2007).

under section 402(b) of the Clean Water Act. In *Defenders of Wildlife*, the state of Arizona applied for a transfer of CWA permitting-authority to it under section 402(b). EPA argued that CWA section 402(b)'s mandatory nature precluded EPA from denying Arizona's application based on ESA considerations. It was not disputed that Arizona had met the nine specified criteria in section 402(b). The issue was whether EPA was required to determine whether under ESA section 7, its transfer decision would jeopardize listed species, thereby in essence adding a tenth required criterion to the list of nine in CWA section 402(b). In a 5–4 decision, the Supreme Court held that EPA was not required to undergo section 7 consultation or otherwise comply with the provisions of section 7 in granting a permit transfer under CWA section 402(b), because the decision to grant such a transfer is not a discretionary one. Of significant importance to the majority was the fact that CWA section 402(b) states that EPA "shall approve" a transfer if each of the nine specified criteria are met.

Because FIFRA section 3(c)(5) provides that EPA "shall register" a pesticide if EPA determines that the specified standards are met, it may appear at first glance that under *Defenders of Wildlife*, compliance with section 7 is not required when EPA makes a registration decision. However, there are significant differences in the requirements of CWA section 402(b) and those of FIFRA section 3(c)(5). CWA section 402(b) specifies an exclusive list of criteria that must be met for EPA to approve a transfer. Each of these criteria relates solely to the issue of whether the state applying for the transfer has the legal authority and other ability to carry out the permitting program. For example, the state must demonstrate that it has the ability to issue fixed-time permits and to ensure compliance with the CWA's substantive criteria, and that it has the ability to provide for public notice and public hearings. These criteria do not in themselves relate to whether a transfer (or permits issued under such a transfer) will jeopardize listed species. Thus, the Court concluded that although EPA could exercise some discretion in applying the criteria, it could impose a completely new criterion addressing listed species impact to the exclusive list of criteria related to legal, administrative, and procedural abilities.

The criteria in FIFRA section 3(c)(5) on the other hand are not limited to determining whether a state has certain legal, administrative and procedural abilities. Instead, this section authorizes EPA to make a determination regarding whether a pesticide will cause "unreasonable adverse effects on the environment," a term which is defined to include "all plants and man and other animals … and the interrelationships which exist among these." Thus, by its very terms, FIFRA section 3(c)(5) authorizes EPA to evaluate risks to "plants and animals," which inherently include threatened and endangered plants and animals. Although FIFRA uses the term "shall," the term is used to mandate that EPA consider, among other things, the impacts on listed species. As the Ninth Circuit stated in *Washington Toxics Coalition*, the mandates of the ESA are complimentary to the mandates of FIFRA. Thus, unlike in *Defenders of Wildlife*, where compliance with the ESA would require that a completely new criterion be added to the statutory list, ESA compliance is complementary to the mandate of FIFRA. Although the ESA is complementary to FIFRA, a FIFRA unreasonable adverse effects determination is not the same as the jeopardy determination that results from the section 7 process. Thus, although EPA is required to consider adverse effects to listed species, its analysis under FIFRA is not a substitute for the analysis under the ESA.

Specifically, under FIFRA, EPA is required to consider a large number of factors, including public health considerations, economics, and social impacts in determining whether a pesticide should be registered. Under section 7 of the ESA, the sole consideration is whether the action will jeopardize the continued existence of the species. In *Washington Toxics Coalition*, the court found that requiring EPA to undergo the section 7 consultation process prior to making an unreasonable adverse effects determination under FIFRA would provide the type of information and Services' expertise that would inform EPA's FIFRA decision making. The court acknowledged that informed decision making is one of the primary purposes of the consultation process. Accordingly, the court's rationale in *Defenders of Wildlife* would not extend to EPA's decisions under FIFRA section 3(c)(5), because unlike EPA's decision making under CWA section 402(b), under FIFRA, EPA is not only authorized to consider effects on plants and animals, it is *required* to do so. The mere fact that FIFRA uses the term "shall" is arguably irrelevant given the dramatically different mandates of CWA section 402(b) and those of FIFRA section 3(c)(5). Thus, it does not appear that *Defenders of Wildlife* in any way obviates or alters EPA's requirement to comply with section 7 of the ESA when making FIFRA registration decisions. Thus, the Ninth Circuit's holding in *Washington Toxics Coalition*, that EPA is required to comply with ESA section 7, is still good law. Further, under the 1989 *Defenders of Wildlife* case,[97] EPA continues to have potential section 9 liability for registering pesticides that actually take listed species.

Despite the overwhelming loss in *Washington Toxics Coalition*, initially the only risk-reduction action that EPA took in response to the court order was to require that a "Point of Sale Notification" be distributed in retail stores that sell the pesticides subject to the Order. This notice is merely a one page flyer with a photograph of salmon, which states: "Salmon Hazard: This product contains pesticides that may harm salmon or steelhead. Use of this product in urban areas can pollute salmon streams. This point of sale notification was produced in compliance with a January 22, 2004 court order, to notify urban users about the potential for **some** pesticides to harm fish" (see Figure 12.1).

After its dramatic loss in *Washington Toxics Coalition*, EPA stopped litigating suits brought to force the agency to comply with section 7 of the ESA in the FIFRA registration process and pursued a policy of settling these cases.[98] Some settlements actually occurred prior to the court decision in *Washington Toxics Coalition*. In

97 882 F.2d 1294 (8th Cir. 1989).

98 *See* Californians for Alternatives to Toxics, *available at* http://www.alternatives2toxics.org (last visited May 26, 2011).

Figure 12.1 EPA's point of sale notification

Californians for Alternatives to Toxics v. *EPA*,[99] EPA agreed, as part of the settlement, to make "effects determinations" for approximately 20 pesticides harmful to dozens of plant and salmon species by specified deadlines. One post-*Washington Toxics* settlement occurred in 2005, when EPA agreed to make "effects determinations" for six pesticides harmful to Barton Springs Salamander within specified time frames in response to a January 26, 2004 lawsuit against the agency brought by the Center for Biological Diversity and the Save Our Springs Alliance (SOSA). The suit, brought in the D.C. Circuit, alleged that EPA violated the anti-take provisions of the ESA when

99 Case No. C00-3150 CW (N.D. Calif.).

it registered six pesticides without reviewing the potential negative effects on the Barton Springs Salamander.[100] The pesticides in question were atrazine, diazinon, carabaryl, prometon, metolachlor and simazine.[101] In the lawsuit, the plaintiffs specifically charged that EPA had failed to comply with sections 7(a)(1) and 7(a)(2) of the Endangered Species Act,[102] which require federal agencies to consult with the Services to guarantee that agency action will not jeopardize the continued existence of any listed endangered or threatened species.[103]

Under the terms of the Settlement Agreement, EPA would make effects determinations relating to the Barton Springs Salamander for the six pesticides according to the following schedule: (1) for atrazine, 12 months from the Settlement Agreement's effective date; (2) for either carbaryl or diazinon, plus one additional pesticide listed in the complaint, 21 months (or 630 days) from the Settlement Agreement's effective date; (3) and for the remaining three pesticides, 25 months (or 760 days) from the Settlement Agreement's effective date.[104] The effects determination could have three potential outcomes for the individual pesticide's effect on the listed species: "no effect," "may affect but is not likely to adversely affect," or "may affect and is likely to adversely affect."[105] If EPA in its effects determination analysis makes a "may affect" or "likely to adversely affect" determination, then it will provide the relevant information to the FWS for formal consultation on that pesticide within 14 days of making that determination.[106] As a result of the settlement, the plaintiffs released all claims pursued in the original lawsuit and agreed to not provide other plaintiffs with information that could lead to similar lawsuits on these same pesticides.[107] Other recent settlements include EPA's 2006 agreement with the Natural Resources Defense Council in which the agency agreed to make "effects determinations" for atrazine's effect on 21 threatened and endangered species within specified time frames,[108] and EPA's 2006 agreement with the Center for Biological Diversity

100 US Environmental Protection Agency, EPA Signs Settlement Agreement Regarding Endangered Species, *available at* http://www.epa.gov/oppfead1/cb/csb_page/updates/es-settlement.htm (last visited May 26, 2011).

101 *Id.*

102 16 U.S.C. §1536(a)(1) and §1536(a)(2) (2011).

103 Center for Biological Diversity vs. Johnson, No. 1:04-cv-00126-CKK, Settlement Agreement, 2–3 (D.C. Cir. 2005), *available at* http://www.epa.gov/oppfead1/cb/csb_page/updates/bartonsprings-agreemt.pdf.

104 *Id.* at 5–6. The effective date, as defined in the Settlement Agreement, is to be the date upon which the "Stipulation and Order of Dismissal with Prejudice," is entered by the court. *Id.* at 4.

105 *Id.* at 5.

106 *Id.* at 6.

107 *Id.* at 10.

108 *See* US Environmental Protection Agency, Natural Resources Defense Council v. EPA, Settlement Agreement, *available at* http://www.epa.gov/espp/litstatus/NRDC settlement_fs.htm (last visited May 26, 2011).

in which the agency agreed to make "effects determinations" for 66 pesticides harmful to the California red-legged frog within specified time frames.[109] More recently, EPA agreed to a stipulated injunction to resolve a lawsuit brought by the Center for Biological Diversity, which establishes a schedule for EPA to review the effects of 75 pesticide active ingredients on 11 federally listed threatened or endangered species.[110] The most recent litigation over EPA's failure to comply with the ESA's section 7 consultation requirements when making regulatory decisions on pesticides under FIFRA is a lawsuit filed on January 20, 2011 by the Center for Biological Diversity and the Pesticide Action Network North America.[111] In this lawsuit, the plaintiffs assert that EPA has illegally failed to consult under section 7 of the ESA "regarding the impacts of hundreds of pesticides known to be harmful to more than 200 endangered and threatened species." This lawsuit, known as the "Mega lawsuit," because it involves hundreds of pesticides, could have significant impacts on farmers and the agricultural industry. Accordingly, the lawsuit has caught the attention of many industry groups who, arguing it could have devastating effect, have brought it to the attention of Congress. Congress has since held hearings on the topic and a National Academies, National Research Council committee is being established to look into the risk-assessment process used by EPA in making its "effects determinations." On June 6, 2011, the court denied motions filed by numerous chemical industry and farming stakeholder groups to intervene in the lawsuit, thereby barring them from becoming parties to the suit.

The Agencies' Regulatory Action and Inaction

EPA's 35-year history with the ESA as applied to pesticide regulation has not been a good one. The Center for Biological Diversity (CBD) has criticized EPA as "display[ing] a stunning lack of initiative in complying with the Endangered Species Act," and having demonstrated a "reckless disregard for the impact of its Pesticide Registration Program on wildlife, most importantly, on endangered species."[112] The primary concern expressed by CBD is EPA's failure to carry out its obligation to conserve endangered species with a program to address pesticide impacts to such species.[113] The affirmative obligation for federal agencies to

109 *See* US Environmental Protection Agency, Center for Biological Diversity v. EPA, Settlement Agreement, *available at* http://www.epa.gov/espp/litstatus/redleg-frog/rlf.htm (last visited May 26, 2011).

110 Order Approving Stipulated Injunction and Order, Center for Biological Diversity v. EPA (case No. 07-2794-JCS)(May 17, 2010).

111 Center for Biological Diversity et al. v. EPA et al, Case No. 3:11-cv-00293 (January 20, 2011).

112 Litmans & Miller, *supra* note 1, at 51; 16 U.S.C. § 1536(a)(1) (2011).

113 *Id.*

conserve threatened and endangered species derives from section 7(a)(1) of the ESA, which provides that all federal agencies shall use their authorities in furtherance of the purposes of the Act by carrying out programs for the conservation of listed species. Prior to 1989, EPA had yet to formulate an effective method for consultation and review for potential pesticide threats to endangered species.[114] The agency's early attempts at meeting its section 7 responsibility consisted of case-by-case pesticide registration reviews for individual species.[115] This process was cumbersome, however, and in 1982 EPA instead initiated the "cluster approach," where all pesticides with similar use patterns—that is all pesticides that were used in a particular agricultural or ecological setting (such as corn farms, forests, rangeland, or areas suited to mosquito larval growth)—would be considered together, and the FWS would prepare one BiOp for all listed species potentially impacted by the pesticides.[116] EPA began implementing this approach, and in the early 1980s consulted with the Services on clusters including the corn cluster, the small grain cluster, the forest cluster, the mosquito larvicide cluster and the rangeland/pastureland cluster. In 1983, the Services issued BiOps for each cluster, making jeopardy determinations for 21 listed species for the corn cluster from 1 or more of 39 pesticides; 21 listed species for the small grain cluster from 1 or more of 58 pesticides; 58 listed species for the forest cluster from 1 or more of 23 pesticides; 77 listed species for the mosquito larvicide cluster from 1 or more of 11 pesticides; and 159 listed species for the rangeland/pastureland cluster from 1 or more of 32 pesticides. This process, while quicker than the case-by-case method, suffered from problems of its own—namely minor uses for pesticides were not reviewed and a final cluster package review would take upwards of two to three years to complete.[117] More importantly, EPA failed to take action to implement any Services recommendations in the 1983 cluster BiOps.

A 1986 independent review of EPA's pesticide program found that the agency did not comply with section 7 of the ESA in one-third of all pesticide cases. Among the problems were registration of pesticides before receiving BiOps from the FWS and a failure to restrict harmful pesticides.[118] In response to the review, in 1987 EPA announced it would seek full compliance with the ESA, by issuing restrictions for pesticides where a determination had been made that usage had been harmful to listed species, and by issuing further restrictions on labeling in conjunction with information bulletins providing use instructions.[119] The proposal faced considerable opposition, and Congress itself intervened, delaying the EPA

114 Jim Serfis, *Pesticide Regulation*, in BALANCING ON THE BRINK OF EXTINCTION, 214–23 (Kathryn A. Kohm, ed., 1991).

115 *Id.* at 216.

116 *Id.* at 216–17.

117 *Id.* at 217.

118 *Id.*

119 *Id.* at 218.

program until 1988.[120] Due to congressional pressure and agricultural lobbying, EPA had not adopted the program as of 1989. At that time, the agency instead proposed a two-prong approach, which consisted of an individual species-based review (rather than pesticide cluster-based) focusing on those species most in need of protection followed by a determination of the highest acceptable rate of pesticide exposure for that species.[121] This approach was included in EPA's 1989 proposed Endangered Species Protection Program (ESPP), which was designed to establish a process for future consultations.[122]

The proposed ESPP attempted to address risks to listed species by requiring a label statement on each pesticide product that would instruct users to obtain and consult with "county bulletins." These county bulletins would be developed for each county in the United States containing listed species habitat. The bulletins would consist of maps showing the location of listed species habitat and would contain instructions on how to properly use pesticides to reduce risks to listed species existing in that particular county. The county bulletin program had many shortcomings. First, it was voluntary and unenforceable. Second, it depended upon pesticide users taking the initiative to obtain county bulletins and then comply with their recommendations. Prior to the easy access to county bulletins via the internet, this was a cumbersome task that few, if any, users would voluntarily have undertaken. Most significantly, however, EPA's progress in developing county-wide bulletins was extremely slow, and out of the more than 1,800 listed species, and thousands of registered pesticides, very few bulletins were developed—with bulletins for only one or two species in very few counties in each state. For example, in the state of Florida, which has more than 108 listed species, bulletins were developed in only three counties for only one species, the Florida Torreya tree.[123] EPA issued a county bulletin for Gadsden County, Florida, that lists 17 active pesticide ingredients that are to be used in accordance with the bulletin's requirements.[124] Users of these pesticides are provided the sample trade names for the active ingredients, as well as the affected listed endangered species, the Florida Torreya, an evergreen tree native to the Apalachicola River Valley. EPA has issued written limitations on the application of the 17 pesticides, providing strict requirements for pesticide application methods depending on the proximity to ravines and bluffs. Within ravines and bluffs, users can only apply pesticides via tree injection; along ravines and bluffs, users may apply pesticides via ground application, but for aerial application of the 17 active ingredients near ravines

120 *Id.* at 219.

121 *Id.* at 220.

122 Endangered Species Protection Program, 54 Fed. Reg. 27984 (July 3, 1989). *See also* Litmans & Miller, *supra* note 1, at 51.

123 *Id.*

124 US Environmental Protection Agency, Pesticides: Endangered Species Protection Program, Gadsden County, Florida, *available at* http://www.epa.gov/espp/florida/gadsd.htm (last visited May 26, 2005).

and bluffs, the user must maintain a certain buffer area dependent on the method of application.[125] Similarly, for the State of Maryland, a bulletin was developed in only one county for one species of fish. Needless to say, the county bulletin program did not meet the goal of either protection of listed species or compliance with the ESA.[126]

The 1986 internal review, together with the 1989 *Defenders of Wildlife* loss, appeared to nudge EPA into action to comply with the ESA consultation provisions. In the late 1980s and early 1990s, EPA consulted with FWS in two substantial consultations, each involving a large number of pesticides and a large number of potentially affected listed species. The first of these was when EPA reinitiated consultation on selected portions of five previous cluster BiOps. EPA's reason for reinitiating consultation was: (1) to reevaluate jeopardy posed to aquatic species based on a new analysis of estimated environmental concentrations; (2) to evaluate pesticides that may affect four bird species listed since the 1983 BiOps were completed; (3) to consider new reasonable and prudent alternatives to avoid jeopardy to species occurring solely or largely on federal lands, and for the red-cockaded woodpecker and the wood stork; (4) to assess the potential of certain pesticides to indirectly harm listed species through their food supply; and (5) to consider withdrawing/canceling jeopardy opinions for pesticides that had been cancelled or suspended. This reinitiated consultation involved 112 pesticides, each potentially affecting one or more of 165 listed species. The consultation resulted in a 1989 BiOp issued by FWS. It took approximately 10 months from the time of the reinitiation of consultation to the time of the issuance of the final BiOp. The 1989 BiOp superseded the 1983 cluster BiOps and made a total of 1,867 jeopardy findings. Thereafter, EPA initiated consultation on 16 vertebrate control pesticides potentially affecting a number of species, including 30 mammal species, 15 bird species, 9 reptile species, and 1 insect species. It took approximately two years from the time of the initiation of the consultation to the issuance of the BiOp. As a result of this consultation, FWS issued a 1993 BiOp, which made 189 total jeopardy findings. Accordingly, in the 1989 and 1993 consultations alone, FWS made 2,056 jeopardy findings.[127] The overwhelming number of jeopardy findings resulting from its first significant consultation efforts seemed to paralyze EPA. The

125 *Id.*

126 For further discussion of EPA's failure to implement the reasonable and prudent alternatives from the 1989 and 1993 BiOps, as well as EPA's failure to develop more than a few county bulletins, *see* Patti A. Goldman, *Protecting Endangered Species From Pesticides: Making the ESA Work or Finding Loopholes*, ALI-ABA Conference on September 18–19, 2003, SJ023 ALI-ABA 31 (discussing the history of litigation seeking to compel EPA consultation on pesticide registrations).

127 Detailed tables developed by the author showing the reasonable and prudent alternatives for each of the 1867 pesticide/species jeopardy findings in the 1989 BiOp and for each of the 189 pesticide/species jeopardy findings in the 1993 BiOp, as well as the incidental take authorizations and reasonable and prudent measures, are on file with the author.

1989 BiOp is 677 pages long and the 1993 BiOp is 189 pages long. In the 1993 BiOp alone, FWS recommended over 165 reasonable and prudent alternatives for the various species/pesticide combinations. In the 1989 BiOp, FWS provided a menu of approximately 27 reasonable alternatives, most of which applied to many of the large numbers of pesticide/species combinations. Examples of just a few of the many reasonable and prudent alternatives from the 1989 BiOp include:

- Prohibit use of the chemical within 100 yards of the water's edge for ground applications and one-quarter mile for aerial applications at sites of known populations or within designated critical habitat, whichever is larger.
- Prohibit use within one-half mile radius of the species' occupied habitat.
- Applicators of the listed jeopardy pesticides must limit their use within all identified wood stork rookeries, including a buffer extending 8–12 miles from the rookery …
- Applicators of the listed forestry-use pesticides will be required to conduct a survey for red-cockaded woodpecker colonies prior to using this pesticide in forests containing pine trees over 30 years old …
- After periods of heavy rains, as measured by surface water (greater than 4 inches) within identified habitat, do not apply chemical within a 100 yards radius of the known breeding sites of the Puerto Rican crested toad. Restrictions shall remain in place for no less than 25 days.

Examples of reasonable and prudent alternatives from the 1993 BiOp include:

- Prohibit the use in occupied habitat.
- Use within occupied habitats only by qualified individuals (e.g., wildlife biologists, certified applicators).
- Prohibit use within 100 yards of occupied habitat.
- A black-tailed dog colony or complex of less than 80 acres having no neighboring prairie dog towns may be treated without a ferret survey.
- A white-tailed prairie dog colony or complex of less than 200 acres having no neighboring prairie dog towns may be treated without a survey.
- Urban situations may be treated without conducting ferret surveys.
- Black-tailed colonies/complexes over 80 acres but less than 1,000 acres, prairie dog control may be allowed after completing a black-footed ferret survey within 30 days of proposed treatments on colonies proposed for treatment, provided no ferrets or their sign are found.
- For complexes over 1,000 acres, no control shall be allowed until the complex has been evaluated for its potential as a recovery site and until the complex has been block cleared.
- EPA shall maintain records, including the amount of acres of prairie dog towns/complexes controlled or the amount of chemical sold, and including application rates.

EPA appeared not to know how to translate information gleaned from the 1989 and 1993 BiOps into regulatory restrictions that would reduce risks to listed species. EPA's limited ability to regulate pesticide use primarily through label directions was a poor vehicle for incorporating the large number of detailed reasonable and prudent alternatives recommended by FWS in the BiOps. Among the questions EPA would have faced in responding to the BiOps were: How would EPA incorporate these into label language? Would pesticide users be able to understand and properly follow these restrictions? Could pesticide users realistically be expected to read potentially dozens of pages of label language and follow the restrictions when applying the pesticides? Could pesticide users realistically be expected to know the locations of the breeding sites of the Puerto Rican crested toad? Was it reasonable to expect a pesticide user to know the location of black-tailed dog colonies of less than 80 acres having no neighboring dog towns? How would EPA enforce these detailed label restrictions?

It is no wonder that EPA struggled with how to carry out its FIFRA mandate, with its limited authority for regulating use, and still comply with the ESA. EPA chose not to attempt to impose these detailed restrictions on pesticide labels, and instead sought to find a more workable solution to the problem. The near impossibility of imposing all reasonable and prudent alternatives for every listed species on each pesticide label led EPA to continue to rely on its county bulletin ESPP program, despite the failure to finalize it, its many shortcomings, and EPA's own admission that it was not up-to-date and far from complete.[128]

Unfortunately, a detailed review of the record reveals that in the ensuing years since the 1989 and 1993 BiOps, in the absence of litigation forcing it to do so, EPA has failed to take any action whatsoever to require pesticide users to comply with any of the reasonable and prudent alternatives recommended by the FWS for the 2,056 jeopardy findings, or any other risk-reduction measures that would address the jeopardy findings. Research into regulatory actions taken on each pesticide for which one of the 2,056 jeopardy opinions were issued did not reveal any regulatory actions taken to reduce risks to any of the affected listed species in response to any of the jeopardy opinions. This conclusion is based in part on a detailed analysis of all of the jeopardy opinions in the 1993 BiOp, tracing the regulatory decisions on each pesticide for which one of the 189 jeopardy opinions were issued from the date of the BiOp to the present. This research investigated: (1) whether EPA had canceled, suspended, or limited any registration in response to the jeopardy opinions; and (2) whether EPA had imposed any of the reasonable and prudent alternatives recommended by the FWS as either a label restriction or any other type of regulatory mechanism. Of the 189 jeopardy opinions and 165 RPAs suggested in the 1993 BiOp, this research did not reveal a single instance in which EPA took any regulatory action in direct response to the findings or suggestions in the BiOp. Moreover, research investigating the pesticides subject to the 1989 BiOp has not revealed any regulatory action taken by EPA to impose

128 Litmans & Miller, *supra* note 1, at 52.

label restrictions or to otherwise reduce risks for the 1867 jeopardy opinions in the 1989 BiOp. Four of the pesticides that were evaluated in the 1989 BiOp have since been severely restricted or banned by EPA: (1) granular carbofuran (severely restricted); (2) endrin (banned); (3) EPN (banned); and (4) mevinphos (banned). However, there is no indication that these regulatory actions resulted from the findings or recommendations in the BiOp. Most significantly, the research revealed no instance in which EPA imposed a reasonable and prudent alternative recommended by the Services as FIFRA label restriction, or any other type of pesticide use restriction.

Instead of imposing the FWS suggested RPAs as label restrictions or taking other regulatory action in response to the 2,056 jeopardy findings in the 1989 and 1993 BiOps, EPA has repeatedly attempted to justify its failure to act by referring to the County Bulletin program initiated in the 1980s. For example, in its reregistration eligibility documents (REDs) for the pesticides found to cause jeopardy in the 1993 BiOp, EPA's sole nod to the endangered species issue was to include in each RED the following statement: "The Agency is not imposing label modifications at this time through the RED. Rather, any requirements for product use modifications will occur in the future under the Endangered Species Protection Program."[129] Unfortunately, as described above, the county bulletin program is virtually non-existent, as it only addresses a very few species in a very few counties of the US, leaves to the pesticide user to decide whether or not to obtain the bulletins and whether or not to follow their instructions, and is completely unenforceable.

It is only fair to note that EPA has taken some very limited regulatory actions based on risks to wildlife, but not directly in response to recommendations of the Services under the ESA. However, out of the many thousands of pesticides and affected species, EPA has only taken a very small number of very limited actions due to wildlife impacts. In fact, the only reported judicial or administrative case in which EPA took regulatory action based primarily on risks to wildlife was *Ciba Geigy* v. *EPA*,[130] in which EPA proposed canceling certain uses of the pesticide diazinon on golf course and turf grass due to the risk the pesticide posed to wild birds. Specifically, the case addressed the question of whether FIFRA requires a precise determination of risk or harm (e.g. the chemical has adverse effects 51 percent of the time it is used) in order to support cancellation of a registration.[131] Another related point of contention was whether a devastating effect on bird

129 The REDs, including those for the pesticides evaluated in the 1993 BiOp (Aluminum and Magnesium Phosphide, Brodifacoum, Bromadiolone, Bromethalin, Chlorophacinone, Diphacinone, Pival, Sodium Cyanide, Sodium Flouroacetate, Warfarin, and Zinc Oxide) can be viewed at http://www.epa.gov/pesticides/pestlabels/index.htm (last visited May 26, 2011).

130 874 F.2d 277 (5th Cir. 1989).

131 *Id.*

populations or merely a significant adverse effect would justify cancellation.[132] In this case, the chemical company's contentions, that there should be more exact thresholds and more significant effects on the overall bird population, were rejected by the court. The Fifth Circuit held that FIFRA gives the Administrator sufficient discretion to conclude that recurring bird kills are an unreasonable adverse environmental effect regardless of whether they significantly reduce bird populations.[133] Ultimately, the case was remanded to the Administrator to rectify the former Administrator's failure to read the word "generally" as meaning "usually," "commonly," or "with considerable frequency." The phrase "generally causes unreasonable adverse effects on the environment" is also read to include any potential general causation of adverse effects.[134]

During the 1980s and 1990s, EPA considered canceling certain other pesticides based on risks to wildlife, but EPA failed to take any significant action to address such risks. In 1991, EPA proposed the cancellation of the pesticide ethyl parathion, due to risks to both humans and wildlife from the high acute toxicity of the pesticide. EPA ultimately accepted a settlement with the manufacturer of ethyl parathion, however, canceling only the ground application uses of the pesticide, which posed significant risks to human farmworkers. The settlement did not address aerial application of the pesticide, which posed the greatest risks to birds and other wildlife, due to spray drift associated with this form of application.[135] Despite the fact that ethyl parathion had been implicated in the deaths of thousands of birds, EPA declined to take regulatory action to address these risks. Ultimately, in 2001, the manufacturer of ethyl parathion voluntarily canceled the remaining uses of the pesticide, after a concerted campaign led by the American Bird Conservancy in partnership with Defenders of Wildlife, the Pesticide Action Network, and the World Wildlife Fund to pressure EPA and the manufacturer of the pesticide to end all uses.[136] Nevertheless, despite the fact that ethyl parathion was considered to be one of the most acutely toxic pesticides and had been documented as the cause of thousands of bird kills, and despite decades of study by EPA, the agency itself failed to take regulatory action to protect wildlife. With regard to the granular form of another pesticide, carbofuran, which was implicated in the deaths of many birds that ingested the granules, presumably believing they were seeds, EPA

---

132 *Id.* at 280.

133 *Id.*

134 *Id.* at 278–80.

135 *See* Ethyl Parathion: Receipt of Requests for Cancellation, Maintenance Fee Cancellation, Cancellation Order, Notification Requirement, Memorandum of Agreement, Request for Comment on Tolerance Reduction/Revocation, 56 Fed. Reg. 65061–01 (December 13, 1991); Ethyl Parathion: Amendment of Cancellation Order, 57 Fed. Reg. 3500–01 (January 29, 1992); and Ethyl Parathion: Correction to the Amended Cancellation Order, 57 Fed. Reg. 6168–01 (February 20, 1992).

136 Ethyl Parathion: Notice of Use Cancellation, 66 Fed. Reg. 47667–01 (September 13, 2001).

entered into a settlement with the manufacturers in 1991 to phase out the use of the granular pesticide. However, EPA continued to allow the use of the liquid form of the pesticide, which has also been implicated in widespread bird mortality.[137]

Equally disturbing as EPA's failure to implement any recommendations from the ESA consultations it did conduct, is that after the 1993 consultation EPA did not engage in consultations with the Services on pesticide regulatory decisions until forced to do so by court order. Clearly, the system was not working. This failure ultimately led to the rash of lawsuits in the early 2000s against EPA for its failure to comply with section 7 of the ESA.

As described above, in the past several years, EPA has come under increasing criticism for its failure to fulfill its obligations under the ESA.[138] EPA makes a large number of regulatory decisions regarding pesticides every year. Currently, there are approximately 20,000 registered pesticide product formulations, containing approximately 675 active ingredients and 1,835 other ingredients. Approximately 470 of the 675 active ingredients are used in agriculture.[139] In a typical year, EPA makes hundreds of significant regulatory decisions regarding pesticide registration. For example, in 2003 alone, EPA registered 31 new pesticide active ingredients, approved the 334 new uses of previously registered active ingredients on over 1,500 different crops, and issued more than 6,500 more minor registrations. During this same time period, EPA also completed reregistration assessments on 28 registered active ingredients, and processed nearly 500 emergency exemption requests.[140] Since the 1993 BiOp, EPA has not initiated any formal consultations, whatsoever, on any of thousands of registrations or other FIFRA regulatory decisions, unless required by court order or settlement agreement. Instead, EPA continued to rely on the never-finalized ESPP program, including the limited voluntary county bulletin program.

In December 2002, EPA revived its ESPP by filing in the Federal Register a notice of its proposed implementation of the ESPP.[141] The 2002 proposed ESPP was

---

137 *See* Pierre Mineau, *Birds and Pesticides: Are Pesticide Regulatory Decisions Consistent with the Protection Afforded Migratory Bird Species Under the Migratory Bird Treaty Act?* 28 Wm. & Mary Envtl. L. & Pol'y Rev. 313, 322 (2004).

138 Goldman, *supra* note 126, at 31; Mineau, *supra* note 137. *See also*, Marcilynn A. Burke, *Klamath Farmers and Cappuccino Cowboys: The Rhetoric of the Endangered Species Act and Why it (Still) Matters*, 14 Duke Envtl. L. & Pol'y F. 441, 487–491 (2004) (discussing a number of regulatory attempts to weaken the consultation process including with regard to pesticide registration).

139 *Endangered and Threatened Species Effects Determination Program:* Overview of the Ecological Risk Assessment Process in the Office of Pesticide Programs, US Environmental Protection Agency, Endangered and Threatened Species Effects Determinations 7 (January 23, 2004), *available at* http://www.epa.gov/oppfead1/endanger/consultation/ecorisk-overview.pdf.

140 Endangered Species Protection Program Implementation, 67 Fed. Reg. 71549–02 (December 2, 2002).

141 *Id.*

in essence a reiteration of the 1989 proposed ESPP, which was never finalized. The notice discussed how, via ESPP, EPA would register pesticides under FIFRA and how the agency would balance the interests of its responsibilities under the ESA and the desire to avoid "unnecessary burden" on farmers and pesticide users.[142] The notice primarily discussed EPA's quantitative testing approaches undertaken in a pesticide registration process, including both exposure tests and toxicity tests on listed species.[143]

Another major focus of the notice was the revival of the County Bulletin Program, in which EPA announced that it would develop and update county bulletins, and would post the bulletins on its website.[144] EPA announced that it would develop bulletins with the assistance of the FWS, NOAA, USDA, states and tribes, and would issue bulletins only for counties in which such measures are considered necessary.[145] Bulletins would specifically identify: (1) the listed species of concern; (2) pesticides that may harm the listed species; (3) the protection measures for that species as well as any habitat information; and (4) a county map indicating where pesticide usage should be modified from its standard use.[146] The county bulletins would be designed to inform the public of pesticide application limitations in their community.[147] An interested pesticide user could review the county bulletins, which are available on the EPA's website, and check for any use restrictions or boundary requirements for pesticide application.[148] In addition to the substantive aspects of the bulletins, EPA proposed modifying pesticide labels to encourage users to follow the information contained within the county bulletin.[149] The modified label would also reference the effect the pesticide could have on listed species and how the user could obtain the relevant county bulletin.[150] Interestingly, label statements that would be amended would not be county-specific, but would simply reference the potential harms to listed species and guide the user to the particularized county bulletin to find information for his county. EPA also proposed the following generic label:

> This product may have effects on federally listed threatened and endangered species or critical habitat in some counties. When using this product, you must follow the measures contained in the County Bulletin for the county in which

142 *Id.*

143 *Id.* at 71553–4.

144 *Id.* at 71558.

145 *Id.*

146 *Id.*

147 US Environmental Protection Agency, Pesticides: Endangered Species Protection Program, "How to Use the County Bulletins," *available at* http://www.epa.gov/espp/how-to.htm (last visited May 26, 2011).

148 *Id.*

149 *Id.*

150 Endangered Species Protection Program Implementation, *supra* note 140, at 71559.

> you are applying the pesticide. To determine whether your County has a Bulletin consult http://www.epa.gov/espp/usa-map.htm. Bulletins also may be available from local pesticide dealers, extension offices, or State pesticide agencies.[151]

Unfortunately, given EPA's poor track record in developing and updating county bulletins over the past few decades, EPA's reiteration of this program in its 2002 proposal did little to comfort those concerned with protecting listed species from pesticides. The CBD Report outlines several other shortcomings of the 2002 ESPP, including EPA's misinterpretation of its duties under the ESA and its general institutional lack of concern for listed species.[152]

The Bush Administration's initial response to the criticisms that EPA failed to comply with the ESA in making pesticide regulatory decisions, and criticism that the voluntary County Bulletin Program was inadequate to comply with the ESA or to protect listed species, as well its response to the recent court losses, was to attempt to amend the joint regulations for consultation under section 7 of the ESA to eliminate the need for EPA to consult with the Services when making regulatory decisions under FIFRA. On August 5, 2004, the Services issued a final rule regarding consultation practices among the Services and EPA for pesticide registrations.[153] The agencies' purported rationale for the rule was to provide a more efficient approach to making decisions on whether new pesticides would "adversely affect" a listed species.[154] Because the "Services believe that EPA's expertise in ecological risk assessments of pesticides, together with the safeguards built into the alternative consultation agreement, make case-by-case discussions … necessary for FIFRA actions," there would be no formal consultation for any FIFRA actions that EPA determines are not likely to adversely affect (NLAA) any endangered species.[155] Under the new rule, EPA would make its own ESA analysis for NLAA determination purposes.[156] Once EPA makes its NLAA determination, the analysis would be complete and there would be no role for the Services to second-guess EPA. If EPA were to conclude that the FIFRA action is likely to jeopardize the continued existence of any listed species or its critical habitat, EPA would prepare an effects determination (which is made with the assistance of a Services Representative); this effects determination would serve as a functional equivalent to the biological opinion that the Services normally provide.[157] At that point the relevant Service would review the determination and may adopt it, modify it, or provide its own biological opinion providing reasonable and prudent

151 *Id.*

152 Litmans & Miller, *supra* note 1, at 52–3.

153 Joint Counterpart Endangered Species Action Section 7 Consultation Regulations, 69 Fed. Reg. 47732 (August 5, 2004) (codified at 50 CFR Part 402.01).

154 *Id.*

155 *Id.*

156 *Id.*

157 *Id.*

alternatives available to the EPA.[158] In effect, the rule would have allowed EPA to bypass consultation if EPA concluded the pesticide regulatory decision was "not likely to adversely affect" a listed species. And if EPA concluded a regulatory action was "likely to effect" a listed species, EPA would in essence write the BiOp which the Services could adopt or modify.

Widely criticized by environmental organizations, not surprisingly, the rule was strongly supported by the FIFRA Endangered Species Task Force, a committee made up of 14 major agro-chemical companies.[159] The primary criticism of the rule was that it would provide EPA with an upfront approval by FWS and NMFS of the EPA's risk assessment procedure while eliminating the oversight of the decision via the removal of post-assessment consultation.[160] Critics argued that EPA staff did not possess the necessary expertise to make effects determinations without input from the Services. Further, allowing EPA to conduct a review of the effects a pesticide would have on listed species was criticized given that history has routinely shown that FWS and NMFS have been critical of EPA's scientific approaches in the consultation process.[161] Environmental organizations feared that the new rule would undercut the ESA and put listed species at greater risk. Consequently, a number of environmental organizations filed suit alleging that the new rule violated the ESA. On September 23, 2004, a coalition of eight environmental groups filed suit challenging the Joint Counterpart Endangered Species Act Regulation. [162]

In 2006, a federal district court ruled that several of the provisions of the new rule were not consistent with the mandates of section 7 of the ESA.[163] Specifically, the court invalidated the provisions of the new rule regarding the process by which EPA would make NLAA determinations, finding these provisions to be arbitrary and capricious.[164] The court found that these portions of the rule were, by their very terms, in conflict with the section 7 statutory mandate, and therefore, they could not survive a *Chevron* step-one test—i.e., Congress has not directly and unambiguously spoken to the issue.[165] Moreover, the court found overwhelming evidence that in promulgating the rule, the Services did not comply with their own ESA section 7 obligations to avoid jeopardy to listed species.[166] Ironically, shortly

---

158 *Id.*

159 Goldman, *supra* note 126, at 31. *See also* Litmans & Miller, *supra* note 1, at 58.

160 *Id.*

161 *Id.*

162 *Id.*

163 Washington Toxics Coalition v. U.S. Fish & Wildlife Serv., 475 F.Supp.2d 1158 (W.D. Wash. 2006).

164 *Id.*

165 Chevron U.S.A., Inc. v. Natural Resources Defense Council, Inc., 467 U.S. 837 (1984).

166 Washington Toxics Coalition v. U.S. Fish & Wildlife Serv., 475 F.Supp. 1158, 1193; Defenders of Wildlife v. Kempthorne, No. 04-1230(GK), 2006 WL 284432 (D.D.C. September 29, 2006). For a complete analysis of both cases and the distinct approaches

after the 2006 *Washington Toxics Coalition* ruling, the federal District Court for the District of Columbia upheld challenges to ESA regulations promulgated by the Services that would establish a similar consultation process for the National Fire Plan regulatory actions.

The recent court losses seem to have prodded EPA reluctantly to initiate at least some ESA consultations. Since 2004, in response to its loss in *Washington Toxics Coalition* and settlement agreements in other cases, EPA has issued numerous "effects determinations," many of which resulted in LAA findings triggering consultation with the Services, in addition to the consultations required by court order.[167] Currently, EPA is engaged in consultation with the Services on a large number of pesticides. Numerous draft BiOps and final BiOps have been issued, including one covering the organophosphate pesticides chlorpyrifos, diazinon, and malathion and one covering the carbamate pesticides carbofuran, carbaryl and methomyl.[168] In the final BiOps on the organophosphate pesticides, NMFS concluded that the continued use of the covered pesticides would jeopardize the continued existence of 27 species of salmon and steelhead listed under the ESA and would adversely affect critical habitat for 25 of the 27 species.[169] In the final BiOp for the carbamate pesticides, NMFS concluded that continued use of the covered pesticides would jeopardize the continued existence of 22 salmonids listed under the ESA and adversely affect the habitat of 20 of these species. In both final BiOps, NMFS set forth a number of reasonable and prudent alternatives (RPAs) to avoid jeopardy and to protect critical habitat. In both BiOps, NMFs also concluded that it would be necessary to cancel the registrations for all of the covered pesticides or to completely prohibit their use in watersheds inhabited by salmonids to avoid any take of the listed fish species. Thus, NMFS included an incidental take statement in each BiOp that authorized some takes provided EPA ensured that the registrations for the pesticides at issue complied with NMFS' RPAs and reasonable and prudent measures (RPMs) contained in the BiOps. On September 11, 2009, EPA

---

taken by each court, *see* Cynthia A. Drew, *Beyond Delegated Authority: The Counterpart Endangered Species Act Consultation Regulations*, 37 Envtl. L. Rep. 10483 (2007).

167 The author has compiled a table of all of the pesticides for which EPA has made, or is in the process of making, effects determinations or consulting with the Services. The table shows the status of the pesticide, the registered use, the toxicity level, whether the pesticide is subject to a court order, the effects determinations made, what the effects determinations are based on and the status of each of the consultations. The table is on file with the author.

168 *See* National Marine Fisheries Service Endangered Species Act Section 7 Consultation Biological Opinion Environmental Protection Agency Registration of Pesticides Containing Chlorpyrifos, Diazinon, and Malathion (November 18, 2008) and National Marine Fisheries Service Endangered Species Act Section 7 Consultation Biological Opinion Environmental Protection Agency Registration of Pesticides Containing Carbaryl, Carbofuran, and Methomyl (April 20, 2009) both *available at* http://www.epa.gov/espp/litstatus/wtc/index.htm (last visited June 16, 2011).

169 *Id.*

announced its plans to request the manufacturers of the pesticides chlorpyrifos, diazinon and malathion to voluntarily impose label restrictions to reduce impacts to listed salmon species. The proposed label restrictions would apply buffer zones around certain water bodies, and would impose application restrictions based on wind speed, soil moisture and weather conditions. Fish mortality reporting would also be required. Similarly, on May 14, 2010, EPA announced that it intended to impose label restrictions on the pesticides carbofuran, carbaryl and methomyl, to reduce impacts to salmon species and would also require fish mortality reporting. These label restrictions would also include spray drift buffers and application limitations based on wind speed, soil moisture, and weather conditions. EPA stated that because these limitations are geographically specific, it would communicate the new limitations through the County Bulletin Program.

Despite its recent progress in conducting effects determinations and initiating consultations with the Services when required to by court order, to date, EPA has not actually implemented any of the RPAs or RPMs suggested by the BiOps. Because of EPA's failure to adopt mitigation measures, on November 29, 2010, the Northwest Coalition for Alternatives to Pesticides and several other organizations filed a "Complaint for Declaratory and Injunctive Relief" requesting that the Court declare that: (1) EPA is in violation of section 7(a)(2) of the ESA by failing to ensure that the use of the pesticides diazinon, malathion, chlorpyrifos, carbaryl, carbofuran and methomyl is not likely to jeopardize the continued existence of listed salmonids and destroy or adversely modify their critical habitat; (2) EPA is in violation of section 9 of the ESA for takes of listed salmonids through the authorized use of the six pesticides listed above without complying with the BiOps' RPAs and RPMs; and (3) to enjoin, vacate and set aside EPA's authorization of the uses of these six pesticides that do not comply with the BiOps' RPAs or RPMs.[170] This case is still pending in the court.

In 2004, EPA also made available a document described as an overview of its Ecological Risk Assessment (ERA) Process.[171] The ERA process begins with what EPA refers to as a Screening-Level Risk Assessment to evaluate a substance's potential impact on non-target organisms, including listed species. Each active ingredient is subject to an individual risk assessment, but EPA does not routinely include an evaluation of mixtures of active ingredients.[172] A suite of toxicity studies performed on a limited number of organisms representing broad groupings, such as birds,

---

170 Complaint for Declaratory and Injunctive Relief, Northwest Coalition for Alternatives to Pesticides, Pacific Coast Federation of Fishermen's Associations, Institute for Fisheries Resources, and Defenders of Wildlife v. EPA 2010 WL 4948598 (W.D. Wash. November 29, 2010).

171 US Environmental Protection Agency, Overview of the Ecological Risk Assessment Process in the Office of Pesticide Programs, U.S. Environmental Protection Agency (January 23, 2004), *available at* http://www.epa.gov/oppfead1/endanger/consultation/ecorisk-overview.pdf.

172 *Id.* at 30.

mammals, freshwater fish, etc., are used in the screening-level risk assessment.[173] In addition to the data submitted by the registrant to support registration, EPA also consults open literature studies for additional toxicity information to be used in the screening-level risk assessment.[174] The analysis phase of the screening-level risk assessment includes characterization of exposure and effects.

The exposure characterization provides "a quantitative analysis of the critical environmental fate and transport properties of the pesticide active ingredient."[175] Potential exposure of plants, wildlife, and aquatic life, to pesticide residues in the environment, are estimated using this information together with the pesticide use and conditions of the use site. The types of data used vary depending on the use site, but include controlled environmental fate and transport laboratory studies, persistence studies, mobility studies, bioconcentration studies, degradation studies, metabolism studies, and field studies.[176] Aquatic systems modeling uses a tiered system in order to allocate resources more efficiently to assessment efforts of varying complexities and potential risks.[177] Terrestrial organism exposure modeling generally emphasizes a dietary exposure route for uptake of pesticide active ingredients and assumes that organisms are only exposed to a single pesticide residue.[178] Estimates for terrestrial exposure are based on the application method: spray applications or granular, bait, and treated seed applications.[179] The effects characterization phase includes "describing the types of effects a pesticide can produce in an organism and how those effects change with varying pesticide exposure levels."[180]

The risk characterization phase integrates the effects and exposure characterization "to evaluate the likelihood of adverse ecological effects on non-target species."[181] The Risk Quotient (RQ) method for expressing risk is used to compare exposure over toxicity. RQs are compared to Levels of Concern (LOCs), which are EPA's interpretive policies used to analyze potential risk to non-target organisms and the need to consider regulatory action. The result of the screening level assessment is an effects determination that the pesticide: (1) will have "no effect"; (2) "may affect but is not likely to adversely affect the species or critical habitat"; or (3) "may adversely affect the species or critical habitat."[182] If the screening-level risk assessment indicates that a pesticide "may potentially impact, either directly or indirectly, listed species or critical habitat," a species-specific

173 *Id.* at 32.
174 *Id.* at 44.
175 *Id.* at 34.
176 *Id.* at 34–5.
177 *Id.* at 37.
178 *Id.* at 39–40.
179 *Id.* at 40.
180 *Id.* at 41.
181 *Id.* at 46.
182 *Id.*

and habitat-specific ecological risk assessment is conducted.[183] A "no effects" determination is reached when it is concluded that "there are no indirect effects and levels of concern (LOCs) for listed species are not exceeded for direct effects during the screening-level assessment."[184] A "may affect" determination is reached when "indirect effects are anticipated or exposure may exceed the LOCs for direct effects." EPA uses "best professional judgment" to distinguish between actions that "may affect but are not likely to adversely affect" and those that are "likely to adversely affect" a listed species or critical habitat.[185]

EPA further refines those aspects of the screening-level assessment for which a "not likely to adversely affect" determination could not be made.[186] EPA decides "whether use of the pesticide 'may affect' a particular listed species and if so, whether it is 'likely to adversely affect' the species."[187] The overall goal for this process is "to protect the listed species and critical habitat by potentially modifying a pesticide's use in a manner that is least disruptive to agriculture and other pesticide users."[188] Data sources used at this level of risk assessment include the best available and current information regarding biological requirements and habits of listed species, sub-county commodity information, geographic features precluding exposure, incident information, sales and use information, local use practices, and monitoring data.[189] The exposure characterization is refined using geographic proximity, specific assessment methodologies, and biological and habitat requirements.[190] Geographic proximity data is used to determine if there is overlap and, thus, potential for a listed species to be exposed to a pesticide. The assessment methodology is reviewed to determine whether the methodology is the most appropriate for the species-specific or habitat-specific assessment.[191] However, it is seldom possible to have a model that exactly fits a particular site. EPA determines whether there is any biological factor that would preclude exposure that may cause direct effects and, in the case of chronic effects, whether there is temporal overlap in pesticide residues and species activities and habits that may result in exposure at a level and duration that produces the effect.[192] During the risk characterization, EPA determines that an action "may affect" a listed species "if the RQ exceeds the endangered species LOC, and a species-specific analysis indicates temporal and spatial overlap between pesticide use and

183 *Id.* at 65.
184 *Id.*
185 *Id.* at 65 and 72.
186 *Id.* at 65–6.
187 *Id.* at 66.
188 *Id*
189 *Id.* at 67–9.
190 *Id.* at 69–70.
191 *Id.* at 70.
192 *Id.* at 70–71.

the species presence."[193] A "no effects" determination is made when "there are no indirect effects nor exposure at levels that may result in direct effects."[194] EPA distinguishes between "likely to adversely affect" determinations and "may affect but is not likely to adversely affect" determinations based on its best professional judgment of the significance and likelihood of effects.[195] Information on incidents, sales and use, local use practices, and monitored levels is used in conjunction with the degree to which LOCs were exceeded to determine whether the predicted effect based on labeled use of the product is likely to occur.

As part of its 2004 ERA process, EPA itself recognizes a number of the assumptions, uncertainties, strengths, and limitations of the screening-level risk assessment.[196] For example, for screening-level risk assessment, the actual habitat requirements of the particular species are not considered. Instead, a maximum level of exposure is based on the assumption that the "species occupy, *exclusively and permanently*, the treatment area being modeled."[197] Another limitation is that only dietary exposure is considered for spray applications, excluding incidental soil ingestion exposure, inhalation exposure, dermal exposure, and drinking water exposure.[198] Additionally, problems arise from the differences between laboratory and field conditions in regards to dietary intake.[199] A two-fold underestimation in exposure potential is created by not accounting for increased energy demands on organisms in the wild when comparing dietary residues to dietary toxicity thresholds.[200] Another uncertainty is the relationship of the listed species' sensitivity to the most sensitive species tested.[201] "[I]t is not likely that tested species represent the most sensitive species within the broad taxonomic groups used."[202]

Another area of concern is that, while the risk-assessment document sets up an extremely complex methodology for assessing effects on wildlife, it does not amend the data requirements to require registrants or applicants to develop or submit more comprehensive or better data on wildlife effects. Instead, it continues to rely on the limited wildlife data requirements currently contained in 40 C.F.R. 158, which, as described previously, are extremely limited and do not address the full suite of risk concerns. Another shortcoming of EPA's ecological risk-assessment process is that it focuses solely on the impacts to the organisms themselves and does not address impacts to habitat that indirectly affect wildlife species.[203]

193 *Id.* at 71.
194 *Id.*
195 *Id.*
196 *Id.* at 51.
197 *Id.* at 57.
198 *Id.* at 57–8.
199 *Id.* at 59.
200 *Id.* at 60.
201 *Id.* at 63.
202 *Id.*
203 Litmans & Miller, *supra* note 1, at 54.

EPA's most recent effort to explain the ESPP occurred on November 2, 2005, when EPA published a notice in the Federal Register describing how the ESPP will be implemented in the field.[204] EPA describes its goal as meeting its responsibilities under FIFRA in compliance with the ESA and without unnecessarily burdening pesticide users. However, EPA's plan is not a legally binding regulation and EPA may decide to change it at any time without notice and comment.[205]

Under the plan, pesticide actions, such as indoor products determinations and emergency exemptions under §18 of FIFRA, are potentially excluded from the scope of the ESPP.[206] The "EPA's overall strategy is to address listed species concerns within the context of the pesticide registration, reregistration, and registration review processes."[207] An effect determination based on "EPA's assessment of a pesticide use's potential effects to listed species" is generally conducted to support the registration status of a pesticide.[208]

Endangered Species Protection Bulletins (Bulletins) will be used to implement changes to a pesticide's use when necessary to protect a listed species in a geographically specific area.[209] Bulletins will be implemented on a county scale. Information provided in the Bulletins includes the identity of the species of concern, the name of the active ingredient(s) to which the limitations apply, a description of the use limitation, a county map showing the specific geographic area to which the use limitations apply, and a picture and description of the species when it would not cause further threat to the species. Unlike the prior voluntary county bulletins, however, EPA will treat the Bulletins as mandatory and enforceable pesticide use limitations[210]

The pesticide label language that will be used when geographically specific use language is necessary to protect listed species will include the following at the beginning of the product's Directions for Use: "ENDANGERED SPECIES PROTECTION REQUIREMENTS."[211] EPA intends to make Bulletins available six months before they go into effect. Applicators are required to use Bulletins in effect the month in which they will be applying the pesticide.[212] EPA intends to treat the Bulletins just as any other label provision in terms of enforcement under FIFRA section 12(a)(2)(G), regardless of whether the failure to follow the Bulletin

---

204 Endangered Species Protection Program Field Implementation, 70 Fed. Reg. 66392 (November 2, 2005).

205 *Id.*

206 *Id.* at 66398.

207 *Id.*

208 *Id.* at 66399.

209 *Id.* at 66400.

210 *Id.* Access to the bulletins will be provided at: US Environmental Protection Agency, Pesticides: Endangered Species Protection Program, Endangered Species Protection Bulletins, *available at* http://www.epa.gov/oppfead1/endanger/bulletins.htm (last visited May 26, 2011).

211 70 Fed. Reg. 66400.

212 *Id.*

results in harm to a listed species.[213] The misuse and misbranding provisions of FIFRA, as well as liability under Section 9 of the ESA, will apply to pesticide users who fail to follow the applicable label provisions. In terms of monitoring, EPA will continue to use existing monitoring data from risk assessments, the US Geological Survey, information provided under the Clean Water and Safe Drinking Water Acts, and information from State or Tribal monitoring programs.[214] The "EPA also intends to develop a process for monitoring the effectiveness of Bulletins after the Program has been in effect for some time."[215]

In the past few years, in response to the *Washington Toxics Coalition* decision, and to comply with various settlement agreements, EPA has undertaken endangered species effects determinations for a number of pesticides as required by court order or settlement agreement. Some of these effects determinations have resulted in a finding of no adverse effect.[216] Many of the effects determinations, however, have resulted in the initiation or reinitiaton of consultation with the Services under section 7. However, environmental groups had to sue EPA to force it to initiate the consultations[217] and had to sue NMFS to force it to complete the consultations when more than five years had passed since EPA initiated the consultations and NMFS had not completed them.[218] Approximately seven years after EPA initiated consultation, NMFS finally issued the final BiOps on the pesticides at issue. Significantly, in the final BiOps, NMFS made jeopardy findings for a number of pesticides for which EPA had initially made "not likely to adversely affect" determinations. In a November 18, 2008 BiOp regarding three pesticides, diazinon, malathion, and chlorpyrifos, NMFS concluded that the pesticides are likely to jeopardize 27 populations of listed salmon and called for buffer zones next to salmon streams where the chemicals are used. In an April 20, 2009 final BiOp on three additional pesticides, carbaryl, carbofuran, and methomyl, NMFS concluded that these pesticides are likely to jeopardize the continued existence

213 *Id.* at 66401.

214 *Id.* at 66402.

215 *Id.*

216 *See, e.g.*, US Environmental Protection Agency, Pesticides: Endangered Species Protection Program, Memorandum: No-Effect Determination for Atrazine for Pacific Anadromous Salmonids, *available at* http://www.epa.gov/oppfead1/endanger/litstatus/effects/atrazine-memo.pdf; US Environmental Protection Agency, Pesticides: Endangered Species Protection Program, Memorandum: "No Effect" Determination for 2,4-D EHE for the Northern California/Southern Oregon Coastal Coho Salmon, *available at* http://www.epa.gov/oppfead1/endanger/litstatus/effects/24dmemo.pdf; and US Environmental Protection Agency, Pesticides: Endangered Species Protection Program, Memorandum: No Effect Determination for Paraquat Dichloride for Pacific Anadromous Salmonids, *available at* http://www.epa.gov/oppfead1/endanger/litstatus/effects/ne_paraquat.pdf.

217 Wash. Toxics Coalition v. EPA, Case No. C01-013132C, Orders issued July 16, 2003 and August 8, 2003.

218 *See* NW Coalition for Alternatives to Pesticides, LLC. v. NMFS, Case No. 07-1791-RSL (Stipulated Settlement Agreement and Order of Dismissal, 2008).

of 22 listed Pacific salmonid species and found that the effects of carbaryl and carbofuran are likely to destroy or adversely modify designated habitat for 20 listed salmonids.[219] Since then, NMFS has issued numerous draft BiOps, many of which contain findings of "likely jeopardy" to a large number of listed species. EPA has consistently responded to these draft BiOps by providing written comments that take issue with many of the RPAs proposed by NMFS.[220]

Despite the recent progress that has been made, EPA has yet to find a way to implement the RPAs and RPMs from the Services BiOps to prevent jeopardy and takes of listed species. The Services, as well as EPA, appear to have become bogged down in the scientific minutia and bureaucracy and have failed to make any meaningful progress in protecting the species themselves. To move forward, the agencies will need some clear Congressional direction on how to proceed in a manner that reconciles the conflicts between the statutes, and sets forth a clear path for agency action.

The Legislative Response

The ongoing 35-year battle between FIFRA and species protection laws, which led to the flood of litigation starting in 2002, drew the attention of members of Congress concerned with what they perceived as the overly broad mandate of the ESA as it relates to private property rights. In response to these concerns,

---

219 National Marine Fisheries Service, Endangered Species Act Section 7 Consultation Biological Opinion, Environmental Protection Agency Registration of Pesticides Containing Carbaryl, Carbofuran, and Methomyl (April 20, 2009), *available at* http://www.nmfs.noaa.gov/pr/pdfs/carbamate.pdf.

220 In a March 1, 2011 draft BiOp, NMFS proposed the following RPA, "[d]o not apply pesticide products containing 2,4-D, diuron, or chlorothalonil when soil moisture is at field capacity, or when a storm event likely to produce runoff from the treated area is forecasted to occur within 48 hours following application by NOAA/NWS (National Weather Service) or other similar forecasting service." National Marine Fisheries Service, Endangered Species Act Section 7 Consultation Draft Biological Opinion, Environmental Protection Agency Registration of Pesticides Containing 2,4-D, Triclopyr BEE, Diuron, Linuron, Captan, and Chlorothalonil (March 1, 2011), *available at* http://www.epa.gov/oppfead1/endanger/litstatus/biop4-march2011.pdf; EPA responded that this RPA would pose difficulties to enforcement personnel due to difficulties determining "what the weather forecast had been prior to the date/time of application, and, if a storm had been predicted, whether that storm could be expected to result in runoff." US Environmental Protection Agency, Response to NMFS March 1, 2011 Draft Biological Opinion, *available at* http://www.epa.gov/espp/litstatus/nmfs-3-11-11-draft-biop.pdf; NMFS responded by amending the RPA to read, "[…] when a storm event likely to produce direct runoff to salmonid-bearing waters from the treated area […]," leaving it in all other respects the same. NMFS Revised Draft Biological Opinion (May 13, 2011), *available at* http://www.epa.gov/oppfead1/endanger/litstatus/5-13-11-2nddraft-biop.pdf.

the US House of Representatives, led by Congressman Richard Pombo (R-CA), passed the Threatened and Endangered Species Recovery Act of 2005 by a vote of 229 to 193 on September 29, 2005.[221] The legislation would have dramatically altered several provisions of the 1973 Act, including a requirement that the government pay private landowners if FWS regulations limit development plans as well as changes in the method of species listing.[222] Most significantly as relates to this chapter, however, section 20 of the House Bill would have provided that any agency action in compliance with FIFRA would also be deemed to be in compliance with the ESA.[223] Such a change at the legislative level would remove all FIFRA-related registration questions from the consultation requirements of section 7(a)(2) and would presumably make the recent rule changes and subsequent lawsuits to section 7(a)(2) moot. Environmental groups were united in their strong opposition to the Bill, which was characterized as an all-out assault on the ESA and an unmitigated disaster for endangered wildlife.[224] After the House passage of the Pombo Bill, it languished in the Senate, where Senate Committee on Environment and Public Works—Fish, Wildlife and Water Subcommittee Chair Lincoln Chafee (R-RI) and minority Committee members including then Senators Hillary Rodham Clinton (D-NY) and Barack Obama (D-IL) opposed the passage of the Bill. With the Democratic takeover of the Congress in the Fall of 2006 and Representative Pombo's failed 2006 re-election campaign, the House Bill appears to be dead, at least for the time being. Nevertheless, controversy over the ESA, as well as a variety of efforts to reauthorize the Act, continues. Recently, Congress has been considering the relationship between the ESA and FIFRA. In March, 2011, the National Academies, National Research Council was asked to conduct a study assessing ecological risk assessment under FIFRA and the ESA. The NRC committee on Ecological Risk Assessment under FIFRA and the ESA was tasked with examining the "scientific and technical issues related to the methods and assumptions used by the [EPA, USFWS and NOAA] to conduct scientific assessments of the ecological risks from pesticides registered by EPA under [FIFRA] to species listed under the [ESA]. Specifically, the NRC statement of task directs the committee to address issues of what constitutes the best available scientific data and information, how to address sublethal, indirect,

221 H.R. 3824 Threatened and Endangered Species Recovery Act of 2005, 109th Congress, 1st Session.

222 *Id.*

223 H.R. 3824, Section 20(a), page 81. A report entitled "Implementation of the Endangered Species Act of 1973," submitted to the House Committee on Resources provides some insight into the various amendments to the ESA. "Implementation of the Endangered Species Act of 1973," Report to the House Committee on Resources, Richard W. Pombo, 109th Congress, May 2005, *available at* http://www.waterchat.com/Features/Archive/050517_ESA_Implementation_Report.pdf.

224 *See* Bob Irvin, *Pombo's "All-Out Assault"*, The Sacramento Bee, Wednesday 22 February 2006, *available at* http://www.klamathbucketbrigade.org/SacBeeOpinion_Pombo'sall-outassault022406.htm (last visited May 26, 2011).

and cumulative effects, how to address mixtures and inert ingredients, the use of models, interpretations of uncertainty, and the use of geospatial information and datasets.[225] At the time of the writing of this book, the NRC Committee had not yet completed its work or released its report.

In May, 2011, a joint hearing of the House Committees of Agriculture and Natural Resources was held to look at the relationship between the ESA and pesticide use. Members of Congress questioned officials of EPA, USDA, FWS and NMFS about the ongoing consultations that have been initiated in response to the various court orders and settlements since the 2004 *Washington Toxics Coalition* case.[226] Some Members suggested that EPA stop participating in the consultation process and that the Services stop issuing BiOps until after the NRC committee has examined the scientific issues that have arisen in some of the consultations.[227] The House of Representatives' versions of the 2012 Farm Bill, under consideration during the summer of 2012, contained provisions that would bar EPA from amending pesticide registrations due to endangered species concerns.[228] On June 16, 2011, more than 130 environmental, public health, food safety and sustainable-farming groups sent a joint letter to EPA asserting that there are several significant flaws in the FIFRA pesticide registration process and arguing that the law needs to be strengthened. The letter requests that EPA immediately initiate formal consultations under the ESA on pesticides "known to be harmful to hundreds of federally threatened and endangered species." The letter also discusses the recent Congressional interest in weakening the law and the substantial influence that the pesticide industry has over EPA's decisions under FIFRA.[229] At the time of this writing, it is difficult to predict what, if any, action Congress may take to address the conflicts between FIFRA and the ESA.

The Sources of Tension in the Law

To fully understand the conflicts between the ESA and FIFRA, it is helpful to consider the political and historical atmosphere in which each statute was enacted. One of the most significant conflicts between the ESA and FIFRA is the completely

225 The National Academies, Committee on Ecological Risk Assessment under FIFRA and ESA: Statement of Task (2011).

226 *See House Considers Pesticide Uses, Endangered Species*, Farm Press Daily (May 9, 2011), *available at* http://westernfarmpress.com/government/house-considers-pesticide-uses-endangered-species (last visited June 18, 2011).

227 *Id.*

228 H.R. 872, Federal Agriculture Reform & Risk Management Act (July 2012).

229 Ensure Pesticide Evaluations Protect Endangered Wildlife and Human Health, letter to Lisa P. Jackson, Administrator of the US Environmental Protection Agency from Center for Biological Diversity, Pesticide Action Network North America and 131 More Organizations and Businesses, *available at* http://www.panna.org/press-release/130-groups-call-epa-stronger-pesticide-protections (last visited June 18, 2011).

different standards that govern regulatory action under the respective statutes. As discussed above, FIFRA, at least in the way EPA has chosen to implement it over the past three or more decades, involves a balancing of the risks associated with the use of the pesticide against the social and economic benefits to society accruing from the use of the pesticide. Thus, even a pesticide that poses high risks to threatened or endangered species in theory could be registered under FIFRA if that pesticide provides economic benefits that outweigh those risks. Section 9 of the ESA, on the other hand, prohibits "takes" of threatened and endangered species. Economic considerations do not come into play under this section. The section 7 consultation mandates ensure that federal agency actions do not jeopardize the continued existence of a threatened and endangered species. Accordingly, the very terms of the statutes have created a catch-22 situation for EPA. If EPA follows the FIFRA cost/benefit standard, it may approve a pesticide that jeopardizes a threatened or endangered species. Accordingly, it may be in violation of the ESA. On the other hand, if EPA chooses to comply with the ESA and deny or severely restrict a registration, EPA could be vulnerable to legal challenges for not properly implementing its FIFRA mandate to consider economics in its registration decisions. Moreover, because pesticides are by their very nature intended to kill organisms in the environment and because there is habitat for the more than 1,800 listed species throughout a wide and vast range of the territory of the United States, strict compliance with the ESA under the existing FIFRA framework likely would result in EPA banning or severely restricting a large majority of registered pesticides. Such an interpretation would lead to the ESA virtually swallowing up FIFRA. This dilemma is likely a large contributor to EPA's ongoing reluctance to comply with the ESA in implementing its pesticide registration program. The only reconcilable approach under the existing laws appears to be to impose detailed label instructions for each pesticide in each geographic location in which that pesticide may adversely affect a listed species through the FIFRA labeling mechanisms. Unfortunately, this approach is extremely unwieldy and as discussed above could result in extremely lengthy and complex label instructions that are unlikely to be complied with. Consequently, the existing FIFRA structure is simply incompatible with the mandates of the ESA.

As with the ESA, the MBTA's standards are not easily reconcilable with those of FIFRA. First, the MBTA imposes a strict liability standard for "takes" of migratory birds.[230] Courts have applied this strict liability standard to pesticide-related bird deaths.[231] This strict liability standard is in direct conflict with the explicit balancing decisions required for FIFRA pesticide registration. As one author has stated, "regular repeated bird kills might … [be] tolerated had the benefits of the pesticide in question been greater."[232] Moreover, as with the ESA,

230 Mineau, *supra* note 137, at 313, 329–31.

231 United Sates v. FMC Corp., 572 F.2d 902 (2d Cir. 1978) & United States v. Corbin, 444 F.Supp. 510 (E.D. Cal. 1978).

232 Mineau, *supra* note 137, at 331–32.

and as others have noted, pesticide labeling under FIFRA does not protect birds from poisoning.[233]

In addition to the conflicting standards of the ESA and FIFRA, the differing focuses of the two statutes create incompatibility. FIFRA creates a national registration process, while the ESA evaluates individual actions' impacts on a specific habitat and species. The ESA is concerned with preventing injury to individual members of each listed species and preventing significant modifications to the habitat of each listed species that would result in injury to the members of the species. Such modifications include habitat modifications that impact breeding and nesting, activities that typically occur in specific geographic locations during specific times of the year for each species. The ESA is also concerned with preventing injury to designated critical habitats, which by their very nature are geographically defined. Accordingly, the ESA is geographically and temporarily focused. On the other hand, under FIFRA, a decision on whether to register or cancel a pesticide is made on a nationwide basis without any detailed consideration of specific geographic or temporal factors. For example, a particular pesticide may easily meet the cost/benefit registration standard because on a nationwide basis the benefits of the pesticide far exceed the environmental or health costs. However, this decision ignores the fact that the pesticide may pose substantial risks to a particular listed species that nests in a particular geographic location during certain times of the year. Although, in theory, such geographic and temporal concerns could be addressed through label restrictions directing users not to use the pesticide in certain geographic locations during certain times of the year, the reality is that they would be extremely unwieldy. It would be extremely unlikely that EPA could require such detailed label restrictions on every pesticide product to address every geographic or temporal restriction needed to protect every listed species in the entire United States. Moreover, even if EPA did require such detailed label restrictions, it is unlikely that a pesticide user would take the time to read these complex restrictions and determine which, if any, restrictions apply to the user's intended use in a given location and at a particular time for each and every listed species that may be affected, let alone actually comply with such restrictions. Moreover, monitoring users to ensure they comply with the label restrictions and enforcing against those who did not would be virtually impossible.

Finally, the ESA and FIFRA are inconsistent in that they provide for very different risk-reduction methods. Under the ESA, the FWS or NMFS will issue, as part of a BiOp, an incidental take statement, which identifies actions which will not be considered to be a prohibited taking under section 9. The incidental take statement specifies the reasonable and prudent measures that must be implemented to minimize risk of takes. Unless these measures are complied with, any resulting

233 *Id.* at 337–8. In this article, the author concludes that because MBTA's provisions relate only to direct, lethal pesticide exposures, they do not fully address the problem. *Id.* at 335. The author concludes that pesticide labeling, on its own, fails to protect migratory birds. *Id.* at 337–8.

takes will be a violation of the Act. These reasonable and prudent measures typically are very detailed and very species-specific, geographically defined and temporally defined. As described above, FIFRA's mechanism for regulating use of pesticides to reduce risk is through label restrictions. Imposition of detailed reasonable and prudent measures set forth in incidental take statements in BiOps is impracticable and unlikely to result in widespread compliance by purchasers and users of pesticides.

Chapter 13

International Perspectives on Pesticide Law and Ecology

Concerns regarding the relationship between pesticides and ecology are not unique to the United States. Pesticide concerns arise anywhere on the globe where crops are grown or where vector-borne diseases exist—in other words, virtually everywhere on earth. Many of the same concerns related to pesticides in the US are shared by other countries. For example, finding appropriate ways to control pests while minimizing harms to humans and the environment from pesticides are worldwide concerns. However, not all parts of the world face all of the same challenges. The developing world in particular faces a number of unique challenges from both a high need for pest control and high ecological and human health risk from pesticide use.

The Green Revolution, which transformed US agriculture to be chemical input reliant, also helped to create a chemical pesticide addiction in the developing world. As in the US, a combination of government policies and agrichemical industry influence promotes chemical pesticide use in the developing world by relying on "prophylactic calendar-based spraying" to achieve maximum yields.[1] Government policies in many developing countries mirror those of the US, with subsidy and crop insurance programs incentivizing industrial agricultural approaches, including heavy reliance on chemical pesticides.[2] However, there is great variation in chemical pesticide use among different regions of the developing world.

Many areas of Asia are major users of chemical pesticides. Van Emden and Peakall explain how pesticide use in Asia skyrocketed during the last few decades of the twentieth century.[3] By 1990, more than 26 percent of the pesticide world market was in Asia, compared to approximately 22 percent in North America, 6 percent in Africa and West Asia and 11 percent in Latin America.[4] Depending on the exact region, most Asian chemical pesticide use is on rice and cotton.

Although chemical pesticides still dominate in most of Asia, government-sponsored Integrated Pest Management (IPM) programs are beginning to play a larger role in many areas. Van Emden and Peakall provide a fascinating example of a shift in government policy in Indonesia that led to a dramatic reduction in

1 Helmut F. van Emden & David B. Peakall, Beyond Silent Spring: Integrated Pest Management and Chemical Safety 168 (1996).

2 *Id.* at 168–9.

3 *Id.* at 169.

4 *Id.*

chemical use and a marked increase in the adoption of IPM techniques.[5] During the Green Revolution, Indonesia was able to dramatically increase its rice production through intensive chemical pesticide use. However, this intensive pesticide use ultimately led to the development of a major pest problem from the brown leafhopper, a previously minor pest. The brown leafhopper pest outbreak led to significant economic loss. Attempts to control the pest with chemical pesticides only exacerbated the problem. In response, during the early to mid-1980s, the Indonesian government adopted IPM and the reduction of chemical pesticide use as the official government policy. Consequently, not only did pesticide use fall by approximately 60 percent, but rice production grew by approximately 15 percent.[6]

As van Emden and Peakall describe it, in contrast to the widespread pesticide use in Asia, the pesticide culture in most parts of Africa and Latin America is quite varied. In most regions of Africa, agriculture is dominated by small-scale farmers. The high cost of chemical pesticides, and, in some areas, the limited availability of these chemicals, limits their use by the small-scale farmers.[7] Nevertheless, pesticide use is widespread on larger farms and commercial cooperatives.[8] Due to the vast geographical, climatic, economic, and cultural diversity in Latin America, as well as the varying access to international markets, pesticide use varies greatly from region to region.[9] Although in recent years there has been some movement toward IPM and even organic farming in the developing world, many farmers in developing countries are subject to even greater constraints than their American counterparts. In many parts of the developing world, basic research on pests, economic thresholds, and IPM practices is inadequate.[10] Even where research exists, there is a great need for better organizations to disseminate technical knowledge and train farmers in IPM, as well as financial support to help facilitate a shift to IPM.[11] IPM requires detailed knowledge and ongoing monitoring, which, although perhaps worthwhile in the long-run, is a major barrier to its initial adoption by the average farmer trying to eke out a living who is comfortable with predictable spraying regimes that do not require ongoing monitoring or specialized knowledge.[12]

One controversial issue with regard to pesticide use in the developing world is whether pesticides that are banned or severely restricted in the US or other developed countries should be exported to countries in the developing world. For example, for decades after the 1972 banning of DDT in the US, American pesticide manufacturing companies continued to manufacture DDT

5 *Id.* at 168, 70.
6 *Id.* at 175–6.
7 *Id.* at 185.
8 *Id.* at 186.
9 *Id.* at 200.
10 *Id.* at 215.
11 *Id.* at 216–17.
12 *Id.*

and export it to the developing world. Oftentimes, in what has come to be known as the "circle-of-poison," pesticides banned for use in the US that are nevertheless exported to developing countries ultimately end up coming back into the US as residues on foods. While pesticide residues on food are not permitted unless in compliance with a tolerance established under the Federal Food, Drug, and Cosmetic Act (FFDCA),[13] testing of imported foods to detect unpermitted pesticide residues is very limited and there is thus certainly some level of illegal residue entering the US on imported food. The larger issue, however, is whether it is appropriate for US manufacturers to sell products deemed to be too risky for the US to other parts of the world. Of the approximately 1.2 billion pounds of pesticides produced in the US per year, approximately 300 million pounds are exported out of the US to other parts of the world. At the same time, approximately 200 million pounds of pesticides are imported into the US from other countries, resulting in a net supply of 1.1 billion pounds per year.[14]

US Law on the Import and Export of Pesticides

FIFRA section 17 expressly addresses requirements for pesticide export and import in the US.[15] EPA regulates the import and export of pesticides under FIFRA. The importation of pesticides is governed by FIFRA section 17(c).[16] All imported pesticides intended for use in the United States must be registered as required by section 3 of FIFRA before being permitted entry into the US.[17] As is required for any pesticide registration, imported pesticides must be properly labeled in accordance with EPA's labeling regulations adopted pursuant to FIFRA, must not be adulterated if intended for animal feed or food use, must have tolerances or exemptions from tolerances under the FFDCA, and must otherwise comply with all other legal requirements under FIFRA.[18] When importing pesticides into the US, the importer is required to provide a "Notice of Arrival (NOA) of Pesticides and Devices" to the appropriate EPA Regional Office.[19] Once the EPA Regional Office has approved the shipment, EPA returns the NOA form to the importer.[20] Upon arrival of a shipment of pesticides or devices, the importer must present

13 21 U.S.C. § 346.

14 US Environmental Protection Agency, About Pesticides, Pesticide Industry Sales and Usage 2006/2007, *available at* http://www.epa.gov/opp00001/pestsales/ (last visited April 13, 2011).

15 7 U.S.C. § 136o (2011).

16 *Id.* at § 136o(c) (2011).

17 US Environmental Protection Agency, Pesticides: International Activities, EPA Import and Export Trade Requirements, *available at* http://www.epa.gov/oppfead1/international/trade/index.html (last visited May 26, 2011).

18 40 C.F.R. § 152.50 (2011).

19 19 C.F.R. § 12.112 (2011).

20 *Id.*

the approved NOA form to the district director of Customs at the port of entry.[21] EPA's Office of Enforcement and Compliance Assurance (OECA) is primarily responsible for enforcement of pesticide importing requirements, with the support of EPA's Regional Offices.[22]

In contrast to the full suite of FIFRA and FFDCA regulations that apply to pesticides imported into the US, pesticides exported from the US to other countries are subject only to very limited regulation under FIFRA.[23] These limited regulations are primarily in the form of notification. Unregistered pesticides—i.e., pesticides that are not approved for use in the US may be manufactured in the US and exported subject to the requirements of FIFRA section 17(a). Specifically, FIFRA section 17(a) provides, "[N]o [unregistered] pesticide ... intended solely for export to any foreign country shall be deemed in violation of this subchapter ... if prior to export, the foreign purchaser has signed a statement acknowledging that the purchaser understands that such pesticide is not registered for use in the United States ...."[24] The requirement is shipment-specific for a particular exporter, product, and purchaser. EPA has stated that it is placing the "highest priority on timely notification for two categories of exported pesticides [. . .]: 1) pesticides on the International list of Prior Informed Consent (PIC), most of which have been banned or are severely restricted in the US; and 2) other pesticides banned and severely restricted in the US for health or environmental reasons, which are not on the PIC list."[25]

While FIFRA section 17(a) addresses requirements for exporting any unregistered pesticide, FIFRA section 17(b) addresses the export of pesticides for which EPA has canceled or suspended the registration. Specifically, FIFRA section 17(b) provides, "[w]henever a registration, or cancellation or suspension of the registration of a pesticide becomes effective, or ceases to be effective, the Administrator shall transmit ... notification thereof to the governments of other countries and to appropriate international agencies."[26] Thus pesticides that EPA has determined pose unreasonable adverse effects on the environment and thus are not permitted to be used in the US, may still be legally exported as long as EPA has provided a one-time notification of its regulatory action to other governmental and international agencies.[27] Unlike the foreign purchaser acknowledgement

21 19 C.F.R. § 12.113 (2011).

22 US Environmental Protection Agency, Pesticides: International Activities, EPA Import and Export Trade Requirements, *available at* http://www.epa.gov/oppfead1/international/trade/index.html (last visited May 26, 2011).

23 *Id.*

24 7 U.S.C. § 136o(a)(2) (2011).

25 US Environmental Protection Agency, Pesticides: International Activities, EPA Import and Export Trade Requirements, Export of Unregistered Pesticides (FIFRA section 17(a)), *available at* http://www.epa.gov/oppfead1/international/trade/index.html (last visited May 26, 2011).

26 7 U.S.C. § 136o(b) (2011).

27 *Id.*

requirement, however, there is no requirement that the foreign government sign any type of acknowledgement statement, and the notification is not provided on a per shipment basis.[28] In other words, FIFRA section 17(b) merely requires a notification of EPA regulatory action and leaves it completely to the discretion of the importing country as to whether they allow the import or impose any additional requirements.

In addition to the foreign purchaser acknowledgement statement and notification requirements of sections 17(a) and (b), as described above, EPA by regulation imposes certain labeling requirements for any pesticide intended solely for export from the US.[29] Specifically, pesticides intended solely for export are not required to be registered provided the pesticides are labeled "Not Registered for Use in the United States," meet certain minimal labeling requirements, are labeled in both English and the language(s) of importing country, and the exporter complies with specified record-keeping requirements. [30] Although EPA's restrictions on pesticide exports are limited, for many years international organizations have made attempts to impose additional requirements on pesticide exports and to impose other international standards to address the risks of pesticides.

International Agreements on Pesticides

International efforts to address the risks of pesticides, primarily focused on the human health effects, date back to the early 1960s when the World Health Organization (WHO) established the WHO Pesticide Evaluation Scheme (WHOPES). This scheme, which has been updated in more recent years, is comprised of a multi-phased testing program, ranging from laboratory studies to large-scale and epidemiological studies to assess the efficacy, human health impacts, and environmental impacts of pesticides.[31] However, international efforts to go beyond research and regulate the sale or use of pesticides did not commence until the mid-1980s, and significant international agreements relating to pesticides were not reached until the 2000s. Many of these agreements still have not been fully implemented. In general, most of the international agreements on pesticides include a push to "harmonize" regulatory standards. One concern with such harmonization is that countries with higher standards will have to lower their standards to harmonize with others.

---

28 *Id.*

29 40 C.F.R. § 168.65 (2011).

30 *Id.* at § 168.65(b) (2011).

31 World Health Organization, WHO Pesticide Evaluation Scheme: WHOPES, *available at* http://www.who.int/whopes/en/ (last visited June 3, 2011).

The United Nations Food and Agriculture Organization (FAO) International Code of Conduct on the Distribution and Use of Pesticides

One of the earliest international attempts to address concerns with the export of pesticides was the United Nations Food and Agriculture Organization (FAO) International Code of Conduct on the Distribution and Use of Pesticides.[32] This is a voluntary code of conduct for pesticide management practices that provides standards for public and private entities engaged in, or associated with, the application or distribution of pesticides.[33] The Code contains provisions for reducing health and environmental risks, and encourages governments and the pesticide industry to cooperate to reduce such risks. First published in 1985, the code was updated in 1989 to include Prior Informed Consent (PIC) procedures. The FAO Code was again revised in 2002 to reflect changes in international policy related to the adoption of the Rotterdam Convention on the Prior Informed Consent procedure in 1998. The most recent, 2005, revision adopts an expanded definition of Integrated Pest Management (IPM) and incorporates a life-cycle approach to pesticide management.[34] The PIC procedures in the FAO Code became the model for the later adoption of mandatory PIC procedures in the 1998 Rotterdam Convention discussed below.

The Rotterdam Convention on the Prior Informed Consent Procedure for Certain Hazardous Chemicals and Pesticides in International Trade

One of the most important international agreements relating to the import and export of pesticides throughout the world is the Rotterdam Convention on the Prior Informed Consent Procedure for Certain Hazardous Chemicals and Pesticides in International Trade.[35] This multilateral treaty seeks to promote shared responsibilities regarding the importation of hazardous chemicals, including pesticides, and to promote the open exchange of information.[36] The treaty encourages exporters of hazardous chemicals to comply with labeling requirements, to include safe handling instructions, and to notify buyers of any known bans or restrictions.[37] Parties to the treaty are free to make their own decisions whether to

---

32 Food and Agriculture Organization of the United Nations, International Code of Conduct on the Distribution and Use of Pesticides (Revised 2005), *available at* ftp://ftp.fao.org/docrep/fao/009/a0220e/a0220e00.pdf.

33 *Id.* at 1.

34 *Id.* at 2.

35 Rotterdam Convention on the Prior Informed Consent Procedure for Certain Hazardous Chemicals and Pesticides in International Trade (2008), *available at* http://www.pic.int/en/ConventionText/RC%20text_2008_E.pdf.

36 *Id.* at 4 (Article 1).

37 Rotterdam Convention, What is RC? Overview, *available at* http://www.pic.int/home.php?type=t&id=5&sid=16 (last visited June 5, 2011).

allow or ban the importation of chemicals listed in the treaty.[38] Exporting countries are required to ensure that producers within their jurisdiction comply with treaty requirements.[39] One of the most significant features of the treaty is that it creates legally binding obligations requiring member countries to implement the Prior Informed Consent (PIC) procedure.[40] The PIC procedure is based on the voluntary PIC procedure initiated by the original FAO International Code of Conduct on the Distribution and Use of Pesticides in 1989.[41] The Rotterdam PIC process requires exporters of pesticides covered under the Convention, which include pesticides that have been banned or severely restricted by Parties and pesticides that Parties have notified for inclusion, to obtain the prior informed consent of an importing country before exporting to that country.[42] A new chemical is considered for inclusion under the Convention when two or more countries in specified regions of the world nominate the chemical for inclusion.[43] The Conference of the Parties (COP) makes decisions regarding the inclusion of new chemicals in Annex III of the Convention.[44] Once a chemical is listed in Annex III, a "decision guidance document" (DGD) containing information about the chemical and any regulatory decisions banning or restricting it is circulated to all Parties.[45] Currently, there are 40 chemicals listed in Annex III of the Convention and therefore subject to the PIC procedure, including 25 pesticides, 4 severely hazardous pesticide formulations, and 11 industrial chemicals.[46] The Rotterdam Convention website provides information on the current substances listed under the Convention.[47] Once the Parties receive the DGD, they have nine months to draft an interim or final response regarding whether to allow future import of the chemical and whether any conditions will apply to its import.[48] Importing countries' decisions must be trade-neutral and therefore must "apply equally to domestic production for domestic use as well as to imports from any source." The PIC Secretariat circulates the import decisions and the Convention requires exporting country Parties "to take appropriate measures to ensure that exporters within its jurisdiction comply with the decisions."[49] While the Convention primarily relates to chemicals and pesticides that Parties have "banned or severely restricted," it is important

38 *Id.*

39 *Id.*

40 Rotterdam Convention, *supra* note 35, at 10–11.

41 Rotterdam Convention, What is RC? Overview, *available at* http://www.pic.int/home.php?type=t&id=5&sid=16 (last visited June 5, 2011).

42 Rotterdam Convention, *supra* note 35, at 12–13.

43 *Id.* at 8.

44 *Id.*

45 *Id.* at 9.

46 What is RC? *supra* note 41.

47 Rotterdam Convention, Chemicals, Annex III, *available at* http://www.pic.int/home.php?type=t&id=29&sid=30 (last visited May 26, 2011).

48 Rotterdam Convention, *supra* note 35, at 10.

49 What is RC? *supra* note 41.

to note that inclusion under the Convention does not represent a global "ban" on the listed substances.[50] The PIC process therefore enables Importing Parties to make informed decisions, based on information provided through the Convention, regarding whether to allow import of the chemical or pesticide in question.[51]

In addition to the PIC procedure itself, the Convention also includes a number of requirements related to pesticide export. A Party that plans to export a chemical that is banned or severely restricted for use within its jurisdiction must, prior to the first shipment and annually thereafter, inform the importing Party that such export will take place.[52] In addition, when exporting chemicals for occupational use, exporting Parties are required to ensure that up-to-date safety information data sheets are sent to the importer.[53]

The Rotterdam PIC process was adopted on September 10, 1998 in Rotterdam, the Netherlands, and entered into force on February 24, 2004.[54] Seventy-two countries including the US have signed the Rotterdam convention, though to date, the United States is a Signatory, but not a Party.[55] For the United States to become a Party, the Senate must ratify the Convention and Congress must pass implementing legislation.[56] It is EPA's stated intention to make the US export notification program compatible with the international process contemplated under the Rotterdam Convention, while still meeting domestic legislative requirements.[57] EPA also states that, "revisions to the US export notification program will be considered in the context of implementation of the PIC Agreement."[58]

The Stockholm Convention on Persistent Organic Pollutants

In addition to the Rotterdam Convention, which imposes certain requirements on the exportation of hazardous chemicals and pesticides, certain substances that have been determined to pose unacceptable environmental risks are subject to this more proscriptive international agreement. The Stockholm Convention on Persistent Organic Pollutants is an international environmental treaty that seeks to eliminate or restrict the production and use of persistent organic pollutants

---

50 *Id.*

51 *Id.*

52 Rotterdam Convention*, supra* note 35, at 12.

53 *Id.* at 14.

54 US Environmental Protection Agency, Pesticides: International Activities, International Agreements and Treaties, *available at* http://www.epa.gov/oppfead1/international/agreements/index.html (last visited May 26, 2011).

55 Rotterdam Convention, Parties, Ratifications, *available at* http://www.pic.int/home.php?type=t&id=63&sid=17 (last visited June 5, 2011).

56 International Activities, *supra* note 54.

57 *Id.*

58 *Id.*

Bridge Document 13.1 The Stockholm Convention on Persistent Organic Pollutants

The Parties to this Convention,

Recognizing that persistent organic pollutants possess toxic properties, resist degradation, bioaccumulate and are transported, through air, water and migratory species, across international boundaries and deposited far from their place of release, where they accumulate in terrestrial and aquatic ecosystems,

Aware of the health concerns, especially in developing countries, resulting from local exposure to persistent organic pollutants, in particular impacts upon women and, through them, upon future generations,

Acknowledging that the Arctic ecosystems and indigenous communities are particularly at risk because of the biomagnification of persistent organic pollutants and that contamination of their traditional foods is a public health issue,

Conscious of the need for global action on persistent organic pollutants, …

(POPs).[59] Under the treaty, countries agree to reduce or eliminate the production, use, and/or release of 12 POPs, and the Convention contains a review process that is used to initiate the addition of other POPs of global concern.[60] The Convention covers pesticides, industrial chemicals, and unintentionally produced POPs.[61] The 12 substances that were the original target of the Stockholm Convention and which are commonly referred to as the "Dirty Dozen" include the pesticides: aldrin, chlordane, DDT, deildrin, endrin, hexachlorobnzene, mirex and toxaphene and the nonpesticidal PCBs, dioxins and furans.[62] In addition, the Convention of the Parties listed nine additional chemicals in 2009, including the pesticides: chlordecone, alpha hexachlorocyclohexane, beta hexachlorocyclohexane, lindane, and pentachlorobenzene.[63]

POPs are singled out in the convention as being of particular concern and requiring action greater than simply the PIC procedure of the Rotterdam Convention. As described above, many of the dirty dozen POPs, such as DDT, were used extensively after World War II, when industry was undergoing a technological

59 Stockholm Convention on Persistent Organic Pollutants (2009), *available at* http://chm.pops.int/Convention/ConventionText/tabid/2232/language/en-GB/Default.aspx (last visited June 5, 2011).

60 International Activities, *supra* note 54.

61 Stockholm Convention on Persistent Organic Pollutants, Convention, The POPs, *available at* http://chm.pops.int/Convention/ThePOPs/tabid/673/language/en-GB/Default.aspx (last visited June 5, 2011).

62 *Id.*

63 *Id.*

boom and the Green Revolution was beginning. As described above, POPs are of particular concern not only because they are toxic to humans and the environment, but because they can persist in the environment for extremely long periods of time, can bioaccumulate in the food chain, and can travel long distances via wind and water.[64] POPs have been found virtually everywhere on earth, including thousands of miles away from any place they have been used in pristine Arctic areas. Many POPs are also believed to be endocrine disruptors in both humans and other species. Of the 12 substances on the Dirty Dozen list, 9 are pesticides and 1, dioxin, can be found as a trace contaminant in certain pesticides.[65] The goal of the Stockholm Convention is to phase out or, at a minimum, reduce the production, use, and release into the environment of the Dirty Dozen POPs.[66] Stockpiles of wastes containing POPs must be managed in a safe, efficient, and environmentally sound manner, taking into account international rules, standards, and guidelines.[67] Parties are required to submit National Implementation Plans (NIPs) within two years after joining the convention.[68] The Convention also establishes a system to provide technical and financial assistance to developing countries to help them to meet their obligations under the agreement.[69] Although the Stockholm Convention initially focused on the Dirty Dozen, the agreement provides a process to add more chemicals to the list.[70] In May of 2009 the Convention began to add other substances to the list. The US is a signatory and the treaty was ratified in 2004.[71]

The Codex Alimentarius Commission

This Commission is part of the UN FAO and the WHO joint food standards program.[72] The Codex Commission sets international standards related to food

---

64 *Id.*

65 US Environmental Protection Agency, International Programs, Toxics, Persistent Organic Pollutants: A Global Issue, A Global Response, *available at* http://www.epa.gov/international/toxics/pop.html (last visited May 26, 2011).

66 Stockholm Convention on Persistent Organic Pollutants, Convention, The POPs, *available at* http://chm.pops.int/Convention/ThePOPs/tabid/673/language/en-GB/Default.aspx (last visited June 5, 2011).

67 Stockholm Convention on Persistent Organic Pollutants, *supra* note 59, at 12–14 (Article 6).

68 *Id.* at 14 (Article 7).

69 *Id.* at 20–22 (Articles 12 & 13).

70 *Id.* at 15-16 (Article 8).

71 US Environmental Protection Agency, Pesticides: International Activities, International Agreements and Treaties, Stockholm Convention on Persistent Organic Pollutants, *available at* http://www.epa.gov/oppfead1/international/agreements/index.html (last visited May 26, 2011).

72 Food and Agriculture Organization/World Health Organization Joint Food Standards Program, Codex Alimentarius, *available at* http://www.codexalimentarius.net/web/index_en.jsp (last visited June 5, 2011).

safety, which are relied upon under other international and trade agreements.[73] The primary mission of Codex with regard to pesticides is to develop Maximum Residue Levels (MRLs) for pesticides in food.[74] Countries that adopt the MRLs developed by Codex are presumed to be in compliance with WTO trade agreement obligations. In the US, in setting tolerances under the FFDCA for pesticides in food, EPA must either adopt the Codex MRLs or provide an explanation of why EPA has determined that the Codex standards are not adequate.[75]

Other International Agreements Relating to Pesticides

The Rio Declaration on Environment and Development

In addition to the international agreements specifically addressing pesticides, a number of other international agreements, including environmental agreements and trade agreements, have significant implications for pesticide trade, use, and disposal. The Rio Declaration on Environment and Development is a document produced at the 1992 UNCED Earth Summit in Rio de Janeiro for the purpose of conserving and promoting the sustainable use of biological diversity, and achieving the fair and equitable sharing of benefits arising from genetic resources designed for commercial use.[76] It consists of 27 principles intended to guide global sustainable development.[77] While the Declaration does not specifically address pesticide concerns, it does contain a number of broad provisions that apply to national decisions regarding pesticides as well as to other environmental decision making. Among other things, the Declaration includes provisions that affirm state sovereignty over natural resources, give priority to the least developed nations, require countries to encourage public awareness and participation in environmental decision making, compensate for victims of pollution and other environmental damage, promote the precautionary principle, and promote environmental impact assessment for proposed activities that are likely to have a significant adverse impact on the environment.[78]

---

73 Codex Alimentarius, Food and Agriculture Organization/World Health Organization Joint Food Standards Program, Understanding the Codex Alimentarius (2006), *available at* ftp://ftp.fao.org/codex/Publications/understanding/Understanding_EN.pdf.

74 *Id.* at 10.

75 US Environmental Protection Agency, Pesticides: Regulating Pesticides, Food Quality Protection Act FFDCA Amendments, *available at* http://www.epa.gov/pesticides/regulating/laws/fqpa/fqpaffdc.htm (last visited June 3, 2011).

76 United Nations Environment Program, Rio Declaration on Environment and Development, *available at* http://www.unep.org/Documents.Multilingual/Default.asp?DocumentID=78&ArticleID=1163&l=en (last visited June 5, 2011).

77 *Id.*

78 *Id.*

The worldwide concern over the safety of biotechnology products is evidenced by the significant role that biotechnology played in the negotiations of the United Nations Convention on Biological Diversity.[79] The agreement that was eventually reached contained a number of provisions that were relevant to biotechnology. Two provisions in particular, Article 8 (In-situ Conservation) and Article 19 (Handling of Biotechnology and Distribution of its Benefits), were directed at addressing international concerns with biotechnology products.[80] The Rio Declaration recognized the need for, and established a process for developing, an international agreement on genetically modified organisms (GMOs). Specifically, Article 8 requires that Contracting Parties establish or maintain means to regulate, manage, or control the risks associated with the use and release of living modified organisms resulting from biotechnology that could affect the conservation and sustainable use of biological diversity.[81] Article 19(3) of the Declaration directed the Parties to "consider the need for and modalities of a protocol setting out appropriate procedures, including, in particular, advance informed agreement, in the field of the safe transfer, handling and use of any living modified organisms resulting from biotechnology that may have adverse effects on the conservation and sustainable use of biological diversity."[82] The resulting agreement on GMOs is referred to as the Cartagena Protocol on Biosafety to the Convention on Biological Diversity.

Cartagena Protocol on Biosafety to the Convention on Biological Diversity

The Cartagena Protocol is an international agreement supplemental to the Convention on Biological Diversity, designed to protect biodiversity from potential risks associated with living modified organisms (LMOs), in other words GMOs that are alive, that result from modern biotechnology.[83] The Protocol applies to the transboundary movement, handling, transport, and use of all living modified organisms that may have potentially adverse effects on biological diversity or human health.[84] Requirements of the Protocol are designed to provide interested parties with information about safety and handling of LMO products so importing parties can make informed decisions about whether to accept LMO imports. The Protocol establishes a "Biosafety Clearing House" to facilitate the exchange of scientific, legal, technical, and environmental information on LMOs and to provide Parties with assistance in implementing the Protocol.[85]

---

79 United Nations, Convention on Biological Diversity (1992), *available at* http://www.cbd.int/doc/legal/cbd-en.pdf.

80 Cartagena Protocol on Biosafety to the Convention on Biological Diversity, January 29, 2000, 39 I.L.M. 1027 (2000).

81 Convention on Biological Diversity, *supra* note 79, at 6.

82 *Id.* at 12.

83 Cartagena Protocol, *supra* note 80, at 1.

84 *Id.* at 5.

85 *Id.* at 15–16.

Significantly, the Protocol broadly applies to all living modified organisms, whether pesticidal or not, but only applies to those that are actually "living."[86] This Protocol emphasizes the use of the precautionary approach in approval of new technologies and seeks to balance public health concerns with economic benefits in approval of new technologies.[87] One of the most specific features of the Protocol is that, similar to the Rotterdam PIC procedure, it contains what is referred to as "Advance Informed Agreement" procedures for LMOs intentionally introduced into the environment, and specific procedures for LMOs that are intended for use in food or during food production.[88] Under this procedure, the Party of export or the exporter must provide notification and the Party of import must acknowledge receipt of such notification.[89] In addition, the Protocol establishes a decision procedure for importing Parties to follow when determining whether to allow the import of a particular LMO, and an opportunity for review of such decisions.[90] Although the Protocol states that it is reaffirming the precautionary approach contained in the Rio Declaration on Environment and Development, it also states that it recognizes that trade and environment agreements should be mutually supportive and that the Protocol shall not be interpreted as implying a change in the rights and obligations of a Party under any existing international agreements or to subordinate this Protocol to other international agreements.[91]

The advance informed agreement procedure applies only prior to the first international transboundary movement of living modified organisms for intentional introduction into the environment of the Party of import.[92] The Party of import has 90 days to inform the notifier in writing as to whether the international transboundary movement may proceed.[93] The Protocol makes clear that in making such a decision, "[l]ack of scientific certainty due to insufficient relevant scientific information and knowledge regarding potential adverse effects of a living modified organism on the conservation and sustainable use of biological diversity in the Party of import … shall not prevent that Party from taking a decision, as appropriate, … in order to avoid or minimize such potential adverse effects."[94] The Cartagena Protocol entered into force on September 11, 2003. The US is not a party to the Protocol.

GMOs provide an interesting example of the difficulties of addressing biodiversity impacts of pesticides on an international level. Neither FIFRA's export provisions nor EPA's plant-incorporated protectant rule address the broader concern

86 *Id.* at 3.
87 *Id.*
88 *Id.* at 6–7.
89 *Id.*
90 *Id.* at 7–8.
91 *Id.* at 2–3.
92 *Id.* at 6.
93 *Id.* at 7.
94 *Id.* at 8.

of risk to biological diversity internationally. It is conceivable that a pesticide, particularly a plant-incorporated protectant, that does not pose an unreasonable risk in the US, could pose an unreasonable risk in another country. As discussed above, many of the risk issues associated with plant-incorporated protectants relate to their ability to outcrop wild relatives. This is a very different situation than that of conventional pesticides. For conventional chemical pesticides, the risks posed by the substance will tend to be similar regardless of the country or part of the world. For example, if a pesticide is highly toxic to humans or other mammals in the United States, it will also be highly toxic to humans or mammals in other countries. The risks associated with plant-incorporated protectants, on the other hand, may vary more with location depending on a number of factors such as the presence of wild relatives. It would be extremely difficult, if not impossible, for EPA to evaluate all risks of a plant-incorporated protectant in every country in the world. Moreover, to conduct the unreasonable adverse effects analysis required by FIFRA, EPA also must look at the societal benefits associated with the pesticide. It would seem infeasible for EPA to conduct such an analysis for every importing country. EPA is not in a position to evaluate, and place value on, the economic and societal benefits that an importing country derives from a particular pesticide or chemical substance.

Another impediment to EPA addressing risks of plant-incorporated protectants in countries other than the United States is that FIFRA provides EPA with very limited authority to regulate exported pesticides. As described above, the export of pesticides is regulated under section 17 of FIFRA.[95] The primary emphasis of this section is on the provision of information by EPA to foreign governments. Section 17 mandates two systems of notification: a notice to the government of an importing country of the export of unregistered pesticides and a notice to all countries of cancellation or suspension actions taken by EPA.[96] Beyond these notification provisions, FIFRA does not provide EPA with the authority to regulate exports.

The Basel Convention

The Basel Convention on the Control of Transboundary Movements of Hazardous Wastes and Their Disposal[97] does not specifically apply to pesticides, but instead broadly applies to "hazardous wastes." This Convention is designed to reduce and prevent the transboundary movement of hazardous wastes, particularly from developed to less developed countries.[98] The Convention is also intended to reduce the amount and toxicity of generated wastes, ensure environmentally

95 *See* 7 U.S.C. § 136o (2011).

96 *Id.* § 136o(a)(2) & 136o(b) (2011).

97 Basel Convention on the Control of Transboundary Movements of Hazardous Wastes and Their Disposal (1989), *available at* http://www.basel.int/text/17Jun2010-conv-e.pdf.

98 *Id.* at 10.

sound generation and management of such wastes, prevent illegal trafficking in waste, and to provide technical and financial assistance to less developed countries to aid them in managing the hazardous and other wastes they generate in an environmentally sound manner.[99] The convention entered into force in May 1992. The US is a signatory but has not ratified the Basel Convention.

International Trade Agreements Relating to Pesticides

In addition to the various international environmental agreements described above, several international trade agreements contain provisions that are relevant to pesticides and GMOs. Generally, trade agreements seek to facilitate free trade and to eliminate trade barriers. Thus, trade agreements are often in tension with international environmental agreements that impose regulations or permit Parties to impose regulatory standards that could hinder free trade.

The WTO Agreement on Technical Barriers to Trade (TBT Agreement)

The TBT Agreement was created to ensure that regulations, standards, testing, and certification procedures do not erect unnecessary barriers to international trade.[100] It prohibits parties from creating technical requirements designed to limit trade, seeks to promote the development of international technical requirements and standards, and encourages members to adopt the international standards as their own technical requirements.[101] The agreement provides,

> [m]embers shall ensure that technical regulations are not prepared, adopted or applied with a view to or with the effect of creating unnecessary obstacles to international trade. For this purpose, technical regulations shall not be more trade-restrictive than necessary to fulfill a legitimate objective, taking account of the risks non-fulfillment would create. Such legitimate objectives are, *inter alia:* national security requirements; the prevention of deceptive practices; protection of human health or safety, animal or plant life or health, or the environment. In assessing such risks, relevant elements of consideration are, *inter alia:* available scientific and technical information, related processing technology or intended end-uses of products.[102]

This agreement entered into force with the establishment of the WTO in 1995.

---

99 *Id.* at 9–11.

100 World Trade Organization, Technical Barriers to Trade Agreement, *available at* http://www.wto.org/english/tratop_e/tbt_e/tbtagr_e.htm (last visited June 5, 2011).

101 *Id.*

102 *Id.* at Article 2.2.

The WTO Agreement on the Application of Sanitary and Phytosanitary Measures (SPS Agreement)

The SPS Agreement establishes restrictions on member parties' food safety policies regarding pesticides, inspection, labeling, and bacterial contaminants, and on members' animal and plant health policies regarding imported pests and diseases.[103] The Agreement entered into force with the establishment of the WTO in 1995. This Agreement contains provisions against allowing regulation of substances to enact technical barriers to trade if there is inadequate scientific certainty about potential threats to human and environmental health.[104] It places the burden of proof on countries to scientifically demonstrate that a product is dangerous before it can be regulated and thus encourages the WTO to override a country's use of the precautionary principle emphasized in Principle 15 of the Convention on Biological Diversity.[105]

WTO Agreement on Agriculture

This WTO international treaty entered into force with the establishment of the WTO in 1995. The Treaty contains provisions structuring domestic support, including agricultural subsidy systems, reducing tariff barriers to trade by WTO member-states, and requiring developed countries to reduce export subsidies.[106] The Treaty has been criticized for reducing tariff protections for small farmers and for allowing developed countries to flood global agricultural markets with heavily subsidized commodities to the detriment of developing nations' economies.

WTO Agreement on Trade-related Aspects of Intellectual Property Rights (TRIPs)

This WTO international agreement establishes minimum standards for various forms of Intellectual Property (IP) regulation.[107] TRIPs contains requirements that member-nations' laws must meet for various forms of IP, including patents and new plant varieties, and includes enforcement procedures, remedies, and dispute resolution procedures.[108] The Agreement requires that in protecting and

103 World Trade Organization, Agreement on the Application of Sanitary and Phytosanitary Measures, *available at* http://www.wto.org/english/tratop_e/sps_e/spsagr_e.htm (last visited June 5, 2011).

104 *Id.*

105 *Id.*

106 World Trade Organization, Agreement on Agriculture (1994), *available at* http://www.wto.org/english/docs_e/legal_e/14-ag.pdf.

107 World Trade Organization, Agreement on Trade-Related Aspects of Intellectual Property Rights, *available at* http://www.wto.org/english/docs_e/legal_e/27-trips.pdf.

108 *Id.*

enforcing IP rights, parties shall contribute to the innovation, promotion, transfer, and dissemination of technology in a manner conducive to social and economic welfare.[109] The US, European Union, Japan, and other developed countries were major proponents of TRIPs during its negotiation in 1994. Ratification of TRIPs is a compulsory requirement of World Trade Organization membership and states can be disciplined for non-compliance through WTO's dispute settlement program.[110]

The North American Free Trade Agreement (NAFTA)

NAFTA established, among other things, the Technical Working Group on Pesticides (TWG) with a primary objective to develop easier and less expensive pesticide regulation and trade among the US, Mexico, and Canada while continuing to meet the environmental, ecological, and human health objectives of NAFTA.[111] TWG partners address "trade issues, national regulatory and scientific capacity, governmental review burden, and coordinate regulatory decision-making and industry burden reduction."[112] The stated mission of the 5-year plan is "to serve as a focal point for addressing pesticide issues arising in the context of liberalized trade among the NAFTA countries, while recognizing the environmental, ecological, and human health objectives of NAFTA." The plan is guided by three strategic objectives: (1) to provide US, Canadian, and Mexican pesticide users with equal access to and same time introduction of pest management tools, including safer alternatives; (2) to work cooperatively on the re-evaluation/registration review of older pesticides using each country's re-evaluation program to the fullest to increase efficiency; and (3) to integrate smart business approaches and practices into NAFTA TWG work, including harmonization of remaining technical data requirements, promotion of environmental protection, consideration of human safety and science issues, use of effective communication and planning, and maintenance of high international standards.[113] TWG is directed to consider not only the Nation States' existing laws and policies, but also existing international agreements designed to reduce the risk associated with pesticide use.[114] In designing its programs, the TWG is directed to consider fora from around the world, including Codex Alimentarius, the Organization for Economic Cooperation

---

109 *Id.* at Article 8.

110 *Id.* at Article 64.

111 US Environmental Protection Agency, Regulating Pesticides, International Activities, NAFTA TWG on Pesticides, *available at* http://www.epa.gov/oppfead1/international/naftatwg/index.html (last visited June 5, 2011).

112 *Id.*

113 North American Free Trade Agreement Technical Working Group on Pesticides, Five Year Strategy 2008–2013, *available at* http://www.epa.gov/oppfead1/international/naftatwg/general/2008–2013strategy.pdf.

114 *Id.*

and Development, and the World Trade Organization.[115] These organizations encourage such cooperation as a mechanism to promote health and environmental safety while facilitating trade.

The preceding description of the international agreements relating to pesticide use demonstrates that ecological concepts play an extremely limited role in these international laws. First, there is ongoing tension between international agreements designed to afford certain environmental protections and those designed to facilitate free trade and penalize countries whose domestic regulations infringe on free trade. Accordingly, at best, most international environmental trade agreements do little more than affirm the right of member nations to regulate to protect human health and the environment provided they do not unduly interfere with free trade. Those international agreements that do expressly address pesticides, are primarily designed to ensure that importing countries have sufficient information to make informed decisions about what products and practices to permit within their boundaries. However, because ecological risk assessments depend on very localized information about the specific species and ecosystems of concern, even with full implementation of PIC procedures, sufficient appropriately tailored information may not be available to allow importing country decision makers to make ecologically sound choices.

115 *Id.*

Chapter 14

Ecologically Based Pest Management

As described in previous chapters, widespread use of conventional synthetic pesticides has significantly impacted ecological resources by killing non-target organisms, contaminating water and soil, and disrupting and destabilizing the natural processes that tend to suppress pest populations. Nevertheless, it is not helpful to focus on the use of these chemicals in isolation. Instead, we must look at the farming system as a whole and recognize that many ecologically detrimental practices go hand in hand. For example, growing monocultures and eliminating crop-patterning practices can increase the need for chemical pesticide inputs. Consequently, to be able to reduce our reliance on chemical pesticides, thereby reducing the adverse ecological impacts resulting from the use of these substances, we will need to consider "whole system" farming practices that utilize and enhance natural processes. Additionally, by employing more ecologically sound farming practices, we will be able to make working farmlands healthy habitats for wildlife that provide of a range of important ecosystem services.

What happens on the farm dramatically impacts what happens to wildlife, biodiversity, and ecosystems throughout the world. If we wish to protect wildlife, biodiversity, and ecosystem services, we cannot simply rely on taking certain lands out of production and putting them in public or private preserves.[1] It is simply impossible to put sufficient land in preservation to adequately protect the natural environment. The statistics regarding current acres of land in preservation as opposed to acres of land in active agriculture are telling. Currently, only approximately 5 percent of land on earth is preserved in parks and other protected areas. In stark contrast, agriculture and livestock production comprise roughly 50 percent of the globe's habitable land.[2] Accordingly, farming practices that provide habitat for wildlife, support biodiversity, and preserve ecosystem services can substantially improve the overall promulgation of these values throughout the globe. In this way, agriculture can not only provide food and fiber, but also significantly contribute to the protection of resources and services critical to society.

For several decades, there have been calls to adopt more ecologically sound farming practices. For example, the 1992 book *Farming in Nature's Image*, introduced the idea of using nature as a model for agricultural production.[3] More

---

1 Jason Clay, World Agriculture and the Environment: A Commodity-by-Commodity Guide to Impacts and Practices 1–2 (2004).

2 *Id.*

3 Judith D. Soule & Jon K. Piper, Farming in Nature's Image: An Ecological Approach to Agriculture 124 (1992). *See also*, Sara J. Scherr & Jeffrey A. McNeely,

recently, much research and discussion has focused on the idea of "sustainable agriculture" or "eco-agriculture." Specifically with regard to pest management, the concept of looking at the whole farming system and using a diverse variety of natural, enhanced, and supplemented control measures has also been researched and discussed for decades, first as Integrated Pest Management (IPM), and more recently as what is known as "Ecologically Based Pest Management" (EBPM).

Some form of the IPM approach is currently used in many agricultural and non-agricultural settings. The adoption of these approaches has resulted from a number of factors. Interestingly, some of the pesticide regulations that have been most effective in encouraging a shift to more ecologically based practices are not directed at ecological risk, but instead are human health and safety based. For example, EPA's Worker Protection Standards,[4] which impose a number of requirements including prescribing how long after pesticide application a worker must wait before re-entering the field (known as "reentry intervals"), and other restrictions to protect the health and safety of farmworkers, have made certain conventional pesticide application practices impracticable and thus have caused some producers to shift to IPM approaches.[5] Other indirect influences that have encouraged some farmers to shift to IPM include other regulations, such as food safety regulations, that are designed to protect human health, rather than the natural environment.[6] Thus, continuing to ensure that human health and safety are adequately addressed under pesticide law will likely have the added ecological benefits that accrue from a switch from conventional pest control to IPM. IPM adoption has also occurred in areas where either private "Crop Advisors" or state extension agents have promoted the concept and provided the information, training, and services necessary to implement them. Unfortunately, this has been done on an ad hoc basis. There is great variability in the level of education and interest in IPM among extension agents. Some have been very active in promoting IPM, while others have been reluctant to move far beyond traditional chemical input approaches. In some areas of the country, private sector "Crop Advisors" have developed IPM systems for particular crops and provide the scouting and monitoring services necessary to properly implement the IPM programs. In addition, the federal government and many state and local governments have adopted IPM approaches in their own pest management practices. IPM is now widespread on military bases, on public lands, in public schools, and in other government-run facilities. Moreover, certain large food retailers, including Wal-Mart, Wegman's, and Gerber are starting to include

eds, Farming with nature: The Science and Practice of Ecoagriculture (2007); and Jeffrey A. McNeely & Sara J. Scherr, Ecoagriculture: Strategies to Feed the World and Save Wild Biodiversity (2003).

4 40 C.F.R. § 170 (2011).

5 Dr. Norman Leppla, University of Florida, Department of Entomology and Nematology, Personal Correspondence (April 20, 2011).

6 *Id.*

in their contracts with food producers requirements that all foods sold to the retailer be grown using organic, IPM or other environmentally sound approaches.[7]

In the mid-1990s, at the request of the USDA and with support from EPA, the National Academies, National Research Council (NRC), convened a committee on "Pest and Pathogen Control through the Management of Biological Control Agents and Enhanced Natural Cycles and Processes." The committee was charged to "assess the status of the knowledge in areas of pesticide application, host resistance, and biological control practices and to chart future direction." In 1996, the NRC published the findings of the committee in a report entitled "Ecologically Based Pest Management: New Solutions for a New Century."[8]

Bridge Document 14.1 Ecologically Based Pest Management: New Solutions for a New Century

Given our charge and the record of history of the application of pesticides, breeding for disease resistance, and integrating biological control practices into production agriculture, my colleagues on the committee and I deliver this report with one key message: In both science and application, researchers, providers of inputs and growers must progress from a product based approach to an ecologically based pest management system identified as EBPM. Management is the key word. In fact, the word control, as in biological control is misleading. Pests in most cases cannot be controlled; pests must be managed with the objectives of a safe, profitable, and durable outcome.

With a better understanding of ecology, the inherent strengths of the managed ecosystem can be used with more modest inputs than in the past. Essentially, the change to EBPM as proposed here will require a substantial change from the primary practice of product input to the primary mind set of information and management. Ultimately, EBPM will help to address ecosystem health not by administering products alone to treat systems, but by integrating components that maximize use of natural processes with minimum development of resistance.

7 Wal-Mart Stores, Inc., Standards for Suppliers Manual, *available at* http://walmartstores.com/download/4216.pdf (setting forth environmental and ethical sourcing requirements for Wal-Mart suppliers); Wegman's, Customer Service, Frequently Asked Questions, How does Wegman's Know that the Produce they Carry is Safe? *available at* http://www.wegmans.com/webapp/wcs/stores/servlet/FAQDetailView?storeId=10052&catalogId=&langId=-1&faqCategory=ProductInformation (last visited June 3, 2011) (indicating that Wegman's works with farmers who employ IPM practices); and Lester Graham, *Apple Growers Trim Pesticides from the Farm* [radio transcript], Great Lakes Radio Consortium, *available at* http://www.agcenter.org/mediaarticleGLRC.html (last visited June 3, 2011) (discussing Gerber Baby Foods' success in influencing its suppliers to adopt IPM practices).

8 National Research Council, Ecologically Based Pest Management: New Solutions for a New Century vii–viii (1996).

> EBPM will require regulatory oversight that matches the level of risk of biological inputs added to the managed ecosystem. For example, synthetic chemicals are new to the biosphere—they have no base of performance in the environment or in relation to human health. However, biologically based organisms, products, and resistant cultivars are inherently different, for the most part, from synthetics. Biological processes, having existed in nature over time, provide a base of experience that is a major resource to evaluate the safe application and establish appropriate oversight of EBPM …

The NRC report focuses on the problem of pests developing resistance to synthetic chemical pesticides, which currently limits the efficacy of many chemical pesticides and has eliminated any viable chemical control for some pests.[9] When natural controls are removed from a system, through growing monocultures and reducing predator and parasite populations with the use of chemical pesticides, farmers become dependent on continual use of chemical pesticides. Over time, pests are likely to develop resistance to the chemicals used, rendering the chemicals ineffective and creating an increasing need for additional chemicals. Thus, the NRC committee concluded, "[t]here is an urgent need for an alternative approach to pest management that can complement and partially replace current chemically based pest-management practices."[10] The report emphasized the need to create "whole-farming" systems, which integrate pest management with fertilization, cultivation, cropping patterns and other ecologically based production practices.[11] The foundation of EBPM, promoted in the report, is using inputs of knowledge of the managed agricultural ecosystem, including knowledge of pests and their natural enemies, with physical, chemical, and biological inputs as supplements. The report articulates the objectives of EBPM as "the safe, profitable, and durable management of pests."[12] The report criticizes conventional chemical input farming as disrupting and destabilizing the natural processes that suppress pests.[13]

Although the term "EBPM" may be relatively new, many of the mechanisms used in EBPM harken back to the pre-chemical history of agriculture, where cultural and biological controls dominated. Thus, cultural practices such as rotating crops, intercropping, choosing pest resistant crop varieties, and promoting the development of healthy soils are key components of EBPM. These practices enhance natural pest control processes. With decades of research on pests and pest management, as well as the development of new technologies, however, EBPM can build on and move beyond the pest management practices relied upon in the pre-chemical era.

To implement widespread EBPM, several things will need to occur. First, EBPM is heavily dependent on detailed knowledge, not only of pest species, but

9 *Id.* at 1.
10 *Id.*
11 *Id.* at 1–2.
12 *Id.* at 2.
13 *Id.*

more broadly of how these species interact with other species such as predators and parasites and how they are affected by physical, chemical, and biological processes in their environment.[14] Although much research on pest biology and IPM systems has already been conducted, much more will be needed to support widespread use of EBPM.[15] EBPM is an approach that involves the complete system and is designed to minimize adverse effects on non-target organisms and the environment. Biological controls are important components of EBPM because they tend to be target-specific, they tend not to persist in the environment and in general they are not as toxic as synthetic chemical pesticides.[16] Of course, EBPM will not be widely accepted and utilized unless it is financially profitable for farmers.[17] The other major hurdle to overcome is to develop a knowledge base of the complexities of the interactive processes of agro-ecosystems; and this will require multidisciplinary research. Finally, the information derived from this research must be disseminated to producers so they can implement EBPM practices.[18]

Many innovative pest management practices, including EBPM, are aimed at working with nature instead of attempting to assert dominance over it. Many approaches that are similar, but vary somewhat from EBPM, have been proposed. For example, in 1995, the Office of Technology Assessment in the US Congress, published a report entitled "Biologically Based Technologies for Pest Control," which contains many parallels to the NRC's EBPM approach.[19] Similarly, the Consumers Union, in its 1996 publication, *Pest Management at the Crossroads*, promoted the idea of "biointensive IPM," which it defined as "a systems approach to pest management that is based on an understanding of pest ecology. It relies on resistant varieties and promoting plant health, crop rotation, disrupting pest reproduction, and the management of biological processes to diversify and build populations of beneficial organisms. Reduced risk pesticides, including biopesticides, are used only as a last resort and only in ways that minimize risks."[20] Countless other publications have supported the use of IPM, biointensive IPM, eco-agriculture and other similar approaches to pest management. While each approach may have a slightly different emphasis, all of them share the common

---

14 The report articulates the objectives of EBPM as "the safe, profitable, and durable management of pests." *Id.*

15 *Id.* at 3. *See also*, US Congress, Office of Technology Assessment, *Biologically Based Technologies for Pest Control*, OTA-ENV-636 (Washington, DC, US Government Printing Office, September 1995) and CHARLES M. BENBROOK, PEST MANAGEMENT AT THE CROSSROADS (1996).

16 *Id.* at 3–4.

17 *Id.* at 4.

18 *Id.* at 5–7.

19 US Congress, Office of Technology Assessment, *Biologically Based Technologies for Pest Control*, OTA-ENV-636 (Washington, DC: US Government Printing Office, September 1995).

20 BENBROOK, *supra* note 15.

foundational principle that pest management is best achieved with a comprehensive understanding of the ecology of the system and the use of a variety of ecologically based approaches to achieving an ecologically resilient and diverse system that keeps pest populations in check.

The innovative practices promoted by EBPM, and related approaches, often rely on a combination of "old" traditional farming approaches such as utilizing cultural controls coupled with new approaches. Finding the right combination of control practices can decrease the need for chemical inputs, thereby decreasing the environmental impacts of farming. Innovation that relies less on chemical inputs is not necessarily a result of altruistic concerns for the environment. More and more farmers are finding that there are sound economic reasons for abandoning the goal of maximum per acre yields through application of expensive chemical inputs; instead they are seeking sustainable crop yields, and therefore profits, by focusing on building soil fertility and vitality. Of course, a "whole-system" approach to pest management cannot be divorced from other aspects of the production process. Thus, EBPM and IPM both should be considered as part of the broader practices of eco-agriculture or sustainable agriculture.

Over the past few decades, scientists, policy makers, farmer organizations, environmentalists, and others have called for new sustainable agricultural approaches to replace industrialized agriculture. What is commonly referred to as "sustainable agriculture" is generally focused on obtaining and maintaining the three objectives of environmental health, economic profitability, and social and economic equity.[21] As the Union of Concerned Scientists explains,

> … [s]ustainable agriculture views a farm as a kind of *ecosystem* – an 'agroecosystem' – made up of elements like soil, plants, insects, and animals. These elements can be enriched and adjusted to solve problems and maximize yields. This integrated approach is both practical and scientific: it relies on modern knowledge about the interactions within natural systems, as well as cutting-edge technologies, to achieve its results. It is a powerful approach that can produce high yields and profits for farmers while protecting human health, animal health and the environment.[22]

---

21 University of California Davis, UC Sustainable Agricultural Research and Education Program, What is Sustainable Agriculture? *available at* http://www.sarep.ucdavis.edu/concept.htm (last visited May 27, 2011).

22 Sustainable Agriculture—A New Vision, *available at* http://www.ucsusa.org/food_and_agriculture/solutions/big_picture_solutions/sustainable-agriculture-a.html (last visited June 17, 2011). For further descriptions of sustainable agriculture, *see* John H. Davidson, *Agriculture*, in Stumbling Toward Sustainability 360–62 (John C. Dernbach, ed., 2002).

Although defined and described in many ways, the underlying principle of sustainability is the desire to meet current needs of society while still preserving sufficient resources for future generations to meet their needs.[23]

The Union of Concerned Scientists identifies five key techniques of sustainable agriculture: crop rotation, cover crops, soil enrichment, natural pest predators, and biointensive integrated pest management.[24] Crop rotation describes the practice of growing different crops in the same field over a period of time.[25] The benefit of this practice is that it discourages the build-up of pests that can occur when one crop is continuously grown in the same location. Sustainable farming uses cover cropping—i.e., planting specific crops on fields between plantings of the primary crop so that fields do not remain bare—which has benefits including reducing soil erosion, reducing weed growth, and enhancing soil nutrients.[26] Rich healthy soil is critical to sustainable farming. Rather than allowing soils to be depleted, and then needing to use intensive synthetic fertilizers, as is done in industrial agriculture, sustainable agriculture seeks to maintain and enhance soil richness by plowing under cover crops and using natural fertilizers, such as composted animal and plant waste.[27] Soil richness is also maintained by abstaining from heavy use of pesticides, which kill the beneficial soil-inhabiting organisms necessary for healthy soils.[28] In lieu of synthetic pesticides, sustainable farming practices are geared toward maintaining a healthy ecosystem, which allows natural predators and parasites of crop pests to thrive and keep pest populations in check.[29] To the extent natural pest control is not adequate, integrated pest management practices, which employ a range of biological and cultural control practices, as well as limited targeted synthetic chemical pest control, are used.[30] To achieve a system of sustainable agriculture, it is necessary to develop regulatory and incentive-based tools that require or promote these practices. The challenge we face is how to revise current policies to meet the goal of ensuring and affording a healthful food supply while moving toward an agricultural system that is environmentally, economically, and socially sustainable.

If we seek to limit the harm to wildlife, biodiversity, and ecosystem services resulting from synthetic pesticide use, as well as from other aspects of unsustainable farming practices, we must start to view the farm as a healthy sustainable living

---

23 *Id.*

24 Union of Concerned Scientists, Food and Agriculture, Sustainable Agriculture Techniques, *available at* http://www.ucsusa.org/food_and_agriculture/science_and_impacts/science/sustainable-agriculture.html (last visited May 27, 2011).

25 *Id. See also* John Boardman, Jean Poesen & Robert Evans, *Socio-economic Factors in Soil Erosion and Conservation*, 6 Envtl. Sci. & Pol'y 1, 1–3 (2003) (discussing the runoff and soil erosion associated with industrial agriculture).

26 *Id.*

27 *Id.*

28 *Id.*

29 *Id.*

30 *Id.*

system, as opposed to an industrial facility. The premise of eco-agriculture is that the farm is a kind of *ecosystem*—an "agro-ecosystem"—made up of elements like soil, plants, insects, and animals. A healthy farm ecosystem will have healthy and fertile soils, healthy populations of natural predators and parasites, pollinators, and other beneficial species. The farm maintains many ecosystem services that provide benefits beyond the farm, and the negative externalities of farming are reduced. Farming practices are geared toward maintaining and enhancing ecosystem health and function, while at the same time maximizing yields within the constraints of maintaining a healthy ecosystem. This reflects a recent shift in thinking away from the idea that human activity and functioning ecosystems are mutually exclusive.

The concept of eco-agriculture relies on modern knowledge about the interactions within natural systems, as well as cutting-edge technologies, to achieve its results. When done properly, it can produce high yields and profits for farmers while protecting human health, animal health, and the environment. Eco-agriculture draws on the key techniques of sustainable agriculture: crop rotation, cover crops, soil enrichment, maintenance and enhancement of natural pest enemies, and bio-intensive integrated pest management. These techniques not only help to build a sustainable and ecologically sound system, but more specifically they help to manage pests without reliance on ecologically harmful synthetic pesticides. Crop rotation discourages the build-up of pests that can occur when one crop is continuously grown in the same location. The pest management benefits of cover crops include reducing weed growth, thereby reducing the need for herbicides. Soil enrichment helps to build healthy crops that are more resistant to disease, and pest damage and thus may not require other means of pest control. Creating conditions conducive to enhancing populations of predators and parasites of pest species is a way to utilize natural pest controls, thereby reducing the need for other pest control interventions. Finally, employing bio-intensive integrated pest management can eliminate or dramatically reduce the need for synthetic pesticide inputs.

Although not inherently obvious, one key to building an ecologically sound agricultural system is developing healthy soils. The greatest invertebrate biodiversity can be found in the soil. However, such biodiversity is greatly impacted by intensive unsustainable farming practices. We now understand that healthy soil is the key to building healthy agro-ecosystems that do not require synthetic fertilizer or pesticide inputs to be productive. Many strategies to build and maintain healthy soils have the added benefit of enhancing predator and parasite populations while simultaneously discouraging the build-up of pest populations. In his book, *World Agriculture and the Environment*,[31] Jason Clay describes a joint venture that the John Deer Company entered into with a Brazilian agricultural company engaged in sustainable farming practices intended to rebuild the soils of previously degraded pasture to grow crops such as soybeans. The soils of the degraded pasturelands were rebuilt by eliminating tillage, and instead, leaving

31 *See generally*, Jason Clay, *supra* note 1.

organic matter on the surface when seeds are planted coupled with planting two or three crops during a year in a particular sequence and implementing three-year crop rotation.

Rather than working with nature to promote biodiversity on working farmlands, most large-scale industrialized agricultural farming operations actually fight against nature and seek to reduce biodiversity. As described above, this tendency is encouraged by the many government subsidy programs that promote high-yield monoculture production.

Working farmlands can reap the benefits provided by nature in the form of pest control through predators and parasites, by promoting biodiversity on the farmlands themselves. Biodiversity on the farm can be enhanced through reduced pesticide use, intercropping, crop rotation, and the creation and maintenance of refugia on or near the farm field. Crops on working farmlands can become more resistant to disease and pests through practices that maintain the health and fertility of soils, such as crop rotation and planting legumes and other cover crops that improve soil fertility.[32]

In recent years, attempts to adopt eco-agricultural approaches have been made throughout the world. A number of examples have been cited as eco-agriculture "success stories." For example, in parts of Central America, farmers have integrated trees into livestock pastures to provide habitat for forest birds. Also in Central America, farmers have had success with planting biodiversity-friendly coffee plantations. In California, a type of eco-agriculture has been implemented by flooding rice fields for birds during fallow seasons. Several examples come from Indonesia where the use of "agro-forests" that integrate agricultural and forestry crops has been successful, as has the integration of rice terraces with fish and vegetables.

Eco-agriculture can be used to achieve more sustainable agricultural production, which not only produces crops and profits, but also protects human health, biodiversity and ecosystem services, and provides viable local livelihoods. The goal of eco-agriculture is broader than merely reducing reliance on pesticides, and instead is to build a healthy, biologically diverse, sustainable, and functioning agro-ecosystem. Nevertheless, a healthy diverse agro-ecosystem will enhance natural pest control and will not depend on chemical pest control as do current industrialized agricultural monocultures. The question then becomes, what legal or policy changes are necessary to encourage the widespread adoption of eco-agricultural approaches.

32 *Id.* at 546.

Chapter 15

Legal and Policy Proposals to Achieve Ecologically Based Pest Management[1]

A variety of regulatory and incentive-based approaches have been proposed to reduce chemical pesticide use. For example, more stringent regulation and/or taxes can be used to internalize the negative environmental externalities of pesticide use. Imposing stringent regulations to minimize the environmental impacts of pesticide use, or taxing pesticides, will discourage farmers from using chemicals unless absolutely necessary; and even when necessary, they will be required to use the pesticides in ways that minimize harms. However, these approaches alone are unlikely to achieve long-term fundamental change. First, as discussed in detail in previous chapters, large-scale monoculture production is conducive to pest outbreaks and is antithetical to encouraging natural predator and parasitic controls. Thus, simply making it difficult or costly for farmers to use pesticides is not a viable solution. Farmers will still need to make money and civilization will still need food no matter how expensive it is to use pesticides. Instead, it will be necessary to find ways to develop and encourage agricultural production approaches that are not as heavily reliant on pesticides, but that still produce food for society and income for farmers. As described in detail in earlier chapters, to reduce dependence on chemical inputs it will be necessary to develop more ecologically based farming systems. Of course, stringent regulatory programs, pesticide taxes and/or pesticide release reporting requirements can all play a role in encouraging a shift. However, our agricultural system is unlikely to undergo major transformation without a number of other societal changes. First, the public must be educated about the shortcomings of our current industrial agriculture system and must be willing to accept certain changes such as, for example, fruits and vegetables that may not be as attractive or as "perfect" in appearance.

---

1 Portions of this chapter are adapted from Mary Jane Angelo, *Corn, Carbon and Conservation: Rethinking U.S. Agricultural Policy in a Changing Global Environment* 17 George Mason Law Review 593 (April 2010); Mary Jane Angelo, *The Killing Fields: Reducing the Casualties in the Battle Between U.S. Endangered Species and Pesticide Law*, 32 Harvard Environmental Law Review 96 (2008); Mary Jane Angelo, *Regulating Evolution for Sale: An Evolutionary Biology Model for Regulating the Risks Posed by Genetically Modified Organisms*, 42 Wake Forest Law Review 93 (2007); and Mary Jane Angelo, *Embracing Uncertainty, Complexity and Change to Protect Ecological Integrity: An Eco-Pragmatic Reinvention of a First Generation Environmental Law*, 33 Ecology Law Quarterly 105 (2006).

Consumers and local communities will have to support local farmers through farmers' markets, community supported agriculture, or other forms of support to shift away from the long-distance shipping and long-term storage of agricultural products that demand increased pest protection. Equally important, more money must be made available for research and development of non-chemical or lower-risk pesticides and sustainable agriculture technologies, including biointensive IPM and eco-agriculture, and to incentivize farmers to switch to these less-pesticide-dependent methods. A revamping of our current subsidy system will be necessary to funnel subsidies away from industrial commodity production to other crops and more sustainable growing practices. As described in more detail below, this could be accomplished through a combination of simply shifting subsidies to other crops, shifting subsidies to crops grown under some type of "sustainable" criteria, compensating farmers for the ecosystem services provided by growing in an ecologically based manner and/or providing additional funding for working land conservation programs that incentivize sustainable growing approaches.

Proposals for Regulatory Changes

Changes to FIFRA

Despite its early successes in protection of ecological resources at the beginning of the environmental movement in the United States, until recently FIFRA has virtually lain dormant with regard to protection of ecological resources. Ironically, if any area of environmental law should be tailored specifically to address ecological concerns, it is pesticide law, where substances that are intended to kill and disrupt species and natural systems are intentionally released into the environment in large quantities. Moreover, pesticide law has failed to keep pace with recent advances in ecological study and the field of conservation biology. Any regulatory or policy system designed to address the ecological impacts of pesticides and pest management must incorporate a number of ecological concepts and ecologically based management tools.

Modification of the Cost/Benefit Standard

Several changes are necessary for FIFRA to become more ecologically based.[2] As described above, one of the shortcomings of FIFRA is that it relies on cost/benefit analysis rather than more protective feasibility or risk-based standards. A more environmentally sustainable approach that would still allow the use of pesticides needed for agriculture and public health protection would be to revise the standard for registering pesticides under FIFRA to make it clear that

2 For detailed discussions of the author's previous proposals for FIFRA reform *see* the author's prior publications cited in note 1.

pesticides that pose significant risk to human health or the environment may only be registered if there are overriding public health, social, or economic benefits that justify registration. Such a revision would in effect force EPA to apply the standard originally contemplated by the Congress in enacting the 1972 FIFRA. Under such an approach, although economic and social costs would be factors to be considered in determining whether to allow a pesticide to be sold or distributed in the US, they would not be used to "balance" away significant human health or environmental risks. Such risks would only be permitted in cases where there were overriding benefits of the pesticide, taking into account a number of specified considerations in the open-ended balancing.

In addition to the economic considerations that the statutory standard directs EPA to consider in conducting this balancing, FIFRA should be revised to explicitly direct EPA to take into consideration additional factors aimed at ensuring that ecological concerns are adequately valued in the balancing. For example, considerations such as the degree of uncertainty regarding risks, the level of probability or risk, the degree of harm that could occur, the likelihood of a pesticide to spread or reproduce in the environment, either through biological means as in the case of GMOs or physical means, are all factors that should be given serious consideration in any open-ended cost/benefit balancing. Physical means of environmental dissemination include the tendency of a pesticide to move great distances through soil, water, or air, and the tendency of a pesticide to bioaccumulate.

On the "ecological costs" side of the equation, it is critical that the regulations and policies are mindful of all ecosystem organization levels including population and ecosystem levels. Thus, ecological risk assessment used to evaluate risks must not be limited to acute toxicological effects on individuals, but must also include assessment of subacute effects, such as reproductive effects, and other physiological and behavioral effects that could impact the ability of individuals to find food, grow, find mates, and successfully reproduce. In other words, all toxicological effects that could impact the fitness of the species should be considered. In particular, special consideration should be afforded to species that are threatened, endangered, or rare, or play a special role in ecosystem function, such as keystone species, umbrella species, and highly interactive species. Moreover, indirect impacts to the food sources, predators and parasites, and habitat requirements of the species must also be taken into account. EPA must not limit its inquiry to the direct impacts to the species of concern, but must also consider indirect impacts to that species that occur because of harm caused to its food sources, predators and parasites, and other species it interacts with.

Moreover, to be able to fully understand and consider the ecological risks associated with pesticides, EPA must engage in, at a minimum, population-level, if not ecosystem-level risk assessment. EPA's current environmental risk-assessment focus on acute and reproductive effects at the individual level fails to reveal the true ecological costs of pesticide use, and thus does not provide a meaningful accounting on the "cost" side of the cost/benefit equation. At the highest level of organization, EPA must look at impacts to ecological resilience of ecosystem to ensure impacts to one or more species or ecological processes won't destabilize the system. The primary way to protect ecosystem resilience is to ensure the maintenance of species diversity. This will be particularly important for the protection of rare or imperiled ecosystems or for those ecosystems that provide important ecosystem services. Regulations and policies should seek to preserve ecological integrity through preservation of ecosystem diversity and resilience. Tools that could be used to accomplish these goals include adaptive management, which can be employed to create a system that can adapt to new information and changed circumstances.

In addition to population and ecosystem level consideration, some species warrant particular protection. For example, threatened, endangered, or rare species may warrant species protection due to their imperiled status. A level of loss of individuals that may be acceptable in a healthy population of a species may be completely unacceptable from both an ecological and a legal standpoint, and may in fact be directly linked to extinction or extirpation of a threatened or endangered species. Similarly, individuals of certain species that play particularly important roles in ecosystem health, such as keystone species, umbrella species, or highly interactive species, may warrant special consideration. Under EPA's current strict cost/benefit balancing analysis, there is no clear mandate to afford special protection or to give more weight to certain species based on the health of their populations or the role they play in overall ecosystem function.

Concerns with rare or sensitive species or ecosystems can be addressed by an open-ended balancing approach, where the agency is directed to consider costs, but not in a strict cost/benefit monetized balancing. Under such an approach, although costs would be taken into consideration, the agency could be directed to afford greater weight in its balancing to other specified factors, such environmental justice concerns, risks to threatened or endangered species, risks to rare species, or risks to vulnerable ecosystems. It should be noted, however, that under this proposal the analysis would not end with the open-ended balancing. A second step, described more fully below, would require the consideration of local factors, such as the presence of threatened or endangered species, which could be affected by the use of the pesticide in that location. Although the possibility of harm to threatened or endangered species should be considered and afforded weight in the open-ended balancing analysis, it should be noted that under the ESA, jeopardy to listed species cannot be "balanced" away by cost considerations.

Although a detailed analysis of discounting environmental benefits is beyond the scope of this book, as described in previous chapters, many risks of pesticides are long-term risks. For example, the tendency of certain pesticides to bioaccumulate as they work their way up the food chain poses risks that may not be realized for many years after the initial release of the pesticide into the environment. Similarly, the harms from certain pesticides that pose reproductive effects, such as the endocrine disruptors, may not be to the animals exposed, but may instead be to the offspring, or even subsequent generations of the animals exposed. Moreover, many of the potential harms to ecological systems and ecological services that pesticides pose, including disruptions of predator/prey relationships, harm to pollinators, and harm to microorganisms that perform decomposition and nutrient cycling services, may not be evident in the short-term, but may have significant long-term impacts. Accordingly, any pesticide regulation that reduces these long-term harms may not demonstrate immediate benefits. Consequently, as Daniel Farber suggests in his book, *Eco-Pragmatism*,[3] it is important to ensure that the long-term benefits of pesticide risk-reduction measures are adequately valued through the use of low discount rates. The idea of intergenerational equity further supports this notion. To ensure that future generations inherit functioning ecosystems, including functioning agricultural systems, it is incumbent upon this generation to ensure that the long-term benefits of pesticide risk-reduction measures are properly valued through appropriate low discounting.

Improved Consideration of Benefits

Another important revision to FIFRA would be to require EPA to consider benefits in its registration decision making. For example, registration applicants should be required to demonstrate that the pesticide they are seeking to register is efficacious and will provide overriding benefits. As discussed above, FIFRA does not mandate, and EPA has opted not to require, that efficacy data be provided when registering a pesticide. As discussed above, FIFRA currently allows EPA to waive efficacy data and allows pesticides to be registered without a showing of necessity or a consideration of whether lower-risk alternatives are available. On the benefits side of the equation, EPA must change its current approach and begin to consider the efficacy of pesticides. However, this consideration must not be limited to merely evaluating the ability of the pesticide to kill the target species, but must also consider efficacy from a broader perspective. For example, if a broad-spectrum pesticide not only kills the target species, but also kills the predators and parasites that otherwise could help to keep the target species population in check, this could result in "flarebacks" of the pest species population, as well as outbreaks of new pests. EPA should consider these secondary effects in its overall efficacy assessment.

Consideration of the availability of lower-risk alternatives should be required not only when deciding whether to cancel a registration, but also at the time of registration and reregistration. EPA's Reduced Risk policies, which are described in previous chapters, do not include the consideration of lower-risk alternatives in the registration decision-making process. Instead, they merely encourage the development of reduced-risk pesticides, by providing an expedited registration process.

The lack of a requirement for efficacy data is in contrast to other licensing statutes, such as the licensing provisions of the FFDCA governing the approval of

---

3 Daniel J. Farber, Eco-Pragmatism: Making Sensible Environmental Decisions in an Uncertain World (1999).

new drugs, which explicitly require a finding that a drug is "effective" as part of the premarket review process.[4] A new drug is considered to be "effective" if there is a general recognition among experts, founded on substantial evidence, that the drug in fact produces the results claimed for it under prescribed conditions.[5]

In addition to requiring efficacy data, EPA should require that applicants for pesticide registration provide data showing that there are no cost-effective alternative pest control methods available. Likewise, EPA should require applicants to provide information demonstrating that their proposed pesticide is relatively beneficial, either environmentally or economically, over other existing pesticides or pest control methods that are available to address the target pest. Currently, EPA merely assumes that a pesticide manufacturer would not incur the costs of developing and marketing a pesticide if it was not efficacious and did not have benefits, and that any pesticides that are not beneficial will be eliminated through market forces. Consequently, a manufacturer could obtain a registration for a pesticide without ever having to show that the pesticide works for its intended purpose, let alone that the pesticide is necessary for combating particular pests or that existing chemical or non-chemical alternatives are not available. Virtually no chemical pesticide is without at least some risk. Thus, it is at least possible, if not likely, that pesticides are being registered that pose some risks, but have not been demonstrated to have any significant environmental, economic or societal benefit.

Technology-based approaches, which have worked well in other areas of pollution control law,[6] may not work as well where the goal is to release substances into the environment with the intent to kill living organisms. Thus, a different approach may be necessary. FIFRA differs from many other environmental laws in that the environmental risks that are sought to be reduced by other such laws are by-products of activities that are intended to produce some other product or service (e.g., air pollution from energy production), whereas, with FIFRA, the pesticide that poses the risk is the intended product. Accordingly, under FIFRA, the imposition of technology to reduce risks would take a different form than with other technological controls that seek, for example, to reduce air pollution emissions. The FIFRA analogue of the technology-based standard, therefore, is an alternative reduced-risk method of pest control. Such an alternative can be either a lower-risk chemical pesticide, or other non-chemical methods of pest control, such as biological control or cultural control.

Adaptive Management through More Localized Decision Making

For a pesticide regulatory system to be ecologically sound, it must be flexible and able to adapt to new information and changed circumstances. One of

4 21 U.S.C. § 355(b)(1) (2011).

5 *Id.* at § 355(d) (2011).

6 *See* Wendy E. Wagner, *The Triumph of Technology-Based Standards*, 2000 U. Ill. L. Rev. 83, 85–86 (2000).

the cornerstones of an ecologically based environmental law is the use of an environmental baseline. A baseline based on ecological integrity can provide a standard by which any action taken under FIFRA can be measured against the reference point of the baseline. As with any baseline, however, such comparisons cannot be made without good data and ongoing monitoring. To ensure that pesticide usage does not undermine ecological integrity, a number of revisions to the FIFRA program are necessary.

Transferring some decision-making authority to a local level is one way to accomplish this goal. An interesting example of turning regulatory control of pesticides over to local governments took place in Canada starting in 1991, when the small town of Hudson, Quebec passed a local law prohibiting the non-essential application of lawn and garden pesticides on public and private property within the town.[7] In 2001, Canada's Supreme Court upheld the municipality's authority to issue local pesticide bans and severe restrictions.[8] That decision opened the door for widespread local government bans of pesticides and since the time of the decision, 171 of Canada's municipalities have passed municipal by-laws restricting the use of pesticides, such that approximately 79 percent of Canada's total population currently benefits from protection under these laws.[9]

Although, pursuant to the *Mortier* decision, the US could follow this model and allow local governments to make decisions about whether to ban some or all pesticide use within their jurisdictional boundaries or whether to impose some type of permitting system on pesticide use, events that have taken place since the decision in *Mortier* suggest this is unlikely to happen, at least without a major cultural shift. Prior to the *Mortier* decision, only a small number of states had in place laws that preempted local governments from regulating pesticide use. After *Mortier*, all but 11 states have laws preempting local regulation of pesticides.[10] In other words, although federal law does not preempt local governments from regulating pesticides use, most states have chosen to preempt local government regulation of pesticides.

7 Sarah B. Pralle, *Timing and Sequence in Agenda-setting and Policy Change: A Comparative Study of Lawn Care Pesticide Politics in Canada and the U.S.*, 13 Journal of European Public Policy 987, 995 (2006).

8 Spraytech v. Town of Hudson, [2001] 2 S.C.R. 241, 2001 SCC 40 (Can.), *available at* http://www.canlii.org/en/ca/scc/doc/2001/2001scc40/2001scc40.html (last visited June 3, 2011).

9 FLORA Community Web, Private Property Pesticide By-laws in Canada, Population Statistics by Municipality, *available at* http://www.flora.org/healthyottawa/BylawList.pdf.

10 Laura A. Haight, *Local Control of Pesticides in New York: Perspectives and Policy Recommendations*, 9 Alb. L. Envtl. Outlook J. 37, 39 (2004). The US Supreme Court recently addressed FIFRA preemption again in Bates v. Dow Agrosciences, LLC., 125 S. Ct. 1788 (2005) (holding that FIFRA does not preempt claims for defective design, defective manufacture, negligent testing, breach of express warranty, and violation of Texas Deceptive Trade Practices Act, and remanding the issue of whether FIFRA preempts fraud and failure-to-warn claims).

The most significant change to FIFRA necessary to be more ecologically protective would be to amend the statute to create a mechanism for localized decision making. Such decision making can take into account geographic factors to protect wildlife, natural resources, and ecosystem services. This could be carried out by a permitting system for large-scale releases of pesticides into the environment wherein permit conditions could be imposed to maximize protection of natural resources and ecosystem services. For example such permit conditions could include buffers around habitat, buffers around water bodies, buffers around nests, restrictions on spraying certain pesticides during certain times of year to avoid species migration, breeding or nesting, restrictions on spraying under certain weather conditions (e.g., high winds or heavy rain), and any other condition that would reduce the risk of harm to listed species or migratory birds.

The consideration of local factors in making the determination of whether or how to use a specific pesticide in a specific location is of particular import. The benefit of local control over pesticide use is that decisions can be made based on local factors. Such factors could include the presence of threatened, endangered, or otherwise rare species, presence of sensitive species, soil conditions, climatic conditions, proximity to environmentally sensitive lands, types of crops grown, types of farming practices used, severity of pest infestations, or other relevant site-specific factors. Because ecological impacts are necessarily contextual and local, whereas human health impacts are not, EPA's historical failure to adequately address local ecological effects of pesticides may be further evidence of an institutional bias in favor of human health protection and away from ecological protection.

In addition to amending FIFRA to include a pesticide application permitting program, there are a variety of other potential mechanisms available for achieving local decision making regarding actual pesticide use. One mechanism is to provide better training to certified applicators in IPM and non-chemical controls and better information regarding endangered species, ecological processes, the role of predators and parasites, and other local environmental conditions. Similarly, better training could be provided to local agricultural extension agents. A variation on this theme would be to empower local officials—whether they be local government officials or extension agent officials—to make case-by-case or season-by-season decisions on the actual use of pesticides. For example, a local official could be required to evaluate the local conditions, including the particular pest concerns, the climatic conditions and a wide variety of local environmental factors, before "prescribing" that a particular pesticide be used.

As described above, one of the cornerstones of ecologically based environmental law is adaptive management. Because of the complexity and changing nature of ecological systems, as well as our ever-increasing understanding of those systems, it is necessary that any system designed to address ecological risks be flexible and able to adapt to changed circumstances and new information. Accordingly, a prescription approach, whereby a prescription for pesticide use is written based on a feasibility analysis for each large-scale application (or perhaps series of applications during a growing season), could serve as an adaptive management approach. Each time a localized decision must be made to determine what pesticide to prescribe, current local conditions can be evaluated to determine the pest control that would maximize ecological protection to the extent feasible. Moreover, as new pesticides and non-chemical pest control techniques are developed they can be considered when determining what to prescribe. For example, if in year two, an American Bald Eagle or other listed species builds a nest near the field to be sprayed, it may not be appropriate to prescribe the same pesticide that was used on the field in year one if that pesticide is toxic to avian species. Likewise, if a listed species is no longer present near the field, it may be appropriate to go back to the more toxic pesticide. The prescription approach is much more amenable to imposing RPAs set forth in BiOps than EPA's current approach of imposing RPAs through a combination of generic labeling language and direction to consult with bulletins on the internet. The prescription approach would allow the specific RPAs that are relevant to the particular pesticide application to be implemented. For example, if the RPA requires that the pesticide not be applied during breeding season, a prescription would not be given for application of the pesticide during that time. Similarly, if the RPA requires a wildlife survey to be conducted, the entity writing the prescription can ensure that such a survey is conducted and take into consideration the results of that survey prior to writing the prescription. Similarly, if new information about the surrounding ecosystem is discovered, it could affect the decision of what pesticide to prescribe. Moreover, as new pesticides and non-chemical pest control techniques are developed, they can be considered when determining what to prescribe.

The prescription idea is similar to that of a medical doctor prescribing that a patient take a particular medication. Prior to issuing such a prescription the doctor would consider a number of factors, such as the patient's overall health, other medical conditions, other medications the patient is taking, any allergies or sensitivities the patient may have to certain types of medications, the patient's age, the patient's health and lifestyle objectives, and the patient's willingness to accept certain risks to achieve such goals. Moreover, the doctor would adjust the type or amount of medication over time to fine-tune the treatment in accordance with changing circumstances or new information. A prescription-type approach to pesticide application could similarly be adjusted over time to take into consideration changed local conditions, or new information about local environmental factors. Of course, the physician-prescribed pharmaceutical system is not without its shortcomings. High pressure sales tactics by representatives of the pharmaceutical industry, glossy advertisements on television and in magazines promising a wonderful life that can be achieved merely by popping a pill, the public's desire for an "easy fix" in the form of a pill, and industry-sponsored research all may contribute to the trend of physicians over-prescribing medications. It is likely that even with a pesticide prescription system in place, pesticide manufacturers will be able to convince decision makers to prescribe their pesticides. Nevertheless, such a system, if properly instituted, could result in at least some level of informed decision making prior to the release of large amounts of pesticides into the environment.

The likely criticisms of such a system would be that to institute it could entail high costs and possibly the creation of a new bureaucracy. However, the possibility exists of relying on existing infrastructure to facilitate such a system without the need for a completely new institution or significant additional personnel. The existing agricultural extension services could potentially be used to administer such a system. Alternatively, existing state requirements for certified applicator training and certification could be expanded to better educate applicators on local environmental factors that should be taken into account and on lower-risk and non-chemical alternative pest control management approaches that are relevant to particular crops in the particular geographic areas. Such an approach might also rely on existing extension infrastructure and resources.

Professor J.B. Ruhl has also noted that one of the most significant shortcomings of FIFRA is its lack of an adequate mechanism for regulating pesticide use.[11] Ruhl describes the regulatory schemes for various environmental laws, and the variety of exemptions and exceptions built into these statutes for agriculture.[12] As a result, Ruhl notes how the core pesticide statute, FIFRA "does little to regulate farm applications of pesticides and leaves fertilizers untouched."[13] FIFRA does not regulate the farm applying the pesticide, but simply how the pesticide itself is made or sold. Contrasting this regulatory system with those found under the Clean Water Act and Clean Air Act, Ruhl argues that the system, with its lack of permits, performance standards, public reporting requirements, or pesticide monitoring system, lacks any comprehensive framework for the regulation of agricultural pesticide use.[14] Ruhl proposes a new statutory scheme, tailored to the specific features of the agricultural industry.[15]

Another component of Ruhl's proposal would be to adopt a "Farm Release Inventory," an approach similar to the Toxics Release Inventory (TRI) under Emergency Planning Community Right-to-Know Act (EPCRA), which would require farms to publicly report releases of agro-chemicals.[16] Experience with the TRI has shown that simply requiring industrial operations to report to the public the types and amount of toxic releases from industrial facilities results in significant reductions of toxic releases, in part because industry will voluntarily reduce its emissions to avoid being seen as the "bad neighbor," and in part because citizens often use the information to put political pressure on industry to find ways

11 *See* J.B. Ruhl, *Farms, Their Environmental Harms, and Environmental Law*, 27 Ecology L.Q. 263, 337–8 (2000).

12 *Id.* at 293–315.

13 *Id.* at 309.

14 *Id.* at 311.

15 *Id.* at 337–8.

16 *Id.*

to reduce releases or substitute less toxic materials.[17] This reporting requirement could serve to discourage excessive pesticide use. Tied into the release report, Ruhl advocates a taxation system that would begin at certain pre-set levels pursuant to the eco-toxicity of the chemical.[18] The author stresses that these reforms alone are not sufficient in the absence of state regulatory vigilance and a continued vigorous federal role in the fight against air and water degradation.

Other Ecological Considerations

Another change necessary to ensure species protection is a reevaluation of pesticide registration data requirements to address more wildlife and ecological effects. FIFRA should also be amended to promote IPM and less risky pesticides. For example, section 11 of FIFRA should be amended to require that certified applicators receive training on lower-risk alternatives to chemical pesticides, IPM and other sustainable pest management approaches. FIFRA should also be amended to make clear that lower-risk and non-chemical pest management alternatives be considered when evaluating the benefits of a pesticide in both the registration process and the cancellation process.

Another change to either FIFRA or EPA's implementation of the statute would be for EPA to require "resistance management" plans to be developed and adopted for all large-scale agricultural pesticide registrations. As described in previous chapters, one of the greatest harms of chemical pesticide use is that it leads to the development of pest resistance, which puts farmers on a "pesticide treadmill," where newer and stronger chemicals are continually needed to suppress pest populations. Currently, EPA requires resistance management plans when it grants FIFRA registrations for Plant Incorporated Protectants (PIPs). EPA also frequently requires resistance management plans when it grants Experimental Use Permits even for chemical pesticides. Resistance management plans typically require untreated refugia near or within the fields in which the pesticides are applied. These refugia provide a safe haven for a population of pest insects, thereby allowing a significant number of individuals of the pest species that are susceptible to the pesticide being used to survive and reproduce. Without such refugia, most susceptible individuals will be killed by the pesticide and eventually only the non-susceptible individuals will survive and reproduce, thereby resulting in resistant pest populations.

Changes to the CWA

A change that could go a long way to reduce pesticide runoff into surface waters would be to impose stormwater treatment requirements on agricultural discharges that are not currently subject to CWA regulation because they are not defined

17 *Id.* at 337.
18 *Id.* at 338–9.

as point sources. As described above, significant groundwater and surface water impacts are directly linked to agricultural activities. Although it may not be possible to impose the same "end-of-pipe" technologies that are typically imposed on traditional point source dischargers, appropriate technology-based standards could be imposed on agricultural stormwater discharges. Many states have voluntary "Best Management Practices" programs that provide assistance and incentives for farmers who agree to build stormwater treatment ponds or employ other practices to reduce pollutant discharges from their operations. These types of BMPs could be imposed through a federal agricultural permitting program as technology-based "best available" technology. Alternatively, additional funding and technical assistance could be provided to encourage more farmers to implement BMPs. Another major tool available to reduce water pollution from farming operations is through the implementation of TMDLs. As described above, although the federal NPDES permitting system does not apply to non-point source agricultural discharges, TMDLs include both point source and non-point source discharges.[19] TMDLs are allocated to agricultural operations as well as to traditional point sources. Unfortunately, under the existing CWA there is not a federal "regulatory hook" to impose TMDLs on non-point sources. Aside from through the NPDES permitting program, TMDLs are implemented primarily by the states. Some states have developed approaches to, in essence, require farmers to comply with BMPs in order to demonstrate compliance with TMDL allocations. In Florida, for example, the Florida Watershed Restoration Act of 1997 establishes a "safe harbor" for agricultural operations that comply with specified BMPs designed to reduce water pollution.[20] If a farmer complies with the BMP, she is considered to be in compliance with TMDLs and Water Quality Standards and has a safe harbor from enforcement action.[21] If the CWA were to be amended to impose regulatory requirements on agricultural discharges, the Florida approach could be used as a model for implementation.

The issue of localized decision making regarding pesticide use is also of significance with regard to pesticide contamination of water. As described above, the issue of whether NPDES permits are required for the application of pesticides to water came to head with the 2002 case of *Headwaters, Inc., v. Talent Irrigation District.*[22] Historically, EPA had not required NPDES permits for such pesticide

19 33 U.S.C. § 1313(d). *See also* Pronsolino v. Nastri, 291 F.3d 1123 (9th Cir. 2002).

20 Section 403.067, Fla. Stat.

21 Section 403.067(7)(c), Fla. Stat.

22 243 F.3d 526 (9th Cir. 2001) (holding that the application of an aquatic pesticide to irrigation canals in compliance with the registration and labeling requirements under FIFRA, did not eliminate the need for an NPDES permit). For a detailed discussion of the judicial decisions and EPA's position on the issue of requiring NPDES permits for aquatic pesticide application, *see* Kelly C. Connelly (case note), *Pesticides and Permits: Clean Water Act v. Federal Insecticide, Fungicide and Rodenticide Act*, 8 Great Plains Nat. Resources J. 35 (2003) and Paul Herran (case note), *Headwaters, Inc. v. Talent Irrigation*

applications. As described above, EPA's proposed Pesticide General Permit for pesticide application to waters of the US is very generic. It merely reiterates that discharges must not cause or contribute to a violation of water quality standards and requires applicators to use IPM and have pesticide plans. To provide the type of localized decision making that is needed to protect water bodies and aquatic life from pesticides, EPA should revise the Pesticide General Permit to exclude large scale applications of pesticides and applications of certain pesticides in geographic areas inhabited by protected species. These types of applications should be subject to individual NPDES permits that take into consideration local ecological concerns.

Other scholars have made similar suggestions for regulatory reform to provide additional environmental protections from agricultural practices, including pesticide use. For example, David Adelman and John Barton have suggested a number of revisions to existing environmental regulatory programs to ensure that agricultural impacts to the environment are regulated "to the same extent and with the same standards as other industrial operations."[23] The authors argue in favor of technology neutral standards that consider the comparable risks and benefits of each technology. J.B. Ruhl has set forth a mix of regulatory, information reporting, tax-based, and incentive-based changes that could help to elevate many of the environmental harms resulting from agriculture.[24] Ruhl's regulatory proposals, while not dramatic, could result in significant improvements. Specifically with regard to water contamination, Ruhl argues in favor of eliminating certain agricultural exemptions from the NPDES program and using a traditional industrial regulatory approach for large agricultural operations.[25]

As can be seen from the discussion above, mere tinkering with existing environmental statutes will not be sufficient to achieve the reductions in synthetic pesticides and changes in agricultural practices necessary for long-term sustainability and minimal ecological harm. To achieve these goals, it will be necessary to transform our agricultural system to be more sustainable and ecologically based. Imposing more stringent or different regulatory requirements in itself will not lead to such a transformation. However, the fact that environmental regulatory changes alone are not sufficient does not mean that there is no role for policy changes and governmental intervention. In fact, governmental intervention will play a critical role in encouraging such transformation.

Although the proposed changes to FIFRA and the CWA could have significant ecological benefits, they are not sufficient to lead to the type of transformation of

---

District: Application of Aquatic Pesticides to Irrigation Canals, a Discharge, Which Requires a Clean Water Act Permit? 25 Haw. L. Rev. 629 (2003).

23 David E. Adelman and John H. Barton, *Environmental Regulation for Agriculture: Towards a Framework to Promote Sustainable Intensive Agriculture*, 21 Stan. Envtl. L.J. 3, 39 (2002).

24 Ruhl, *supra* note 11, at 334.

25 *Id.* at 335.

farming practices that is called for in the previous chapter. Even if environmental laws were modified to eliminate agriculture exemptions and impose more stringent standards, this would not be enough, because these laws are not designed to discourage industrial agriculture or to promote more sustainable forms of agriculture.

Although the types of regulatory changes described above can help to minimize the harms caused from pollution from agricultural operations and wetlands impacts, they do not go far enough to address the fundamental systemic problem with modern industrial commodity production. To address fully the energy and environmental implications of modern agricultural practices, we must make a dramatic shift to a more sustainable system of agriculture. To accomplish such a transformative shift, mere tinkering with existing regulatory regimes will not be sufficient. A complete overhaul of existing agricultural policy is warranted; and a significant component of such an overhaul would be a complete rethinking of commodity subsidy programs.

Proposals for Changes to Subsidy Programs

The regulatory changes proposed above would result in some environmental benefits. However, simply bringing the current agricultural system more squarely into the realm of environmental regulation will not address the root of the problem—i.e., that our entire system of industrial agriculture is fundamentally flawed and unsustainable. Current environmental statutes, even if they were to be amended to address more comprehensively the environmental impacts of agriculture, simply are not designed to promote the changes in agricultural practices that are needed. For example, FIFRA is not designed to eliminate or reduce pesticide usage, or even to encourage that higher-risk pesticides be replaced with lower-risk pesticides. Instead, the statute merely provides a licensing program that ensures pesticides meet a cost/benefit standard prior to commercialization. Neither FIFRA nor the Clean Water Act contain any mechanisms or authority to require or encourage more sustainable forms of agriculture. Even if CWA were to be amended to remove current agricultural exemptions, existing industrial agricultural practices could continue to be employed, provided sufficient engineering solutions to reduce pollutant discharges or provide wetlands mitigation are implemented. Although perhaps yielding environmental benefits, such an approach would do nothing to address the intensive use of fossil fuel and water inputs of industrial agriculture or to encourage more sustainable practices such as crop rotation, diversification via intercropping, Integrated Pest Management, cultural pest control, biological pest control, efficient irrigation practices, water conservation practices and non-synthetic fertilizers. Moreover, none of the existing environmental regulatory programs provides any authority to provide incentives to influence the types of crops that are grown in particular geographic locales. Thus, in addition to improving environmental regulatory programs, we must look to ways to improve the vast agricultural subsidy system that serves as a major driver of agriculture.

Just as existing regulatory programs could be modified to promote more environmentally friendly practices, current subsidy programs could be modified to do the same. An evaluation of current agricultural subsidy programs requires that we ask ourselves the following questions: Are the goals of the legislation passed in the 1930s, in response to the Great Depression and the Dust Bowl, goals that are most salient today? If not, what other goals must be considered? Even if our goals remain largely the same, is the current labyrinthine agricultural subsidy system really meeting the stated goals? If our goal is to have a safe, secure, abundant, and reasonably priced food supply, it is far from clear that our current system is helping to meet these goals. When the vast majority of our subsidies go to growing crops that are not "foods" without substantial processing and when there is no market for these crops, it is very questionable how these subsidies are helping to meet our societal goals. Moreover, with the more recent crises in the areas of climate change, energy security, and energy scarcity, we must reevaluate whether subsidizing high-energy high fossil fuel input industrialized agriculture is serving the best interests of our nation.

Through a variety of forces (including federal legislation such as Farm Bills), the agricultural system that has arisen in the United States since the 1930s is one composed primarily of "industrialized monocultures controlled by a handful of transnational corporations."[26] The consequences of this agricultural system include a broad range of economic, social, and ecological costs. Economically, for example, many rural communities have been destroyed, and the small farmers that do remain have been saddled with debt and dependence on government subsidies.[27] Socially, the concentration of agricultural ownership and intensification of production has led to agricultural jobs that are often dangerous, low paying, and demeaning.[28] Ecologically, the agricultural industry's reliance on intensive production methods and chemical inputs (in the form of pesticides, fertilizers, antibiotics, and hormones) has led to threats to air, water, and soil quality, stresses on supplies of natural resources such as water, soil, fuel, and land, and decreases in the overall genetic diversity of plants and animals.[29] By supporting certain types of production and commodities over others, the 2008 Farm Bill contributes to this state of affairs. Ultimately, subsidies should be shifted away from industrialized agriculture and toward more sustainable agriculture. Just as subsidies are currently used to promote certain types of agricultural behavior, subsidies may be used as incentives to farmers who engage in sustainable farming practices that protect

26 Deven G. Peña, *Environmental Justice and Sustainable Agriculture: Linking Ecological and Social Sides of Sustainability*, Resource Paper Series, Second National People of Color Environmental Leadership Summit—Summit II 3 (2002), *available at* http://www.ejrc.cau.edu/summit2/SustainableAg.pdf.

27 National Resource Council, Alternative Agriculture 90–93 (1989).

28 Eric Schlosser, *Hog Hell: Smithfields's Workers Face a Modern-Day Jungle*, The Nation, September 11, 2006, at 28.

29 National Resource Council, *supra* note 27, at 97–130.

ecological resources and services. Compensation to those parts of the agricultural industry that engage in sustainable practices is justified by the same rationale that justified the New Deal policy of paying farmers to produce less: the welfare of the nation depends on it.

There are several justifications for a shift away from current subsidy programs to a subsidy program that affirmatively promotes sustainable agriculture. First, the societal goals that led to the creation of the dustbowl-era subsidies and the subsidies promoting high yield in the 1970s are not the same goals that we have today. The public has consistently favored environmentally sound practices and has become increasingly concerned with purchasing healthy "clean" food. Moreover, the climate change crisis forces us to examine our use of fossil fuels. Industrialized agriculture that relies on large amounts of fossil fuel inputs that contribute to the climate change crisis and furthers our dependence on foreign oil cannot be sustained. Finally, as Michael Pollan says, we must stop viewing farming as a zero sum matter where either we completely preserve land from agriculture or we allow agriculture to seriously degrade the biodiversity and other ecological benefits provided by the land that is farmed.[30] For thousands of years humans have engaged in agriculture, much of which was carried out in ways that preserved a significant portion of the environmental resources and ecosystem services. Much has been written in recent years about compensating landowners for ecosystem services that they protect on their land but that benefit all of us. To pay farmers to protect ecosystem services while continuing to be able to farm seems to be a win-win situation. People need food, and thus we all have an interest in ensuring a robust agricultural system. However, people also need and want a healthy environment. Perhaps it makes more sense to pay farmers who engage in practices that meet both of these needs rather than forcing farmers to engage in environmentally harmful unsustainable practices that merely provide industrial feedstocks because it is the only way they can stay afloat financially.

One of the most obvious subsidy changes that could help to facilitate a shift away from industrial agriculture would be to eliminate the subsidies for corn ethanol production. Corn ethanol does not appear to have any net energy or environmental gain and the subsidy programs aimed at encouraging its development only serve to exacerbate the problems associated with industrial production of corn. In contrast, the subsidies for certain other biofuel development, including cellulosic biofuels and other renewable alternative energy sources, may be warranted. Some of the money that is currently spent on corn ethanol subsidization could be shifted to some of these more promising alternative energy development programs.[31] More significantly, however, shifting a substantial portion of the money currently spent on corn ethanol to ecosystem payments for sustainable agriculture could provide

30 Michael Pollan, *Farmer in Chief*, N.Y. Times Mag., October 12, 2008.

31 For a discussion of alternative sustainable energy, *see* Lynn Price & Mark D. Levine, *Production Consumption of Energy*, *in* Stumbling Toward Sustainability 79–98 (John C. Dernbach, ed., 2002).

more bang for the buck in terms of making progress toward energy independence, reducing greenhouse gas emissions, and reducing other environmental degradation and public health problems. Industrial corn production relies on large quantities of fossil fuel inputs. In an era where energy independence and climate change are two of the greatest problems facing our nation, it may be worth spending money to wean ourselves off of energy-intensive farming rather than spending money to promote more energy-intensive farming for a questionable fuel source.

The Farm Bill's large commodity subsidy programs should be evaluated for potential overhaul. Historically, critics have pointed to the Farm Bill's primary purpose of price stabilization through subsidization as being outdated at best and, at worst, a handout to large corporations at the expense of most farmers, the environment, and the overall health of the nation.[32] Economic studies demonstrate that, at least in the developed world, the primary driver of farmers' choices in what crops to grow and what agricultural practices to employ are economic incentives, including price supports and other economic subsidies.[33] Agricultural producers react rapidly to changed economic incentives.[34] Other studies have demonstrated a positive correlation between farm sector assistance, including subsidies, and per hectare use of chemical fertilizers.[35]

To meet this challenge of moving away from an industrial agriculture dominated system to a more sustainable system, it will be necessary to rethink the entire Farm Bill subsidy framework. Just as US agricultural subsidy programs have shaped agriculture in the past by rewarding growers for what are now seen as unsustainable practices, US agricultural subsidy programs can be adapted to promote practices that society now finds more desirable because they conserve resources for future generations. It is certainly worth considering whether a major shift in subsidy programs away from supporting high-yield monoculture industrial commodity production towards promoting a less energy and water intensive, more environmentally friendly and diverse form of agriculture is warranted.

---

32 For representative criticism of the farm bills *see* Brian M. Riedl, The Heritage Foundation, Top 10 Reasons to Veto the Farm Bill, *available at* http://www.heritage.org/Research/Reports/2002/04/Top-10-Reasons-to-Veto-the-Farm-Bill (last visited May 27, 2011); Defenders of Wildlife, Top Ten Major Problems with the House Farm Bill, *available at* http://www.familyfarmer.org/sections/tenproblems.pdf (last visited May 27, 2011); Michael Pollan, *You Are What You Grow*, N.Y. Times Mag., April 22, 2007, *available at* http://www.nytimes.com/2007/04/22/magazine/22wwlnlede.t.html?ei=5090&en=e8328c69f0b3f4be&ex=1334894400&partner=rssuserland&emc=rss&pagewanted=all (last visited May 27, 2011).

33 Johan Boardman, Jean Poesen & Robert Evans, *Socio-economic Factors in Soil Erosion and Conservation*, 6 Envtl. Sci. & Pol'y 1–3 (2003).

34 *Id.*

35 Jan Lewandrowski, James Tobey & Zena Cook, *The Interface Between Agricultural Assistance and the Environment: Chemical Fertilizer Consumption and Area Expansion*, 73 Land Econ. 404 (1997).

Legal scholars have only recently begun to weigh in on ways to achieve a more sustainable agricultural system. For example, Neil Hamilton has asserted that a new sustainable agriculture could, and perhaps should, play an important role in the new "green industry" movement.[36] As Hamilton puts it, agriculture was the "original green activit[y]" and "is inherently part of any sustainable future."[37] Of course, for agriculture to play a role in the green industry movement, it must be transformed from its current incarnation. Hamilton argues that sustainable agriculture is tied to the recent interest in buying local or organic food, joining community supported agriculture and the "clean food" movement. The challenge is how to move away from current practices to promote and encourage a sustainable agricultural system.[38]

By far, the lion's share of farm subsidies under the Farm Bill goes to farmers with large industrialized monocultures of commodity crops using significant pesticide inputs. Recent iterations of the Farm Bill, including the 2008 Farm Bill, however, include a number of conservation programs. Many of these programs, such as the Conservation Reserve Program (CRP) and the Wetland Reserve Program (WRP) involve taking land out of production to preserve it for conservation purposes. While such conservation programs certainly help to achieve certain conservation goals, they do nothing to change the way working lands are farmed to reduce pesticide input or to become generally more sustainable.

Since 1996, two significant working lands conservation programs have been added to the Farm Bill, Environmental Quality Incentives Program (EQIP), which was added in 1996 and revised in 2002 and 2008, and the Conservation Stewardship Program (CSP),[39] which was originally called the Conservation Security Program in 2002 and modified as the CSP in the 2008 Farm Bill. The CSP provides financial and technical assistance to farmers and ranchers who implement conservation measures on their actively farmed land.[40] Assistance is available for a number of different types of conservation practices including watershed protection, soil and water and air quality improvement, carbon sequestration and water and energy conservation.[41] While the CSP is not directly geared toward encouraging reductions in pesticide use or shifts to lower-risk pesticides, assistance is available for practices that increase biodiversity and wildlife, and pollinator habitat, which presumably would include eliminating or reducing harm to biodiversity, wildlife, and pollinators that result from pesticide use. Even more importantly, however,

---

36 Neil D. Hamilton, *Feeding our Green Future: Legal Responsibilities and Sustainable Agricultural Land Tenure*, 13 Drake J. Agric. L. 377 (2008).

37 *Id.*

38 *Id.*

39 16 U.S.C. §§ 3838d, 3838e.

40 Tadlock Cowan, Cong. Research Serv., Conservation Provisions of the 2008 Farm Bill, 8 (2008), *available at* http://www.ncseonline.org/NLE/CRSreports/08Aug/RL34557.pdf.

41 *Id.*

by encouraging increased biodiversity on the working farm, the CSP encourages the types of ecologically based production practices that ultimately reduce the need for pesticide inputs, relying on more ecologically based pest management from predators and parasites. The CSP provides five-year contracts with up to $40,000 payments to eligible farms.[42] The CSP program has a total funding pool of 12 million dollars for 10 years, which is sufficient to enroll approximately 13 million acres per year into the program.[43] One of the most significant features of the 2008 CSP is that it mandates that USDA take steps to tailor the program for organic farms and ranches and to provide outreach and technical assistance to enroll organic producers in the program.[44]

EQIP provides financial cost-share assistance and technical assistance to participating farmers and ranchers who use conservation practices on their working lands.[45] Contracts under EQIP range from one to ten years.[46] Farmers can receive up to 75 percent of the costs incurred in implementing conservation practices as a cost-share and up to 100 percent of any income forgone due to the installation or adoption of the conservation practice.[47] Total payments under EQIP generally may not exceed $300,000 per person or legal entity.[48] Incentives under EQIP are used to encourage energy conservation, water conservation, and conservation planning.[49] As is the case with CSP, EQIP encourages a broad range of conservation practices and is not specifically directed toward reducing harm from pesticide use. However, some of the environmental benefits that EQIP seeks to promote include soil, water and air quality, and wildlife habitat, all of which are impacted by pesticide use.[50] The 2008 Farm Bill provides $7.325 billion in funding for EQIP for 2008–2012.[51]

Although EQIP funding is available to most types of agricultural operations, the 2008 Farm Bill makes clear that one of the purposes of EQIP is to encourage the conversion from conventional farming to organic farming.[52] Prior to the 2008 Farm Bill, some states used their EQIP programs specifically to benefit organic producers. The 2008 Farm Bill expressly allows EQIP funding to be used for organic production and transition to organic production.[53] Farmers who agree to

42 *Id.*

43 *Id.*

44 *Id.* at 9.

45 *Id.* at 7.

46 16 U.S.C. § 3839 (2011).

47 Cowan, CRS, *supra* note 40, at 8.

48 *Id.* at 7.

49 *Id.*

50 *Id.*

51 *Id.* at 8.

52 Renee Johnson, Cong. Research Serv., Organic Agriculture in the United States: Program and Policy Issues, 9 (2008), *available at* http://www.fas.org/sgp/crs/misc/RL31595.pdf.

53 *Id.*

transition to organic production and to seek USDA certification under the National Organic Program are eligible to receive up to $20,000 per year for financial assistance.[54] EQIP does not set aside a specific dollar amount for organic funding. Instead, funding for organic growers and transition to organic is included in the overall EQIP budget.[55]

The 2008 Farm Bill contains additional programs that specifically benefit organic producers. Organic agriculture is one of the fastest growing agricultural sectors in the US and demand has been exceeding supply, resulting in many organic products being imported from other parts of the world. The failure of domestic organic production to keep pace with demand is attributed at least in part to the high costs associated with converting to organic production and obtaining USDA organic certification. The 2008 Farm Bill seeks to encourage conversion to organic production through a number of different programs. As discussed above, EQIP funding is now explicitly authorized for organic production and conversion to organic agriculture. In addition, the 2008 Farm Bill significantly expands funding available for organic producers in the National Organic Certification Cost Share Program, a program initially established in the 2002 Farm Bill. This program allows producers and handlers to recoup up to 75 percent of their annual costs of obtaining organic certification from USDA.[56] The maximum annual payment per farmer is capped at $750.[57] The 2008 Farm Bill increased funding from $5 million to $22 million.[58]

The 2008 Farm Bill took a number of important steps to encourage more environmentally friendly agriculture, including Ecologically Based Pest Management (EBPM). Nevertheless, the amount of funding earmarked for working lands conservation programs is extremely modest in comparison to the massive subsidies available for industrialized commodity production. While the current Farm Bill contains several conservation programs such as the CSP, EQIP, WRP, and CRP,[59] and contains provisions which promote the accessibility of organic and local foods,[60] many critics argue that any reform that the 2008 Act contains is superficial, and that the status quo of major subsidies to large-scale, industrialized producers of a few commodities was preserved to the detriment of most farmers,

54 US Department of Agriculture, Natural Resources Conservation Service, Environmental Quality Incentives Program (EQIP) Support for Organic Growers, *available at* http://www.nrcs.usda.gov/programs/eqip/organic/index.html (last visited June 5, 2011).

55 *Id.*

56 Johnson, CRS, *supra* note 52.

57 *Id.*

58 *Id.*

59 For a complete list of conservation programs in the 2008 farm bill, *see* USDA, 2008 NRCS Farm Bill Conservation Programs, *available at* http://www.nrcs.usda.gov/PROGRAMS/farmbill/2008/ (last visited May 27, 2011).

60 *Id.*

the environment, and the overall health of the nation.[61] While the conservation programs and supports for more sustainable foods that do exist should be commended, they should also be greatly expanded.

For example, the Union of Concerned Scientists recommends shifting EQIP livestock funds to sustainable livestock and poultry production rather than using the funding to subsidize large CAFOs.[62] It suggests using the funding for managed rotational grazing, pasture and range management, and other environmentally sound practices. Likewise, they argue in favor of new funding and technical assistance to encourage a shift from grain-fed feedlot animal operations to grass-fed operations.[63] Other recommendations include providing more subsidies for small and medium farms on par with the funding currently provided primarily to the largest agricultural operations.[64] As with the regulatory modifications described above, these types of alterations to current subsidy programs could have significant environmental benefits. However, the approach of simply modifying existing subsidy programs to promote more environmentally and socially sound farming practices may not be sufficiently transformative to get to the core of the problem—i.e., we are using vital environmental and energy resources to grow surpluses of industrial feedstocks that do not contribute to the need for nutritious foods. Merely providing subsidies for industrialized agriculture that makes some improvements in its practices is not enough. As Pollan has argued,[65] we need to find a way to move our current fossil-fuel-based industrialized agriculture to a solar-based sustainable agriculture. To achieve the agricultural transformation that is needed to have a truly sustainable food production system, a more dramatic shift in policy may be warranted.

Some commentators have argued in favor of dramatically reducing or even eliminating agricultural subsidies. However, it is important to recognize that such a move in itself may not significantly reduce the environmental degradation associated with agriculture. For example, some studies have shown that decreasing subsidies may not reduce the overall pollution associated with fertilizer use because, although per hectare amount of fertilizer used might be reduced, lower subsidy levels may actually increase the number of hectares farmed.[66]

---

61 *See, e.g.*, Sustainable Agriculture Coalition, Grassroots Guide to the 2008 Farm Bill 3 (October 2008), *available at* http://sustainableagriculturecoalition.org/wp-content/uploads/2008/11/sac-farm-bill-guide.pdf.

62 Union of Concerned Scientists, Food and Agriculture, Food and Agriculture Solutions, Big Picture Solutions, New Policies for a New Agriculture, *available at* http://www.ucsusa.org/food_and_agriculture/solutions/big_picture_solutions/new-policies-for-a-new.html (last visited May 27, 2011).

63 *Id.*

64 *Id.*

65 Pollan, *Farmer in Chief*, *supra* note 30, at 64.

66 Lewandrowski et al., *supra* note 35.

One of the most significant market-distorting flaws with subsidy programs is that historically they were, and in some cases still are, coupled with production levels—i.e., the more a farmer grows, the more government money she receives. Consequently, growers have a financial incentive to produce the highest yields possible, which requires high water and fossil fuel inputs and has significant environmental consequences. One option to shift growing practices away from this type of agriculture is to decouple subsidies from production levels. In 1996 the US took a major step toward decoupling by instituting income support systems that are not based on production levels. As described above, direct payments, countercyclical payments, and ACRE payments are made regardless of whether the producer actually grows the crop or how much of the crop is produced. The European Union took a similar step in 2003 when it decoupled its subsidies from production. In addition, as part of the EU's 2003 overhaul of agricultural policy, it imposed requirements that growers must comply with certain specified environmental practices in order to receive subsidies.[67] However, although these programs have decoupled payment from actual production amount, they are only partially decoupled. For example, countercyclical payments are tied to historical base acres and fixed program yields. Consequently, farms do not receive larger payments for larger yields. However, these payments are still linked to market conditions and payments increase in response to lower national average prices.[68] Thus, on a national basis, if a commodity is produced at such a high rate that there is a large surplus, and prices are consequently low, farmers will receive a larger subsidy than they would in the absence of such surpluses. Unfortunately, these attempts to partially decouple subsidies from production levels do not appear to have significantly reduced high-yield production of commodity crops. The reasons for the failure of decoupling to impact agricultural practices are unclear and most likely multifaceted. For example, one study suggests that the US's attempts at decoupling have not alleviated the problem of production distortion because decoupled lump payments increase farmers' household income, which enables increased investment into farming operations and which in turn can lead to increased production levels.[69] One study found that farmers typically spend a large percentage of the decoupled subsidy payments they receive to invest in their farming operations.[70]

67 Daniel Bianchi, *Cross Compliance: The New Frontier in Granting Subsidies to the Agricultural Sector in the European Union*, 19 Geo. Int'l Envtl. L. Rev. 817 (2007).

68 US Department of Agriculture, Economic Research Service, Briefing Rooms, Farm and Commodity Policy, What is Meant by Decoupling? *available at* http://www.ers.usda.gov/Briefing/FarmPolicy/decoupling.htm (last visited May 27, 2011).

69 *Id.* (citing US Department of Agriculture, Economic Research Service, AmberWaves February 2003, Farm Payments: Decoupled Payments Increase Households' Well-Being, Not Production, *available at* http://www.ers.usda.gov/AmberWaves/Feb03/Features/FarmPayments.htm (last visited May 27, 2011)).

70 *Id.*

Another significant problem that remains after US income supports were decoupled from production levels is that other programs such as price supports were not decoupled. Thus, although there may not be a financial incentive for a farmer who historically grew corn to switch to another crop under the income support programs, a strong incentive continues to exist for that farmer to grow as much corn as possible under the price support programs. As described above, the non-recourse loan program in essence directs the federal government to buy commodities at higher prices than could be obtained on the open market. Consequently, farmers continue to have a strong incentive to produce as much as possible. Income supports will not be affected one way or the other by production levels, and price support levels will rise in direct proportion to the amount produced. Decoupling one program without decoupling the others does not achieve the goal of allowing market forces, rather than government subsidies, to dictate what crops to grow and at what production levels. Adelman and Barton have noted that even with the US's attempts at decoupling, "loan-deficiency subsidies continue to discriminate against farmers who employ environmentally sound crop rotations and, despite record surpluses, induce farmers to maximize their yields through overuse of agrichemicals and irrigation."[71]

The USDA has recognized that the US's decoupling approach has not achieved its intended goal in part because, when the 1996 decoupling occurred, the decoupled income subsidies were intended to be the primary commodity subsidies. However, declines in market prices for many years after the decoupling led to an increase in subsidy payments from price support programs that were not decoupled, such as the non-recourse marketing loan program.[72]

Another potential explanation for the failure of decoupling is that, because income supports are only available to farmers who historically grew large commodity crops, even with decoupling these farmers may continue to grow those commodity crops because that is what they do. A farmer who has grown corn all of her life, who knows how to grow corn, who has invested large amounts of money into equipment to grow corn, and who lives in a community that has a strong corn-growing culture, may choose to continue to grow corn even though she could receive the same income supports if she were to grow other crops. Thus, decoupling of income support programs in itself is not enough.

Decoupling price supports as well as income supports could prove to be more effective at promoting a shift away from overproduction of large commodity crops. Of course, to encourage such a shift, price supports would have to be available for crops other than the major commodity crops. In an era of surpluses of commodity crops such as corn that are primarily used in large animal feedlot operations or as

71 Adelman and Barton, *supra* note 23, at 33.

72 US Department of Agriculture, Economic Research Service, AmberWaves February 2003, Farm Payments: Decoupled Payments Increase Households' Well-Being, Not Production, *available at* http://www.ers.usda.gov/AmberWaves/Feb03/Features/FarmPayments.htm (last visited May 27, 2011).

industrial feedstocks for processed foods, it simply makes no sense from a public policy standpoint to continue to subsidize only these commodities, and to fail to afford the same subsidies to producers of non-commodity food crops. If public policy concerns argue in favor of any subsidies, it would be better to subsidize a broad array of healthy food crops grown in an environmentally sound manner, without the heavy dependence on fossil fuel inputs, and to decouple all subsidies from high-yield unsustainable commodity production. To achieve this, not only must subsidy programs be modified to apply to a broad variety of crops rather than a few commodity crops, but also a mechanism must be in place to tie subsidies to environmental protection.

The uncertainty regarding the extent to which decoupling changes farmer behavior is only half of the equation. Perhaps the more important question is whether there will be resultant environmental benefits even if a farmer plants less acreage in a decoupled system. Planting fewer acres may result in more environmental benefits on the land that is left fallow, but if the farmer compensates for this by more intensely farming the planted acres, there may not be a net environmental benefit. For example, one study analyzing the effects of subsidies based on production levels and decoupled income payments on species richness, supports the conclusion that both types of subsidies contribute to loss of species richness.[73] The two different subsidy programs, however, appear to influence different species in different ways, with decoupling appearing to have more influence on avian richness than insect richness.[74] This study suggests that production-dependent subsidies, such as price supports, tend to lead to increases in the area of arable land and consequent decreases in the area of fallow land. Income supports that are decoupled from production levels, on the other hand, tend to induce a shift to more fallow land and less actively farmed land. This trend supports the idea that decoupling can result in increased biodiversity due to less land being farmed.[75] However, the study shows that the impacts to particular species or groups of organisms from such changes are not well understood, and there may be substantial differences among different taxa.[76] Thus, decoupling alone may not be sufficient to achieve a more sustainable agricultural system.

Another option for encouraging such a shift to sustainable agriculture is to make a more dramatic change to subsidy programs. For example, a system could be created in which subsidy levels are tied to the adoption of varying levels of sustainable practices. Perhaps a tiered system could be created where growers who continue to grow in a monoculturistic industrial fashion, but reduce their use of fertilizers, pesticides, and water and employ certain best management practices to limit erosion, depletion of organic matter in soils, and contamination of groundwater

73 Thomas K. Gottschalk et al., *Impact of Agricultural Subsidies on Biodiversity at the Landscape Level*, 22 Landscape Ecology 643 (2007).

74 *Id.*

75 *Id.*

76 *Id.*

and surface water and harm to surrounding biodiversity, receive a tier-one level of subsidy. Of course, this subsidy should be available for a variety of crops—not merely for commodity grain crops. Another tier could be a higher level of subsidy provided to growers who meet existing USDA organic certification growing standards. This subsidy would serve to reward organic growers, encourage more growers to go organic, and lower consumer prices for organic products, thereby increasing consumer demand for such products. However, although USDA organic certification standards call for the use of certain sustainable practices, they are primarily targeted at prohibiting the use of certain substances such as synthetic pesticides and fertilizers, genetically modified organisms, antibiotics and pesticides rather than requiring widespread implementation of sustainable practices. Thus, perhaps a third tier is also warranted. The third tier could be comprised of growers who, while not going as far as to meet organic standards, engage in a set of identified sustainable practices, which would require a very different approach to farming than large-scale monoculture industrialized production.

Ecosystem Services Payments

Perhaps the most useful model for this proposed shift in the subsidization of agricultural production is ecosystem services payments. In general, an ecosystem-services approach recognizes, protects, and restores the multitude of services provided by ecosystems—services such as purifying air and water, detoxifying and decomposing waste, renewing soil fertility, regulating climate, mitigating droughts and floods, controlling pests, and pollinating plants.[77] Ecosystem services payments are one of several instruments that can be used to protect ecosystem services, and they usually take "the form of a subsidy, either as a direct payment or tax break."[78] They are justified by a public goods argument: "society at large benefits from these activities but because of market failures does not pay for them."[79] Moreover, converting to and implementing more sustainable agricultural practices will undoubtedly cost farmers money, at least in the short run.[80] Despite criticism from some policy analysts that they are wasteful or inefficient subsidies, or that they are paying people to do what they "should" be doing anyway, ecosystem services payments can be found in many places throughout environmental law and policy

---

77 *See* James Salzman, Barton H. Thompson, Jr., & Gretchen C. Daily, *Protecting Ecosystem Services: Science, Economics, and Law*, 20 STAN. ENVTL. L.J. 309 (May 2001).

78 James Salzman, *Creating Markets for Ecosystem Services: Notes from the Field*, 80 N.Y.U. L. REV. 870, 886 (June 2005).

79 *Id.*

80 *See* Linda K. Lee, *Groundwater Quality and Farm Income: What Have We Learned?* 20 REV. OF AGRICULTURAL ECON. 168 (evaluating 21 studies which suggest farm income decreases as measures are taken to reduce water pollution from farms).

around the world.[81] Although the existing conservation programs, such as the CRP and WRP, could be considered a form of ecosystem services payments—i.e., paying farmers for the ecological values protected by taking certain lands out of production—to accomplish the goal of truly sustainable agriculture, farmers must be paid, not just for the land they do not farm, but more importantly for the ecosystem services provided by the land that they do farm. By compensating farmers for preserving certain ecological functions such as habitat, carbon sequestration, and use of animal waste on the actively farmed lands themselves, farmers will have economic incentives to adopt more sustainable approaches to farming.

One commentator describes the rationale for providing ecosystem services payments to farmers as follows: "Why not treat farmers' provision of ecosystem services as no different from their provision of other marketable goods? Farmers are certainly well accustomed to contractual arrangements for their agricultural products. Dairy farmers sign contracts to sell their milk; potato farmers do the same for their spuds. Why not treat the provision of water filtration services as a similar business transaction, where farmers manage their land through riparian buffers and grass swales to 'grow the crop of water quality' much the same as dairy and potato farmers do for their cash crops? In many respects, provision of ecosystem services would be no different than supplying traditional farm produce, with the level of compensation dependent on the quality and level of services provided."[82]

J.B. Ruhl has argued that tax-based and incentive-based approaches should be employed to encourage farmers to engage in more environmentally friendly practices.[83] Ruhl's suggestions for incentive-based approaches focus primarily on expanding the existing WRP and CRP "green payment" programs.[84]

Another commentator suggests the following measures that Congress should take in order to fully integrate ecosystem services payments into federal farm policy:[85] funding of research to determine how to calibrate farm practices with ecosystem service delivery at local scales (similar to what is being done with the Florida Ranchland Environmental Services Project discussed below); development of national standards for quantifying ecosystem service values associated with agricultural lands, including the development of proxies that can inexpensively be measured to estimate service delivery potential; preferential treatment in federal "green subsidy" payments programs for farms that would actually deliver ecosystem service values to identifiable local and regional populations; and funding of pilot and permanent demand-based state and local farm multi-functionality programs (such as the Florida Ranchlands Environmental Services Project).

---

81 Salzman, *supra* note 78, at 870.

82 *Id.* at 888.

83 Ruhl, *supra* note 11, at 339.

84 *Id.* at 340–41.

85 J.B. Ruhl, *Agriculture and Ecosystem Services: Strategies for State and Local Governments*, 17 N.Y.U. Envtl. L.J. 424, 458 (2008).

A compelling example of ecosystem services payments in practice, the Costa Rican Payments for Environmental Services Program, is a government-run scheme that pays private landowners direct payments for the ecological services that their lands produce.[86] Implemented by the Sistema Nacional de Areas de Conservacion and the Fondo Nacional de Financiamiento Forestal, the Program encourages the adoption of land uses and forest management techniques that protect primary and secondary forest and encourage forest plantations to meet industrial demands for lumber and other wood products.[87] The Program enters into site-specific contracts with individual small- and medium-sized farmers.[88] Three types of contracts exist: (1) Forest conservation contracts, in which US$210 per hectare is paid over a 5-year period for forest conservation easements; (2) Sustainable forest management contracts, in which US$327 per hectare is paid over a 5-year period for 15-year sustainable forest management easements; and (3) Reforestation contracts, in which US$537 per hectare is paid over a 5-year period for 15- to 20-year reforestation easements.[89]

An example of environmental services payments closer to home, the Florida Ranchlands Environmental Services Project, is "a payment for ecosystem services pilot program designed to test whether cattle ranchers near Lake Okeechobee can provide ecosystem services more cost effectively than by building new public works projects."[90] A collaboration between South Florida cattle ranchers, Florida state agencies, USDA NRCS, researchers, and environmental groups, the eventual aim of the Project is to "allow ranchers to compete to provide environmental services of water storage, phosphorus retention and wetland habitat enhancement in the Northern Everglades ecosystem."[91]

As described above, one of the important components of an ecosystem services payment program, or any other agricultural sustainability program, is to fund additional research, education and training to support the development and widespread use of such a program. Leading agricultural law expert, Johan Davidson, has opined that in "American agricultural research, particularly in the research stations and the land-grant colleges, there is a notable absence of serious and

---

86 Edgar Ortiz Malavasi & John Kellenberg, Program of Payments for Ecological Services in Costa Rica 1, *available at* http://ecosystemmarketplace.com/documents/cms_documents/lr_ortiz_kellenberg_ext.pdf.

87 *Id.*

88 *Id.* at 4.

89 *Id.* at 4–5.

90 Jacob T. Cremer, *Tractors versus Bulldozers: Integrating Growth Management and Ecosystem Services to Conserve Agriculture*, 39 ENVTL. L. REP. NEWS & ANALYSIS 10541, 10546 (June 2009) (citing Sarah Lynch & Leonard Shabman, *The Florida Ranchlands Environmental Services Project: Field Testing a Pay-for-Environmental Services Program*, RESOURCES, Spring 2007, at 17).

91 University of Florida Water Institute, Valuing Ecosystem Services on Florida Ranchlands: Lessons Learned, *available at* http://www.waterinstitute.ufl.edu/symposium/downloads/92718559.pdf.

continuing collaboration with ecologists."[92] Because an ecosystem services-based model of agriculture is such a new and different approach, and because it relies heavily on an understanding of the functioning of ecosystems, and how to integrate agricultural production and ecosystem services, additional research will be needed.

In addition to continued support for programs such as the Conservation Reserve Program and the Environmental Quality Incentives Program (which themselves contain components of environmental services payments), the 2008 Farm Bill, in section 2709, contains promising language for the prospect of environmental services payments being fully integrated into agricultural policy. That section requires the Department of Agriculture to "establish technical guidelines that outline science-based methods to measure the environmental services benefits from conservation and land management activities in order to facilitate the participation of farmers, ranchers, and forest landowners in emerging environmental services markets," and to establish guidelines to develop a procedure to measure environmental services benefits, a protocol to report environmental services benefits, and a registry to collect, record and maintain the benefits measured.[93]

In order to implement these provisions, the USDA created two new entities in December 2008.[94] The Office of Ecosystem Services and Markets and the Conservation and Land Management Environmental Services Board were established in order to "assist the Secretary of Agriculture in the development of new technical guidelines and science-based methods to assess environmental service benefits which will in turn promote markets for ecosystem services including carbon trading to mitigate climate change."[95] It is not yet clear what type of ecosystem services payment programs will result from this new legislation. However, it is important to note that these new provisions do not alter existing large-scale commodity subsidy programs. Although the new provisions may reflect a potential evolution toward a new ecosystem services payment program, a large-scale shift to a sustainable agricultural system will require a more comprehensive overhaul of the current system, rather than the piecemeal addition of a few new programs.

Finally, any proposal to change the Farm Bill Subsidy programs must take into consideration how the various subsidy programs can work together. For example, although the CRP and other conservation provisions in the Farm Bill have been successful in conserving certain environmentally sensitive lands, they

---

92 John H. Davidson, *Agriculture*, *in* Stumbling Toward Sustainability 360, 366 (John C. Dernbach, ed., 2002).

93 *Id.* at 459 (citing Food, Conservation & Energy Act of 2008, H.R. 2419, 110th Cong. § 2709 (2008)).

94 USDA, USDA Announces New Office of Ecosystem Services and Markets, News Release No. 0307.08, *available at* http://www.usda.gov/wps/portal/!ut/p/_s.7_0_A/7_0_2KD?contentidonly=true&navid=farmbill2008&contentid=2008/12/0307.xml (last visited May 27, 2011).

95 *Id.*

may unintentionally also adversely impact some ecosystem services and natural resources. When farmers are paid to set aside acreage for conservation, their natural inclination, as encouraged by price supports, may be to increase yield levels on the remaining lands that they are free to continue farming by farming more intensively. Such increased intensive farming may result in even more environmental degradation and the need for more fossil fuel and water inputs than may have been previously used. Thus, the current conservation programs in the Farm Bill are double-edged swords, protecting certain environmental values while at the same time encouraging the degradation of others. To avoid this problem it may be necessary to integrate the conservation subsidy programs into the price and income support subsidy programs to establish one comprehensive subsidy program that maximizes the promotion of environmentally sound and sustainable practices both on and off of the actively farmed fields.

Development of a robust ecosystem services program for agriculture could go a long way toward promoting a more sustainable agriculture system. However, merely establishing yet another subsidy program will not be sufficient. To achieve the level of change that is warranted to squarely address climate change, dependence on foreign energy, and environmental degradation from agriculture, the conservation subsidy programs should be integrated into a newly constructed version of the current commodity subsidies programs of Title I.[96] Such a "merged" subsidy system could induce the shift to sustainable, ecologically functioning cropland ecosystems, rather than dividing crop land into two categories: conserved fallow land and industrialized monocultures with little or no ecological value on actively farmed land.[97] Under this approach, Title I subsidies would be paid to producers of a large variety of crops (not just the few large commodity crops), but only if they implement practices geared toward low fossil fuel inputs and general environmental sustainability. Rather than tying subsidy levels merely to historic or current production levels, subsidies would be tied to the level of ecosystem services provided. Ecosystem services that could be included as compensable could include—to name a few of the many—the following: natural predator and parasite control of pests, pesticide resistance management, biodiversity, carbon offsets, soil conservation, soil enhancement, water conservation, habitat protection, protection of important nutrient cycling microorganisms, pollinator protection, nutrient retention, flood protection, aquifer recharge, erosion protection, animal waste disposal (by employing it as fertilizer), and a variety of other important services that might be provided depending on the crop and the manner in which

96 Others have proposed similar, but less ambitious, merged "green payments" programs. *See, e.g.*, Katherine R. Smith, *Time to "Green" U.S. Farm Policy (Environmental Impact of Farm Subsidy Programs)*, 11 Issues in Sci. & Tech. 71 (1995).

97 For a discussion of how farms can be made to be functioning ecosystems, *see* Wes Jackson, *Farming in Nature's Image*, *in* Fatal Harvest: The Tragedy of Industrial Agriculture 47 (Andrew Kimbrell, ed., 2002) and Daniel Imhoff, *Faring with the Wild: A Conservation Approach to Agriculture*, *in id.* 367.

it is produced. Under a merged approach certain environmentally sensitive lands could continue to be set aside, as they are in the current CRP and WRP. Lands that are completely protected would most likely provide greater ecosystem services than lands that are actively farmed—albeit in an environmentally sustainable manner. Thus, lands chosen for conservation would receive higher subsidies than working lands, but working lands would still receive subsidies based on the type and level of ecosystem services provided.

Consumer Awareness through Labeling and Certification

One of the key components to shifting toward an ecologically based low pesticide input agricultural system is through consumer choice. Consumers have become much more concerned about ensuring their food is safe and that it is grown in an environmentally sound manner. As can be seen from the recent upward trend in the organic foods market, consumers are becoming more concerned with how their foods are grown and many more consumers are willing to pay a premium for pesticide-free food and other products today than 20 years ago. Although much of this consumer preference may be driven by health concerns regarding pesticide residue on food, the benefits of pesticide-free farming accrue to the natural environment as a side benefit.

One way to educate consumers and encourage them to purchase food and other products produced organically or in environmentally sensitive ways is through certification and labeling programs. The USDA organic labeling program is already in place and is effective in the sense that it assures that purchasers can identify products produced meeting the USDA "organic" standards. However, this program has serious limitations in that the standards are limited to primarily telling farmers which chemicals they cannot use to be considered "organic." While these standards ensure that synthetic pesticides are not used on products labeled organic or on farms certified as organic, they may not be sufficient to achieve widespread transformation in farming systems.

While the existing organic standards and labeling are one way to attract consumers desiring healthier and more environmentally friendly agricultural products, other labeling or certification programs could be used to capture additional environmental benefits. For example, even where organic farming is not possible or profitable, more environmentally friendly farming practices with low synthetic pesticide inputs, including EBPM approaches, may be appropriate. Labeling or certification programs that inform the consumer of the benefits of this type of farming may be appropriate even if "organic" standards are not met. Professor Jason J. Czarnezki has chronicled a number of food labeling programs including organic, carbon footprint, and environmental life-cycle analysis labeling programs, and has discussed how these programs can both respond to consumer demand and influence consumers to choose eco-friendly food products.[98]

Further, if the goal is not to merely reduce synthetic pesticide inputs, but to transform our agricultural system into a more ecologically based system that preserves and enhances natural pest control, there could be a way to certify and label products grown in this fashion to attract consumers who desire to encourage a transformed agricultural system.

Increased Funding for EBPM Research

Research institutions have the ability to experiment and develop non-chemical or low-chemical input approaches to pest management, such research then could be disseminated to farmers.[99] Unfortunately, low chemical input pest control practices tend not to generate profits, and thus, there is no monetary incentive for the agri-business interests that frequently fund pest control research to sponsor such research. To research, develop, disseminate, and market new low-input pest control techniques, it will be necessary to decouple pest control research from the funding of input suppliers.

In addition to the incentive and assistance programs described above, the 2008 Farm Bill provides increased funding for sustainable and organic research. Historically, USDA's research and extension budget for sustainable and organic agriculture was woefully deficient, and the vast majority of the USDA's $2.5 billion research and extension annual budget was for conventional farming. While funding for sustainable and organic agriculture remains relatively small, the 2008 Farm Bill provides for a five-fold increase in the Organic Agriculture Research and Extension Initiative. The goal of this initiative is to facilitate the development of organic production and markets for organic products. The 2008 Farm Bill added two new grant purposes: examining optimal conservation outcomes related to organic production and developing new seed varieties that are well suited for organic production. Funding for this research and extension initiative is $78 million over a several year period.

Beyond Farm Bill funding for sustainable and organic farming research and extension, EPA administers a new funding program specifically geared toward research, public education, training and monitoring related to IPM. Since its inception in 2008, the Pesticide Registration Improvement Renewal Act has

98 Jason J. Czarnezki, *The Future of Food Eco-Labeling: Organic, Carbon Footprint, and Environmental Life-Cycle Analysis*, 30 Stan. Envtl. L.J. 3 (2011).

99 International Assessment of Agricultural Knowledge, Science and Technology for Development, Global Report, Agriculture at a Crossroads: International Assessment of Agricultural Knowledge, Science and Technology for Development (2009), *available at* http://www.agassessment.org/reports/IAASTD/EN/Agriculture%20at%20a%20Crossroads_Global%20Report%20(English).pdf.

provided $2.3 million in grants to advance public–private pesticide stewardship projects that focus on pesticide risk management with emphasis on IPM. The program made $500,000 available in grants for 2011. Pursuant to the program, EPA solicited proposals that address a number of issues related to advancing IPM. EPA has expressly sought proposals seeking reduced-risk management alternatives to certain specified agricultural pesticides. This program indicates a renewed interest in developing IPM approaches that can reduce or eliminate the need for synthetic chemical pesticides. However, the funding of only $500,000 for 2011, suggests that the encouragement of IPM clearly has not risen to be a high priority. One way to encourage the widespread use of IPM or EBPM would be to dramatically increase the funding available for research and development of IPM/EBPM systems that could reduce the need for conventional chemical pesticides. The current funding level of $500,000 per year is miniscule when compared to the billions of dollars that are spent annually on commodity crop subsidies that encourage the use of pesticide and other fossil fuel inputs.

Of course, one of the deterrents to a less input-intensive form of agriculture is that there is money to be made by developing and selling inputs such as chemical pesticides, whereas there is not an obvious way to profit from showing farmers how to achieve production goals without the need to purchase chemical inputs. Accordingly, the pesticide manufacturing industry has a large financial incentive to convince farmers that they will benefit from using their products. While there may be a market for expert IPM consulting services, the market is probably quite limited as compared to a market for chemical products. Moreover, it will not be possible to achieve the necessary transformation of agriculture away from complete reliance on chemical pesticides without taking into consideration the needs and perceptions of farmers and the economies of farming communities.[100] Any new system must not only be profitable for farmers, but must also be able to protect farmers from the risks of significant pest outbreaks. Accordingly, any changes in law or policy that result in reduced chemical pesticide use must be accompanied by laws or policies that improve the availability of alternative non-chemical means of pest management such as through EBPM. Increased funding for research, development and dissemination of EBPM and other ecologically sound pest management practices will be needed to ensure that growers have viable alternative pest management options. Moreover, the alternative pest management approaches must be economically sound for growers. Changes to subsidy programs, working lands conservation programs and ecosystem management can help to ensure that farmers who adopt ecologically sound practices are not economically disadvantaged and perhaps even enjoy increased financial benefits.

100 David Morley & Beth Franklin, *Future Searching for New Opportunities Involving the Pesticide Industry and Sustainable Agriculture*, *in* Bugs in the System: Redesigning the Pesticide Industry for Sustainable Agriculture 169, 177 (William Vorely & Dennis Keeny, eds, 1998).

Conclusion

To achieve the shift away from heavy reliance on synthetic chemical pesticide inputs to a more ecologically based system of pest management, a multifaceted approach is warranted. As described in this chapter, changes to FIFRA and CWA, as well as to EPA's implementation of both of these statutes, will be necessary to provide the type of localized decision making that is necessary to ensure that the use of a pesticide in a particular place at a particular time will not result in unacceptable ecological harms.

Another component of the needed multifaceted approach is the provision of economic incentives in the form of subsidies or ecosystem services payments to producers who adopt organic, EBPM, or other ecologically based pest management systems. To encourage consumers to purchase products grown with EBPM approaches, a certification and labeling program for EBPM, similar to that of the existing organic certification and labeling program, should be developed. Finally, a significant increase in funding for research on EBPM is warranted. Funding should also be made available to fund the dissemination of the EBPM approaches that come out of the research, and to provide training and assistance to growers who choose to implement such approaches.

Index

Figures, Photographs and Bridge Documents are noted in **bold**.